A LIVING PAST

The Environment in History: International Perspectives

Series Editors: Dolly Jørgensen, *University of Stavanger;* David Moon, *University of York;* Christof Mauch, *LMU Munich;* Helmuth Trischler, *Deutsches Museum, Munich*

eseh european society for environmental history

Rachel Carson Center
ENVIRONMENT AND SOCIETY

Volume 1
Civilizing Nature: National Parks in Global Historical Perspective
Edited by Bernhard Gissibl, Sabine Höhler, and Patrick Kupper

Volume 2
Powerless Science? Science and Politics in a Toxic World
Edited by Soraya Boudia and Natalie Jas

Volume 3
Managing the Unknown: Essays on Environmental Ignorance
Edited by Frank Uekötter and Uwe Lübken

Volume 4
Creating Wildnerness: A Transnational History of the Swiss National Park
Patrick Kupper
Translated by Giselle Weiss

Volume 5
Rivers, Memory, and Nation-Building: A History of the Volga and Mississippi Rivers
Dorothy Zeisler-Vralsted

Volume 6
Fault Lines: Earthquakes and Urbanism in Modern Italy
Giacomo Parrinello

Volume 7
Cycling and Recycling: Histories of Sustainable Practices
Edited by Ruth Oldenziel and Helmuth Trischler

Volume 8
Disrupted Landscapes: State, Peasants and the Politics of Land in Postsocialist Romania
Stefan Dorondel

Volume 9
The Nature of German Imperialism: Conservation and the Politics of Wildlife in Colonial East Africa
Bernhard Gissibl

Volume 10
In the Name of the Great Work: Stalin's Plan for the Transformation of Nature and Its Impact in Eastern Europe
Doubravka Olšáková

Volume 11
International Organizations and Environmental Protection: Conservation and Globalization in the Twentieth Century
Edited by Wolfram Kaiser and Jan-Henrik Meyer

Volume 12
State Forestry in Northern Europe: Histories from the Age of Improvement to the Age of Ecology
Edited by K. Jan Oosthoek and Richard Hölzl

Volume 13
A Living Past: Environmental Histories of Modern Latin America
Edited by John Soluri, Claudia Leal, and José Augusto Pádua

A Living Past

Environmental Histories of
Modern Latin America

Edited by
John Soluri, Claudia Leal, and José Augusto Pádua

berghahn
NEW YORK · OXFORD
www.berghahnbooks.com

First published in 2018 by
Berghahn Books
www.berghahnbooks.com

Library of Congress Cataloging-in-Publication Data

Names: Soluri, John, editor. | Leal, Claudia, 1970- editor. | Pádua, José Augusto, editor.
Title: A Living Past: Environmental Histories of Modern Latin America / edited by
  Claudia Leal, John Soluri, and José Augusto Pádua.
Description: New York: Berghahn, 2018. | Includes bibliographical references and index.
Identifiers: LCCN 2017051497 (print) | LCCN 2017059196 (ebook) | ISBN
  9781785333910 (eBook) | ISBN 9781785333903 (hardback: alk. paper)
Subjects: LCSH: Human ecology—Latin America—History. | Nature—Effect of human
  beings on—Latin America—History. | Latin America—Environmental conditions.
Classification: LCC GF514 (ebook) | LCC GF514 .L58 2018 (print) | DDC 304.2098—dc23
LC record available at https://lccn.loc.gov/2017051497

British Library Cataloguing in Publication Data

A catalogue record for this book is available from the British Library

ISBN 978–1-78533-390-3 hardback
ISBN 978–1-78533-391-0 ebook

Contents

List of Illustrations, Tables, and Figures   vii

List of Maps   ix

Preface   x

Introduction. Finding the "Latin American" in Latin American
Environmental History   1
John Soluri, Claudia Leal, and José Augusto Pádua

Chapter 1. Mexico's Ecological Revolutions   23
Chris Boyer and Martha Micheline Cariño Olvera

Chapter 2. The Greater Caribbean and the Transformation
of Tropicality   45
Reinaldo Funes Monzote

Chapter 3. Indigenous Imprints and Remnants in the Tropical Andes   67
Nicolás Cuvi

Chapter 4. The Dilemma of the "Splendid Cradle": Nature and
Territory in the Construction of Brazil   91
José Augusto Pádua

Chapter 5. From Threatening to Threatened Jungles   115
Claudia Leal

Chapter 6. The Ivy and the Wall: Environmental Narratives from
an Urban Continent   138
Lise Sedrez and Regina Horta Duarte

Chapter 7. Home Cooking: Campesinos, Cuisine, and Agrodiversity   163
John Soluri

Chapter 8. Hoofprints: Cattle Ranching and Landscape
Transformation   183
Shawn Van Ausdal and Robert W. Wilcox

Chapter 9. Extraction Stories: Workers, Nature, and Communities
in the Mining and Oil Industries   205
   Myrna I. Santiago

Chapter 10. Prodigality and Sustainability: The Environmental
Sciences and the Quest for Development   226
   Stuart McCook

Chapter 11. A Panorama of Parks: Deep Nature, Depopulation,
and the Cadence of Conserving Nature   246
   Emily Wakild

Epilogue. Latin American Environmental History in
Global Perspective   266
   J. R. McNeill

Selected Bibliography   277

Index   288

Illustrations, Tables, and Figures

Illustrations

1.1. "Corn Patch Fringed with Maquay [*sic*]," Toluca, Mexico, 1907.　32

2.1. Sugar Mill "Tinguaro", established in 1840, in Colon Plains, Matanzas, Cuba.　49

2.2. Sugar Central España Republicana (former España), in Colón municipality, Matanzas, Cuba.　54

3.1. Gathering the coca plant in Bolivia. Wood engraving, 1867.　76

5.1. Village on Napi River, Pacific coast of Colombia, photographed by Robert C. West, circa 1955.　125

6.1. Smog around the Pico de Orizaba, in Mexico City, the second largest city of the Americas, 2015.　139

6.2. New Year's Eve Celebration, Copacabana Beach, 2012.　158

7.1. Construction workers gathered around a large comal with tortillas, Mexico City, 1953.　167

7.2. Ezequiel Arce and his potato harvest, Cusco, Peru, 1934.　170

8.1. Guatemalan cowboy in a field of guinea grass, 1917.　186

9.1. Mexican workers at an earthen oil dam at Potrero del Llano, northern Veracruz, circa 1914.　215

9.2. Ermenegildo Criollo, Cofán leader, in Amazonia, Ecuador.　219

10.1. A coffee farm near Sasaima, Colombia, 2013.　235

10.2. Sign from the Araucárias Municipal Natural Park near Guarapuava, Brazil, 2015.　239

11.1. Families enjoying the central plaza of Huaraz, Peru, during the International Glacier Forum in June 2013.　252

Tables

3.1. Indigenous population of the Andean highlands, circa 2009–2015.   68

Figures

11.1. Total conservation areas created per year in Latin America.   254

Maps

1.1. Mexico's main biomes   25

3.1. The Cinchona regions of South America   77

4.1. Anthropization of Brazilian biomes, 1960   104

4.2. Anthropization of Brazilian biomes, 2000   108

5.1. Tropical rainforest cover   118

8.1. Ranching biomes of Latin America   185

11.1. Designated conservation areas in Latin America, 2015   247

Preface

This set of interpretive essays blends environmental history with cultural, economic, political, and social histories of Latin America and the Caribbean from the nineteenth century to the early twenty-first century. Together they provide a synthesis of the work that scholars have been developing in recent years to reinterpret the region's past. But they also try to go beyond what has been done by suggesting new topics and fresh interpretations. Furthermore, they provide vital historical and geographical contexts for ongoing debates related to sustainable cities, tropical deforestation, mining, conservation, agrodiversity, and tourism, among other topics.

In designing and writing this book, we confronted several epistemological and logistical hurdles. We knew from the start that it would not be a "handbook" or an "encyclopedia", and that entire regions and periods would have to be excluded. Ecosystems and processes that some readers associate with the region's past may be absent or not be covered to their satisfaction. We decided to focus largely on the national period due to "resource constraints" (page limits) and a desire to concentrate on political structures—nation-states—that remain vital today. However, legacies and continuities from the pre-colonial and colonial pasts are integrated into most authors' analyses; these are modern histories that strive to avoid being overly presentist.

We decided from the outset that instead of providing a single narrative, we would offer multiple voices and approaches, exposing some of the seams and unresolved tensions in the practice of Latin American and Caribbean environmental history. In an attempt to explore the region's history from various angles, this volume includes geographically oriented chapters that focus on specific nations (Brazil and Mexico) or macro-regions (Caribbean and tropical Andes), along with chapters that are more thematic in scope, looking at socio-ecosystems (tropical forests and urban areas) and human activities that transform nature (agriculture, conservation, mining, ranching, and technoscience).

In both theory and practice, a cosmopolitan ethos has guided this endeavor, which includes authors based in Brazil, Canada, Colombia, Cuba, Ecuador, Mexico, and the United States. This initiative received German financial support through the Rachel Carson Center for Environment and Society (RCC) in

Munich, where co-editors Claudia Leal and José Augusto Pádua were fellows. Christof Mauch, Director of the RCC, helped to conceptualize this volume and gave it continuous and enthusiastic encouragement for which we are eternally grateful. We are also indebted to the Carson Center's Katie Ritson who edited an issue of the RCC's *Perspectives* that included condensed, preliminary versions of the chapters contained in this volume. Peter Coates, Stefania Gallini, and Jane Carruthers graciously provided comments on those exploratory pieces. We convened workshops, funded in part by the Carson Center, in Sasaima, Colombia (2013) and Chascomús, Argentina (2014), where we shared drafts, aspirations, frustrations, and not a few *tragos*. The three editors also met in Bogotá one last time (2015) thanks to the hospitality of Universidad de los Andes. Adrián Gustavo Zarrilli organized the workshop in Chacomús, and we thank him deeply for taking on that responsibility at the same time that he was organizing the VII Symposium of the Sociedad Latinoamericana y Caribeña de Historia Ambiental (SOLCHA).

The project owes a large intellectual debt to a growing international community of scholars who form SOLCHA. We believe that both the process and the product of this project will strengthen this community. In that regard, we owe our greatest debts to the fantastic group of scholars who have contributed to this volume; it would not have happened without their encouragement, erudition, and patience. Finally, we are grateful for John McNeill's willingness to contribute the essay that concludes this volume.

Often, the greatest value of a work of synthesis lies in the gaps, silences, and distortions identified by engaged readers. For the moment, our work is done; the task of advancing Latin American and Caribbean environmental history now lies in the hands of many scholars interested in this fascinating and promising field.

Finding the "Latin American" in Latin American Environmental History

John Soluri, Claudia Leal, and José Augusto Pádua

"Sometimes you can't be outside, the odor stinks, your throat stings.
It smells of gas. Even if we close our doors, it smells."[1]

This is how residents of Villa Inflamable, a shantytown in Greater Buenos Aires, described their neighborhood in 2004. Villa Inflamable is surrounded by one of Argentina's largest petrochemical complexes, in addition to a hazardous waste incinerator and an unmonitored landfill. The land, air, and water are contaminated with heavy metals; children suffer health problems linked to elevated levels of lead in their blood. Situated on the bank of the highly polluted Matanza-Riachuelo River, the inhabitants of Villa Inflamable also endure frequent flooding. In 2004, a group of residents filed a lawsuit against both the Argentine government and private companies seeking compensation for harm suffered from pollution. The case reached Argentina's Supreme Court, which ruled in 2008 that decisive action needed to be taken by government authorities including a plan to relocate the petrochemical complex. More than a decade later, the people of Villa Inflamable continue to wait.[2]

The story of Villa Inflamable highlights the environmental dimensions of social problems associated with neoliberal policies: since the 1980s, both urban poverty and economic inequalities have skyrocketed in Argentina, compelling poor people to live in undesirable places. But the causes of environmental degradation and pollution have deeper roots: Shell Oil opened its first refinery in 1931 with the blessings of Argentine military ruler General Uriburu, who, like his civilian predecessors, was eager to harness petroleum resources first discovered in Patagonia in 1907.[3] The name of the sixty-four-kilometer waterway—"Matanza-Riachuelo"—alludes to a still older history: in the early nineteenth century, the banks of the "little stream" (*riachuelo*) became the site of slaughterhouses (*matanza*), leather tanneries, and other factories that needed water in order to transform animals raised on the fertile grasslands of the Pampas into commercial goods.[4] The lead and cadmium that course through the blood of residents in Villa Inflamable then, are part of Argentina's national heritage—an entangled web of historical relationships that extend far beyond neoliberal Buenos Aires.

Environmental history seeks to describe and analyze the dynamic relationships between human societies and nonhuman nature that can help us to explain not only industrial places like Villa Inflamable but also seemingly pristine areas like Amazonian rainforests, Caribbean beaches and Andean glaciers. Environmental historians strive not only to introduce new questions, actors, and explanatory frameworks, but also to shed fresh light on familiar topics such as the rise of nation-states, social inequalities, and technological change.[5] This volume, comprised of original contributions researched and written by fifteen scholars coming from diverse academic and geographical backgrounds, is both synthetic and topical. Some of its chapters are organized around nations or regions such as Brazil, Mexico, the Greater Caribbean, and the tropical Andes. Others focus on broader themes not limited to a specific location or region: environments such as urban areas and jungles, or transformative processes like agriculture, conservation, mining, ranching, and scientific research.

Well-established in the United States and parts of Europe, environmental history is still taking shape in Latin America. By no later than the 1930s, notable scholars demonstrated that the region's human history is best understood when not viewed in isolation from the material environments in which it is entangled.[6] In the 1980s and 1990s, a handful of researchers from Latin America and the United States planted the proverbial seeds for the field by publishing interpretive essays and narrative histories about human-driven environmental changes in countries including Argentina, Brazil, and Mexico.[7] Latin American environmental history began to blossom at the outset of the twenty-first century in the forms of scholarly publications, conferences, and graduate programs. Communication and networking among a growing group of scholars studying Latin American and world environmental history greatly increased as reflected and fostered by the Latin American and Caribbean Society for Environmental History (SOLCHA), formally established in 2006.[8]

Nevertheless, the circulation of research findings across this vast, diverse region continues to face barriers posed by geography, academic cultures, language, and technological divides; as a result, the historiography remains fragmented and uneven. Furthermore, the vast majority of published works, with a few notable exceptions, study specific regions or nation-states, so that a general picture is still emerging.[9] In this introductory chapter, we identify some of the central themes that characterize recent scholarship in the field, before exploring four features—colonial legacies, nation-states, transoceanic connections, and tropicality—which, taken together, are crucial for identifying the "Latin American" in the environmental histories that follow.

The theme that most strongly marks the first wave of Latin American environmental history is a preoccupation with the fate of forests. Warren Dean's sweeping history of Brazil's Atlantic Forest (1994) and Reinaldo Funes's study of the Cuban sugarcane industry (2004) both reinterpret the histories of those countries by documenting the centrality of long-term processes of forest removal.[10] The centrality of deforestation is further revealed by studies focused on agriculture, human diseases, mining, and politics, in which the felling of forests serves as a key trigger for other kinds of socioecological change.[11] Other researchers working on case studies of Brazil, Chile, Colombia, Costa Rica, and Mexico have sharpened their focus to explain the logic of logging industries and the policies that regulated them.[12]

This emphasis on forests is understandable in light of the diversity and geographical expanse of forested ecosystems in Latin America. In some instances, the motivation to write forest histories results from an undeniable reality that many forests have disappeared or become highly fragmented over the past two hundred years. In contrast to the environmental historiography on Europe, South Asia, and the United States, state forestry institutions have not figured prominently in Latin American histories with the exception of Chile and Mexico, whose governments enlisted French or German-trained foresters to oversee the management of temperate forests in the early twentieth century.[13] That said, recent scholarship on the contested politics of forest management resonates with studies of regions outside of Latin America that examine the viewpoints of the communities who inhabited forests and their efforts to defend their territories and sylvan resources from states and other actors.[14]

Environmental historians of Latin America have also devoted considerable energy to studying rural landscapes and livelihoods, an outgrowth of the longstanding scholarly interest in the region's export economies. They have privileged analysis of the agroecological dynamics of plantation production of tropical commodities, highlighting widespread deforestation, the exploitation of local people, and the role of plant pathogens.[15] Mining and petroleum have received less attention to date, a curious imbalance given the importance of past and present mineral extraction to the political economies of most Latin American countries. The same slow start holds true for environmental histories of food, energy, and goods produced primarily for internal consumption.[16] Domestic use of natural resources looms large for environmental histories of the late twentieth century, when import substitution industrialization (ISI), large dam construction, and transportation infrastructure expanded, along with public investments in education, health, and housing for a burgeoning population.[17]

Conservation history is another popular topic among environmental historians of Latin America. The findings of this scholarship demonstrates that the establishment of protected areas in the region cannot be understood as the mere importation of "America's best idea," but rather as a result of multiple and sometimes contradictory dynamics that need to be explained. These include a desire to control frontier areas in Brazil, Argentina, and Chile; the building of rural justice in revolutionary Mexico; and the promotion of international tourism in Costa Rica.[18] Historians of science have shown that scientists and naturalists have played critical roles in "nationalizing" nature via natural resource inventories and taxonomies, and advocating for government-led conservation projects.[19] The contributions from this related field also include works on the development of the environmental sciences.[20]

Latin America is widely regarded as the most urbanized region in the contemporary world. Although "megacities" only emerged in the past few decades, urban centers are hardly new phenomena in Latin America, as visitors to Mexico City are able to appreciate when encountering the jumbled architecture of Aztec, Spanish, and Mexican societies. Fortunately, environmental historians are increasingly studying urban water and waterworks, among other topics related to the spaces that determine the experiences of most of the region's inhabitants.[21] In addition, because cities rely on connections to other places for food, energy, water, and building materials, environmental histories of cities promise to reveal the historical importance of regional, rural-urban connections that often lie in the shadows of international ties in Latin American historiography.

There are many forms of socioecological relationships and systems in Latin America whose histories have yet to be fully examined. Histories of watery environments, including rivers, estuaries, reefs, ocean littorals, and even glaciers, are few and far between.[22] Grasslands have not received attention commensurate with their historical importance for the production of grains and livestock.[23] Deserts and semiarid regions, with a few notable exceptions, have yet to be embraced by environmental historians with the same degree of enthusiasm as forests.[24]

Latin American environmental histories are heavily concentrated on the nineteenth and twentieth centuries. This "recentism"—not without its critics—is partly attributable to mounting evidence showing that planetary environmental transformations accelerated dramatically since 1945.[25] The fact that many of us understand environmental history as a form of knowledge production that can generate "tools" for shaping contemporary policy and politics also helps to explain the bias for the recent past. However, as we acknowledge below, the legacies of the precolonial and colonial eras are crucial for explaining both the changes and continuities of the nineteenth and twentieth centuries. Moreover, recent works on sixteenth-century New Spain (Mexico) demonstrate the vitality of early modern environmental histories.[26] Perhaps what is most worthy of critical reflection is that few environmental histories cut across the colonial/republican divide or advance new periodizations.

In sum, Latin American environmental history has taken great strides in recent years as measured by the growing number of practitioners in the field and the expanding geographical, temporal, and thematic scope of research.

That said, there is no shortage of tasks at hand requiring additional empirical and conceptual work. However, instead of providing a "to do" list, we devote the rest of this chapter to exploring four interrelated features that are crucial for understanding the "Latin American" in Latin American environmental history: the enduring legacies of Iberian colonialism, the nineteenth-century formation and persistence of nation-states, transoceanic exchanges, and tropicality.

Enduring Colonial Legacies

A human presence in what is today Latin America began no later than fourteen thousand years ago; since that time, people have modified a wide range of environments via foraging, hunting, fishing, fire setting, farming, water management, mining, and building settlements. The domestication of plants began at least ten thousand years ago in southern Mexico, the tropical Andes, and Amazonia, giving rise to crops such as maize, potatoes, cocoa, and manioc. In some parts of Latin America, indigenous societies' agricultural prowess supported the formation of populous, urbanized societies. However, indigenous cultures left imprints nearly everywhere, including ecosystems still perceived to be wildernesses such as the Amazon River basin and the wind-swept Patagonia steppe lands. By the time Europeans arrived in the late fifteenth and early sixteenth centuries, tens of millions of people lived in Latin America and the Caribbean, with particularly dense populations in Mesoamerica and the Andes. Far from pristine, the Americas circa 1492 consisted of humanized land and waterscapes.[27]

The arrival of Iberian colonizers had uneven effects on Latin American cultures and environments. In the century following initial contacts with people from Europe and Africa, indigenous populations succumbed to introduced pathogens whose devastating effects helped to ensure that short-term military victories, achieved via political alliances, warfare, and enslavement, would become enduring colonies. In contrast to later European imperial ventures in Africa and Asia, the main consequence of colonial rule in the Americas—a demographic collapse—was not the result of explicit state policies aimed at controlling resources (such as forestry in India or game conservation in Africa) but rather an unintended, contingent outcome.[28] Exposure to introduced pathogens would continue to affect indigenous groups in Latin America in the twentieth century, when relatively isolated groups in Amazonia and Tierra del Fuego confronted new diseases and violence in processes that paralleled those in North America, Australia, and New Zealand.[29]

The demographic collapse and ensuing three centuries of Iberian colonialism transformed Latin America and the Caribbean but did not homogenize it. In mining regions (including Central Mexico, highland Peru, and Minas Gerais), human migration and the relentless extraction of silver and gold, but also to a lesser extent of mercury, diamonds, and other minerals, resulted in significant levels of deforestation, pollution, and human health risks. One of the first documented instances of large-scale toxic contamination took place in the famed silver-mining center of Potosí when in 1626 a reservoir burst, creating a surge of water that destroyed many mills and released an estimated nineteen tons of mercury (used for amalgamation) into the Pilcomayo river system, a toxic legacy that remains in the soils of the area.[30] In New Spain, centuries of charcoal production for silver smelting consumed vast quantities of forests.[31] In lowland tropical regions of northeast Brazil and the Caribbean, the introduction of sugarcane and millions of African slaves resulted in regional deforestation and, particularly on Caribbean sugar islands, the extirpation of fauna.[32]

Iberian imperialism also bequeathed to modern Latin America a number of urban centers, many of which would become national capitals (e.g., Mexico City, Lima, Havana, Bogotá, Caracas, Guatemala, Buenos Aires, Santiago de Chile, and Rio de Janeiro). In Mexico City, the center of Spain's American empire, colonial officials initiated a long-term process of environmental change in the form of the *Desagüe*, a massive infrastructure project that eventually drained most of the lakes that surrounded the city in the early sixteenth century.[33] Along with libraries, cathedrals, and convents, Spaniards in Mexico City established Chapultepec, a public park that remains one of the largest urban green spaces on the planet. In their contribution to this volume, Lise Sedrez and Regina Horta Duarte explain how many contemporary urban environmental issues are rooted in decisions made during the colonial period, including fundamental ones such as location.

Perhaps the most taken-for-granted colonial legacy are the countless cattle, horses, pigs, goats, sheep, oxen, and hens that roamed from California to Tierra del Fuego. European colonialism in Africa and Asia did not have nearly as large of an impact in altering the species composition of domesticated animals as it had on the Americas, where large animal domesticates were few in number and there was much room for the newcomers.[34] Only Australia experienced a comparable degree and scale of transformation.[35] As Shawn Van Ausdal and Robert Wilcox demonstrate in this volume, cattle ranching and horsemanship outlived Iberian rule and continue to be a major influence on Latin American ecologies and cultures.

But, there was another, less obvious legacy of Iberian colonialism: an expansion of forests and other native ecosystems due to the abandonment of indigenous agricultural fields, a reduction in intentional burnings, and the decline in human population that would not begin to recover until the late eighteenth century. Determining the extent and composition of forest cover is fraught with uncertainty due to major gaps in evidence and challenges associated with interpreting the data that do exist.[36] Nevertheless, evidence suggests that forests covered as much as 68 percent of Latin America and the Caribbean in the early nineteenth century.[37] Republican discourses and metaphors of an abundant nature, such as the Brazilian republic's birth in a "splendid cradle" (see chapter 4, by José Augusto Pádua), were as much a product of colonialism as they were of any intrinsic prodigality of tropical nature. Nationalist writers sometimes viewed these postcolonial environments as hostile, at other times beautiful, but they nearly always described them as vast and empty, an understanding that drove—and justified—wasteful production processes, a problem that persists to this day.

Finally, Iberian colonialism brought new languages, ontologies, and epistemologies to the Americas.[38] These included not only Christian doctrines and rituals but also Enlightenment ideas and practices about how to order and represent nature, as well as legal codes that redefined property, sovereignty, and social relations. The incorporation—albeit partial—of new ways of knowing and valuing life-forms, time, and human labor is vitally important for thinking about the forces driving ecological change, including the conservation of biocultural resources. Current political debates, including those that have invoked the indigenous Andean notion of *Sumak Kawsay* (good living) for rethinking development, demonstrate the high stakes involved in the ways we interpret the complicated and connected histories of indigenous, African, and European ideas and practices toward nonhuman natures, a topic that Nicolás Cuvi addresses in this volume.[39]

States from Nature

As political revolutions spread across the Atlantic world, the Spanish Empire crumbled in the Americas and gave way in the early nineteenth century to sovereign nation-states in all but a few places. Between the 1830s and the 1950s, when many Africans and Asians fought against European colonial bureaucracies and armies to maintain control over land and resources, Latin Americans participated in novel and fractious experiments in state-making and nation-building in which they enlisted nature.[40] The environment became national patrimony, not imperial plunder as was the case throughout most of Africa and Asia.

Ruling elites—who often violently disagreed about forms of governance—struggled to define state territories by drawing borders that nationalized natures. As Pádua and Claudia Leal argue in their contributions to this volume, states claimed sovereignty over vast, uncontrolled areas—the deserts of northern Mexico, the forests and steppes of Patagonia, the jungles of the Amazon and Orinoco basins—via maps that invariably overlapped, creating disputed territories on paper. In many cases, it took decades for states to reach agreement on the exact boundaries that defined their respective dominions. As Stuart McCook explains, inventories of plants and animals were a fundamental tool in this process of nationalizing nature, effectively extending passports to birds, trees, and other organisms. Governments enlisted scientists who traveled widely, identifying and counting species, illustrating specimens, and compiling long lists of plants and animals that claimed nature as a means to proclaim national grandeur.[41] This incipient form of nationalism did not need effective occupation, and thus came in handy for fledgling states that had trouble imposing their rule over diverse and extensive geographies.

The republican imperative to transform "wastelands" (in Spanish often expressed as *desiertos*) coincided with the expansion of industrializing economies in the North Atlantic. Paradoxically, cash-strapped governments turned increasingly to international investors and markets in order to assert their territorial sovereignty and further nationalize nature by turning "natural resources" into export products. Nineteenth-century Latin American states were quite literally hewn *from* nature—republics built on bird shit and bananas, cattle hides and sheep fleece, coffee and copper. The wealth acquired by extracting minerals and nutrients from the lithosphere, or making commodities from the biosphere, allowed states to increase their tax revenue and enlarge meager bureaucracies. This wealth did not flow spontaneously; metaphors of "open veins" or "commodity lotteries" fail to convey the sustained labor of millions of people and the state policies that directly and indirectly enabled the harvest of rents from the forests, grasslands, mountains, and waterways that comprised the "splendid cradle" in which Latin American nation-states were born.

As Chris Boyer and Micheline Cariño demonstrate for the case of Mexico, revenue-poor states took advantage of the existence of ample "public" lands to grant concessions in exchange for state-building projects, including the measurement of lands and the construction of roads. The prospects of turning forests into wealth via commodification or simply speculation motivated concessionarees. But once states grew stronger, especially after the economic turmoil of the 1930s, they moved in the opposite direction, returning some natural resources to the public domain, a process that arguably had its most forceful expression in post-revolution Mexico. As Myrna Santiago documents in this volume, organized labor often supported states' efforts to claim strategic resources for the nation. Militant miners and oil workers pressured governments to nationalize industries in hope of securing greater control over labor processes and the subsoil resources that they viewed as national patrimony. For example, Mexican oil workers influenced the government of Lázaro Cárdenas to nationalize Mexico's oil industry.[42] Many Latin American states subsequently nationalized key mining and energy sectors, setting examples that would be emulated elsewhere in the world.

The nationalization of key industries was part of a more active state role in environmental management, which after the mid-twentieth century strongly contributed to significant landscape transformations. The rise of populism in the 1930s, along with the geopolitics of World War II and the Cold War, accelerated the creation of government-sponsored, technoscientific "development" programs whose broad outlines would be pursued in similar fashion in many parts of Europe, Asia, and Africa: land reform and agricultural "modernization" via the Green Revolution, dams for irrigation and electricity, highways and mass transit systems, mass education, and public health measures changed the lives of people and altered ecosystems throughout the world. If true that state-led development frequently failed to reduce social inequalities or to promote democracy in Latin America, it almost always resulted in accelerating rates of resource use, linked in part to an enormous increase in human population that often overwhelmed the capacities of government programs.

But not all twentieth-century state policies and projects promoted resource consumption; as Emily Wakild's chapter reveals, Latin American governments also created national parks, established forestry departments, and passed laws protecting fauna. In contrast to colonial Africa or South Asia, elites viewed protected areas as forms of national patrimony rather than foreign impositions. These measures contributed, as early as the 1930s in Argentina, but more commonly in the 1960s, to the building of "nature states"—distinctive spheres of government activity devoted to caring for nature—a phenomenon observable in other parts of the world.[43]

Transoceanic Trade and Ecological Exchanges

Many scholars have documented the movements of people, plants, animals, and pathogens across oceans and their effects on Latin America's biocultural diversity. Although transoceanic movements have connected and transformed other large regions, such as the Indian Ocean, the magnitude and significance of exchanges between the Americas and Africa, Eurasia, and the Pacific are unique. The Columbian Exchange marked only the beginning of a process of biological exchange that has yet to cease; in fact, the potential for more frequent and multidirectional exchanges increased dramatically in the nineteenth century, a period when African slaves, indentured labor from China, India, and the Pacific Islands, along with immigrants from Europe, all converged on the region. These massive flows of people brought with them a menagerie of domesticated animals and crop plants, along with a slew of "hitchhiking" biota, including rats and weeds that have fared all too well in their new homes.[44] More importantly, the flows from Latin America to the rest of the world multiplied and diversified with far-reaching environmental consequences.

The rise of agroexport industries characterized by international investors and markets linked by telegraphs, railroads, and steamships, strengthened and expanded the networks by which people and nonhumans moved across oceans and continents. The production of bananas, cattle, coffee, sugarcane, sheep, wheat, and, more recently, palm oil, grapes, salmon, and soybeans, has led to an influx of organisms from across oceans. Fodder grasses, crop plants, and trees remade terrestrial landscapes, while the introduction of northern hemisphere fish, including trout, salmon and bass, remade freshwater ecologies. The introduction and propagation of African grasses and South Asian cattle breeds transformed tropical cattle ranching, an economic sector often perceived as static if not backward. In Argentina, the displacement of sheep from the fertile Pampas to the steppe lands of Patagonia compelled ranchers to import sheep breeds from Australia and New Zealand that would subsequently spread from Patagonia to the Andes.

The export economies created new agroecological dynamics marked increasingly by "commodity diseases"—outbreaks of plant and animal diseases whose intensity, spread, and meanings were tightly linked to processes of commodification.[45] Agribusiness and governments took initiatives to eradicate or at least limit the spread of organisms that damaged crops and animals. These initiatives included an ever-widening search for varieties and breeds that possessed both disease tolerance and marketability. Plant diseases, herbivores, and nutrient loss also contributed to a rise in the use of agrochemicals and synthetic fertilizers. Commodity diseases in key export crops, including bananas, cacao, coffee, and sugar, prompted planter associations, agribusinesses, and governments to enlist the support of university-trained scientists, and to establish research centers dedicated to plant pathology and breeding.[46]

Export economies were by no means limited to products from farms and ranches. People extracted wealth from forests, including timber but also products such as rubber, *chicle*, quinine, tagua seeds, coconuts, and Brazil nuts that generated wealth from *living* forest species. The strategic importance of tree and palm products such as natural rubber and quinine would compel powerful states, including the British Empire and the United States, to seek to control their production. Forests, along with other kinds of ecosystems, also served as an important source of feathers, skins, and furs from wild fauna. Jaguars, whales, chinchillas, turtles, fur seals, foxes, and rheas were just a few of dozens of animal species hunted extensively for commercial purposes. However, commerce in wildlife also compelled some Latin American governments to regulate hunting as early as the late nineteenth century, in the name of preserving national patrimony, not global biodiversity.

Modern Latin America's involvement with interoceanic trade and exchanges was not limited to biological commodities. Some of the first export markets to emerge in the nineteenth century included mined fertilizers (guano and nitrates). As Santiago explains in her chapter, mining activities increased during the twentieth century: the extraction of bauxite, copper, gold, iron, and tin, along with petroleum and gas drilling, created new kinds of environmental hazards. The transnational character of Latin America's twentieth- and twenty-first-century mining and petroleum industries raises new questions for environmental historians that have less to do with biological exchanges than with energy exchanges: Mexico and Venezuela, along with Brazil, Colombia, and Ecuador, have transferred massive amounts of stored energy (and heat-trapping carbon) to the United States and elsewhere. The extraction and export of copper, principally from Peru and Chile, has helped to enable the large-scale transmission of electricity, transforming many aspects of daily life for billions of people and contributing to localized and planetary ecological changes.[47]

Transoceanic movements continue to be an important source of environmental change in much of Latin America even as new markets (e.g., China) and new products (e.g., soy and lithium) emerge. As Reinaldo Funes observes, the movement of international tourists on cruise ships and airplanes has largely replaced that of agroexports in much of the Greater Caribbean. Illegal drug trades also fuel border-crossing activities with ecological consequences that have yet to be fully assessed. Of course, intercontinental exchanges are by no means the only ones with lasting ecological consequences; indeed, the rights and restrictions associated with citizenship in sovereign nation-states make regional movements far more common than international ones. Nevertheless, centuries of intercontinental exchanges of biophysical matter and energy are—somewhat paradoxically—central to the "Latin America" in Latin American environmental history.

Tropicality: Confronting Diversity

Nearly three quarters of the South American landmass lies within the tropics, in addition to the entire Central American isthmus, the insular Caribbean, and central and southern Mexico.[48] Contemporary estimates indicate that rainforests occupy an astounding 44 percent of Latin America's landmass. Other extensive tropical ecosystems include grasslands and savannas (16.4 percent) and dry tropical forests (8.8 percent). Nature in these tropical ecosystems is extremely diverse: Latin America and the Caribbean, which occupy less than 10 percent of the earth's land mass, possess nearly one-third of the world's documented species of vascular plants (far more than found in Africa or Asia), half of all species of amphibians, and forty percent of Earth's reptiles and birds.[49]

However, even naturalists such as Alexander von Humboldt, who marveled at the American tropics' biological diversity, considered that its heat and fecundity sapped the energy and motivation from human societies. European and U.S.-based writers, artists, and scientists believed that the abundance of nature anywhere in the tropics "overwhelmed the human endeavor" and reduced place to nature itself.[50] This ahistorical thinking about the tropics served to depict the Caribbean region as an area fit only for "inferior races," deemed closer to nature, thus justifying the continued enslavement of Africans and the post-emancipation use of indentured workers from Asia. In fact, the sugar, rum, and slave trades radically altered Caribbean environments while introducing yellow fever and its mosquito vector to the region, a historical contingency that many medical doctors and others would interpret as evidence of the inherent dangers of the tropics for "whites" (see Reinaldo Funes's chapter).[51] This thinking persisted into the early twentieth century and served as a justification for importing black workers from the Caribbean to labor on the Panama Canal and also banana plantations.[52] The "civilizing mission" espoused by both early twentieth-century U.S. government officials in Cuba and Panama and corporate businesses resonated forcefully with the discourses of European colonial officers in tropical Asia and Africa.

The history of labor subjugation in many parts of tropical Latin America reminds us that political elites and investors considered the tropics to be not only unruly and dangerous but also a source of riches. The rainfall and heat that came to embody the notion of tropicality (there are also dry and cold areas in the tropics) served to produce sugar and bananas for export. Tropical plantations efficiently transformed solar energy—along with forests and soils—into caloric energy, while extractive economies, as in the case of rubber, commodified useful elements found in the diversity of life-forms. The tropics, then, were not just imagined spaces, but also very physical places that got transformed as they joined networks of global exchange.

Latin American elites viewed the tropics with deep ambivalence. In the early twentieth century, writers such as Horacio Quiroga, Rómulo Gallegos, José Eustacio Rivera, and Alejo Carpentier presented tropical forests in a vein similar to that of Joseph Conrad's portrayal of the Congo River basin in *Heart of Darkness*. As Rivera's novel *The Vortex* (1925) poignantly showed, the jungle disoriented and ultimately defeated those who sought to subdue or domesticate it. The notion that untamed tropical forests posed a threat to civilization informed Latin American governments' efforts to colonize tropical lowlands through the 1970s.[53] However, elites in early twentieth-century Brazil and postrevolutionary Mexico increasingly rejected theories of racial "degeneracy" and embraced visions of *mestizaje* (racial mixing)—and whitening—that tried to reimagine the tropics as a place of great vitality and potential.

As Leal and Funes argue in their chapters, both elite ideas about tropical nature and economic activities underwent a dramatic shift in the second half of the twentieth century. The development of public health measures to control yellow fever in late nineteenth-century Cuba, along with a decline in demand for sugarcane and the formation of close ties with the United States, created conditions that gave rise to a robust tourist trade beginning in the 1930s when droves of (mostly U.S.) tourists came, lured by desires for Cuban sun, rum, and son. By the end of the twentieth century, the islands of the Caribbean—including Fidel Castro's Cuba—were better known for their sandy beaches than for tropical agriculture. European and North American tourists viewed Caribbean islands as places of rejuvenation, not illness. In a rather bitter irony, many residents of the Caribbean left their "island paradise" to seek jobs in the United States and elsewhere as local agricultural economies collapsed.

Perceptions of tropical forests in continental Latin America also transformed. In the 1980s, formerly menacing jungles became "hot spots" of biological diversity that drew the attention of international conservation movements, political leaders, and pop stars. No longer threatening, jungles discursively transformed into "rainforests" that were now threatened with extinction due to human action. The United Nations recognized Brazil, Colombia, Ecuador, Mexico, Peru, and Venezuela as "megadiverse" nations possessing high rates of endemic species. International conservation networks formed alliances with local organized people to pressure national governments to protect rain forests and their human inhabitants. These movements resulted in the creation of novel forms of conservation, such as extractive reserves, that protected forests without displacing forest residents. In the early twenty-first century, deforestation rates in Amazonia started to decline.

But, rainforests and sandy beaches constitute only a small part of most Latin Americans' lived experiences, even if the vast majority of them live in the tropics, a remarkable continuity that stretches across the pre-Columbian, colonial, and modern eras. Most of these tropical people have lived in agrarian settlements or urban areas. As the chapters by John Soluri and Cuvi demonstrate, highland tropical regions have been centers of agricultural diversity that sustain indigenous communities and supply urban markets. The highland tropics are also places where alternative cosmologies and visions of human-nature relationships continue to germinate. Finally, the Andes are home to nearly all of the world's tropical glaciers, which represent, among other things, important sources of water for rural and urban populations. Many glaciers are likely to disappear long before tropical rainforests do.

There is a diversity within the diversity of Latin America's tropical regions that scholars and others have to confront in order to move beyond stereotypes or deterministic understandings of tropicality. In ecological terms, the productivity and biotic diversity of tropical ecosystems distinguish them from temperate latitudes. However, the productivity of a sugarcane plantation in Brazil, the diversity of potatoes in highland Ecuador, or the variety of bats found in a Central American coffee farm are largely outcomes of historical processes.

Tropicality—the cultural meanings of tropical places and the living things that inhabit them—is also deeply historical, contested, and consequential.

Convergences: Latin America and a Global Environment

There is no single factor that distinguishes Latin American environmental history from other world regions, but rather a conjuncture of multiple socioenvironmental processes that include: Iberian imperialism, early but rather weak state formation, sustained intercontinental material exchanges, and tropicality. Of course, these are broad generalizations that run the risk of blocking from view the rich diversity that lends the geohistorical idea of "Latin America" its material beauty and complexity—its enormous palette of colors, tastes, textures, sounds, smells, and "structures of feeling," that condition and give meaning to everyday life. The chapters that follow collectively reveal some of the seams, tensions, diversity, and silences that characterize environmental histories of the region.

At the same time, the volume also points toward a convergence of Latin American environmental history with that of several other world regions characterized by: population growth, particularly in urban areas, increasing commodification of life-forms and processes, governments that rely on resource rents to maintain political legitimacy, the rise of consumer-driven (and meat-eating) middle classes, and energy-intensive technological systems. One outcome of this convergence has been a tremendous increase or "Great Acceleration" in energy and resource use since ca. 1940 that has created an unprecedented era of human-induced environmental change that some observers have labeled the "Anthropocene."[54]

"Latin American" environmental histories, therefore, should continue to be written as connected histories whose specificity is not derived exclusively or even primarily by the boundaries of its constituent nation-states. Moreover, as many parts of Latin America become linked in new ways to Africa and Asia, environmental historians should be looking in new directions when drawing connections and comparisons. At the same time, our narrative and analytical lenses have to be constantly zooming in and out so as not to lose the ability to focus on the localities where most of life, human and otherwise, plays out.

We end with a small caveat: this volume does not pretend to provide answers for contemporary environmental problems: reading this book will do little to lower the lead levels of children in Villa Inflamable, conserve the Andean flamingo, or reduce atmospheric levels of carbon dioxide. However, by calling attention to the more-than-human processes that have constituted Latin America, we hope to spark conversations about the region's present and possible futures that are informed by notions of sustainability, diversity, and resiliency, rather than growth, purity, and stability. Ultimately, the only sustainable past is one that remains open to interpretation.

John Soluri is associate professor and director of global studies in the Department of History at Carnegie Mellon University. He has published *Banana Cultures: Agriculture, Consumption and Environmental Change in Honduras and the United States* (Texas, 2005), winner of the *George Perkins Marsh Award (2007)*. His research and teaching focus on transnational environmental histories of agriculture, food, energy, and the commodification of nonhumans in Latin America. In a humbling effort to practice what he preaches, he is a longtime board member of Building New Hope, an NGO that works in solidarity with urban and rural people in Central America.

Claudia Leal holds a Ph.D. in geography from the University of California at Berkeley and is associate professor in the department of history at the Universidad de los Andes in Bogotá. She has been a fellow of the Rachel Carson Center for Environment and Society in Munich and visiting professor at Stanford University and Universidad Católica de Chile. She is the author of *Landscapes of Freedom: The Building of a Postemancipation Society in Western Colombia*, and coeditor (with Carl Langebaek) of *Historias de Raza y Nación en América Latina* and *The Nature State: Rethinking the History of Conservation* (with Wilko Graf von Hardenberg, Mathew Kelly, and Emily Wakild).

José Augusto Pádua is a professor of Brazilian environmental history at the Institute of History of the Federal University of Rio de Janeiro, where he also codirects the Laboratory of History and Nature. From 2010 to 2015, he was president of the Brazilian Association of Research and Graduate Studies on Environment and Society. As a specialist on environmental history and politics, he has lectured, taught, and done fieldwork in more than forty countries. He has published many books and articles in Brazil and abroad, including *Environmental History: As If Nature Existed* (edited with John R. McNeill and Mahesh Rangarajan).

Notes

1. Javier Auyero and Déborah Swistun, "Flammable: Environmental Suffering in an Argentine Shantytown," *American Sociological Review* 73, no. 3 (2008): 357–379, 366.
2. Javier Auyero and Deborah Alejandra Swistun, *Flammable: Environmental Suffering in an Argentine Shantytown* (Oxford: Oxford University Press, 2009).
3. Nicolás Gadano, *Historia del petróleo en la Argentina, 1907–1955: Desde los inicios hasta la caída de Perón* (Buenos Aires: Edhasa, 2006).
4. María Alejandra Cousido, "Contaminación de cuencas con residuos industriales: Estudio del caso Matanza Riachuelo, Argentina," *Revista CENIC. Ciencias Químicas* 41 (2010): 1–11.

5. José Augusto Pádua, "As bases teóricas da história ambiental," *Estudos Avançados* 24 (2010): 81–101 (Available in English on the website of the same journal); John R. Mc-Neill "The State of the Field of Environmental History," *The Annual Review of Environment and Resources* 35 (2010): 345–374.

6. Examples include Sérgio Buarque de Holanda, *Visão do paraíso: Os motivos edênicos no descobrimento e colonização do Brasil* (Rio de Janeiro: Editora José Olympio, 1959) and Gilberto Freyre, *Nordeste: Aspectos da influência da cana sobre a vida e a paisagem do nordeste do Brasil* (Rio de Janeiro: Editora José Olympio, 1937). Key contributions from the United States include Alfred Crosby, *The Columbian Exchange: Biological and Cultural Consequences of 1492* (Westport, CT: Greenwood Press, 1972) and studies completed by the Berkeley School of Geography: see Kent Mathewson and Martin S. Kenzer, eds., *Culture, Land, and Legacy: Perspectives on Carl Sauer and Berkeley School Geography* (Baton Rouge: Louisiana State University Geoscience Publications, 2003) and Claudia Leal, "Robert West: un geógrafo de la escuela de Berkeley," prologue to Robert West, *Las tierras bajas del Pacífico colombiano* (Bogotá: Instituto Colombiano de Antropología e Historia, 2000).

7. Noteworthy examples include Nicolo Gligo and Jorge Morello, "Notas para una historia ecológica de América Latina," in *Estilos de Desarrollo y Medioambiente en América Latina*, ed. Nicolo Gligo and Osvaldo Sunkel (Mexico City: Fondo de Cultura Económica, 1980); B. Mateo Martinic, "La ocupación y el impacto del hombre sobre el territorio," in *Transecta botánica de la Patagonia Austral*, ed. O. Boelcke, David M. Moore, and F. A. Roig (Buenos Aires: Consejo Nacional de Investigaciones Científicas y Técnicas, 1985); Fernando Ortiz Monasterio, *Tierra profanada. Historia ambiental de México* (Mexico City: Instituto Nacional de Antropología e Historia, Secretaría de Desarrollo Urbano y Ecología, 1987); José Augusto Pádua, ed., *Ecologia e política no Brasil* (Rio de Janeiro: Espaço & Tempo/IUPERJ, 1987); Fernando Mires, *El discurso de la naturaleza: Ecología y política en América Latina* (San José: Departamento Ecuménico de Investigaciones, 1990); Antonio E. Brailovsky and Dina Foguelman, *Memoria verde: Historia ecológica de la Argentina* (Buenos Aires: Editorial Sudamericana, 1991); Elinor Melville, *A Plague of Sheep: Environmental Consequences of the Conquest of Mexico* (Cambridge: Cambridge University Press, 1994); Guillermo Castro Herrera, *Los trabajos de ajuste y combate: Naturaleza y sociedad en la historia de América Latina* (La Habana: Casa de las Américas, Colcultura, 1995); Warren Dean, *With Broadax and Firebrand: The Destruction of Brazil's Atlantic Forest* (Berkeley: University of California Press, 1995); and Alejandro Tortolero Villaseñor, ed., *Tierra, agua y bosques: Historia y medio ambiente en el México Central* (Mexico City: Potrerillos Editores, 1996).

8. See the webpage of the Latin American and Caribbean Society for Environmental History (SOLCHA) for an overview of such developments: http://solcha.org/. Examples of publications that helped shape this community include a few edited volumes: *Estudios sobre historia y ambiente en América*, vol. 1, *Argentina, Bolivia, México, Paraguay*, ed. Bernardo García Martínez and Alba González Jácome (Mexico City: Instituto Panamericano de Geografía e Historia, El Colegio de México, 1999); *Estudios sobre historia y ambiente en América*, vol. 2, *Norteamérica, Sudamérica y el Pacífico*, ed. Bernardo García Martínez and María del Rosario Prieto (Mexico City: El Colegio de México, 2002); *Naturaleza en declive: Miradas a la historia ambiental de América Latina y el Caribe*, ed. Reinaldo Funes Monzote (Valencia: Centro Francisco Tomás y Valiente

UNED Alzira-Valencia, Fundación Historia Social, Colección Biblioteca Historia Social, 2008). See also, special issues of several Latin American and U.S. journals: *Revista Theomai* 1 (Quilmes, Argentina, 2000); *Varia Historia* 26, edited by Regina Horta Duarte (Belo Horizonte, Brazil, 2002); *Diálogos: Revista Electrónica de Historia* 4, edited by Rony Viales H. (San José, Costa Rica): 2003–2004; *Nómadas* 22, edited by Stefania Gallini (Bogotá, Colombia, 2005); *Historia Crítica* 30, edited by Claudia Leal (Bogotá, Colombia, July–Dec. 2005); *Revista de Historia* 59–60, edited by Carlos Hernández Rodríguez and Anthony Goebel McDermott (Heredia, Costa Rica, 2009); *Latin American Research Review* 46 (USA, 2011); *Hispanic American Historical Review* 92 (Durham, NC, 2012); *Revista de Historia Iberoamericana* 7 (Santiago, Chile, 2014).

9. For a concise synthesis that is particularly strong on the colonial period, see Shawn Miller, *An Environmental History of Latin America* (Cambridge: Cambridge University Press, 2007). For overviews of the field, see Mark Carey, "Latin American Environmental History: Current Trends, Interdisciplinary Insights, and Future Directions," *Environmental History* 14 (April, 2009): 221–252; Stefania Gallini, "Historia, ambiente, política: El camino de la historia ambiental en América Latina," *Nómadas* 30 (2009): 92–102; Patricia Clare, "Un balance de la historia ambiental latinoamericana," *Revista de Historia* (2009): 185–201; Lise Sedrez, "Latin American Environmental History: A Shifting Old/New Field," in *The Environment and World History*, ed. Edmund Burke III and Edward Pomeranz (Berkeley: University of California Press, 2009).

10. Dean, *With Broadax and Firebrand*; Reinaldo Funes Monzote, *From Rainforest to Cane Field in Cuba: An Environmental History since 1492* (Chapel Hill: University of North California Press, 2008 [2004]).

11. José Augusto Pádua, *Um sopro de destruição: Pensamento político e crítica ambiental no Brasil escravista, 1786–1888* (Rio de Janeiro: Jorge Zahar, 2002); John McNeill, *Mosquito Empires: Ecology and War in the Greater Caribbean, 1620–1914* (Cambridge: Cambridge University Press, 2010); Germán Palacio, *Fiebre de tierra caliente: Una historia ambiental de Colombia 1850–1930* (Bogotá: Universidad Nacional de Colombia, 2006); Eunice Sueli Nodari and João Klug, eds., *História ambiental e migrações* (São Leopoldo: Oikos, 2012); John Soluri, *Banana Cultures: Agriculture, Consumption, and Environmental Change in Honduras and the United States* (Austin: University of Texas Press, 2005); Daviken Studnicki-Gizbert and David Schecter, "The Environmental Dynamics of a Colonial Fuel-Rush: Silver Mining and Deforestation in New Spain 1522–1810," *Environmental History* 15 (January 2010): 94–119.

12. Shawn Miller, *Fruitless Trees: Portuguese Conservation and Brazil's Colonial Timber* (Stanford: Stanford University Press, 2000); Claudia Leal and Eduardo Restrepo, *Unos bosques sembrados de aserríos: Historia de la extracción maderera en el Pacífico colombiano* (Medellín: Universidad de Antioquia, Universidad Nacional sede Medellín, Instituto Colombiano de Antropología e Historia, 2003); Pablo Camus, *Ambiente, bosques y gestión forestal en Chile, 1541–2005* (Santiago: Centro de Investigaciones Barros Arana de la Dirección de Bibliotecas, Archivos y Museos, Lom Editores, 2006); Anthony Goebel, "Los bosques del 'progreso': Explotación forestal y régimen ambiental en Costa Rica: 1883–1955" (Ph.D. diss., University of Costa Rica, 2013); Adrián Gustavo Zarrilli, "El oro rojo. La industria del tanino en la Argentina (1890–1950)," *Silva Lusitana* 16 (2008): 239–259.

13. For a study of a prominent European forester in Chile, see Fernando C. Hartwig, *Federico Albert: Pionero del desarrollo forestal en Chile* (Talca: Editorial Universidad de Talca, 1999). There are many historical studies of Indian forestry; important works include S. Ravi Rajan, *Modernizing Nature: Forestry and Imperial Eco-Development, 1800–1950* (Oxford: Oxford University Press, 2006) and K. Sivaramakrishnan, *Modern Forests: Statemaking and Environmental Change in Colonial Eastern India* (Stanford: Stanford University Press, 1999). The literature on U.S. forestry is also extensive; some key works include Thomas R. Cox, The Lumberman's Frontier: Three Centuries of Land Use, Society, and Change in America's Forests (Corvallis, OR: University of Oregon Press, 2010) and Char Miller, *Gifford Pinchot and the Making of Modern Environmentalism* (Washington, DC: Island Press, 2004).

14. Christopher Boyer, *Political Landscapes: Forests, Community, and Conservation in Mexico* (Durham: Duke University Press, 2015) and Thomas Klubock Miller, *La Frontera: Forests and Ecological Conflict in Chile's Frontier Territory* (Durham: Duke University Press, 2014). See also Lise Sedrez, "Rubber, Trees and Communities: Rubber Tappers in the Brazilian Amazon in the Twentieth Century," in *A History of Environmentalism*, ed. Marco Armiero and Lise Sedrez (London: Bloomsbury Academic, 2014) and Claudia Leal, *Landscapes of Freedom: Building a Postemancipation Society in the Rainforests of Western Colombia* (Tucson, AZ: University of Arizona Press, 2018).

15. Funes Monzote, *From Rainforest*; Soluri, *Banana Cultures*; Adrián Gustavo Zarrilli, "Capitalism, Ecology and Agrarian Expansion in the Pampean Region (1890–1950)," *Environmental History* 6 (2002): 560–583; Sterling Evans, *Bound in Twine: The History and Ecology of the Henequen-Wheat Complex for Mexico and the American and Canadian Plains, 1880–1950* (College Station: Texas A&M University Press, 2007); Stefania Gallini, *Una historia ambiental del café en Guatemala: La Costa Cuca entre 1830 y 1902* (Guatemala: Asociación para el Avance de las Ciencias Sociales en Guatemala, 2009); Rony Viales Hurtado and Andrea Montero Mora, "Una aproximación al impacto ambiental del cultivo del banano en el Atlántico/Caribe de Costa Rica, (1870–1930)," in *Costa Rica: Cuatro ensayos de historia ambiental*, ed. Ronny J. Viales Hurtado and Anthony Goebel McDermott (Costa Rica: Sociedad Editora Alquimia 2000, 2011), 85–124; Stuart McCook, "Las epidemias liberales: Agricultura, ambiente y globalización en Ecuador, 1790–1930," in *Estudios sobre historia y ambiente en América Latina*, vol. 2, *Norteamérica, Sudamérica, y el Pacífico*, ed. Bernardo García Martínez and María del Rosario Prieto (Mexico City: El Colegio de México/Instituto Panamericano de Geografía e Historia, 2002), 223–246.

16. Myrna Santiago, *The Ecology of Oil: Environment, Labor, and the Mexican Revolution, 1900–1938* (Cambridge: Cambridge University Press, 2006); Gregory Cushman, *Guano and the Opening of the Pacific World* (Cambridge: Cambridge University Press, 2013); Jó Klanovicz, "Artificial Apple Production in Fraiburgo, Brazil, 1958–1989," *Global Environment* 5 (2010): 39–70; Shawn Van Ausdal, "Pasture, Power, and Profit: An Environmental History of Cattle Ranching in Colombia, 1850–1950," *Geoforum* 40 (2009): 707–719; idem, "Productivity Gains and the Limits of Tropical Ranching in Colombia, 1850–1950," *Agricultural History* 86, no. 3 (2012): 1–32; Robert Wilcox, *Cattle in the Backlands: Matto Grosso and the Evolution of Ranching in the Brazilian Tropics* (Austin: University of Texas Press, 2017); Daniel Renfrew, "New Hazards and Old Disease: Lead Contamination and the Uruguayan Battery Industry," in *Dangerous*

Trade: Histories of Industrial Hazard Across a Globalizing World, ed. Christopher Sellers and Joseph Melling (Philadelphia: Temple University Press, 2012): 99–111.

17. For examples of new directions in the field related to these changes, see Natalia Milanesio, "Liberating the Flame: Natural Gas Production in Peronist Argentina," *Environmental History* 18, no. 3 (2013): 499–522; and Mikael Wolfe, *Watering the Revolution: An Environmental and Technological History of Agrarian Reform in Mexico* (Durham: Duke University Press, 2017).

18. Lane Simonian, *Defending the Land of the Jaguar: A History of Conservation in Mexico* (Austin: University of Texas Press, 1995); Sterling Evans, *The Green Republic, A Conservation History of Costa Rica* (Austin: University of Texas Press, 1999); Seth Garfield, "A Nationalist Environment: Indians, Nature, and the Construction of the Xingu National Park in Brazil," *Luso-Brazilian Review* 41, no. 1 (2004): 139–167; Gregory Cushman, "'The Most Valuable Birds in the World': International Conservation Science and the Revival of Perú's Guano Industry," *Environmental History* 10 (2005): 477–509; José Augusto Drummond and José Luiz Andrade Franco, *Proteção à natureza e identidade nacional no Brasil, anos 1920–1940* (Rio de Janeiro: Ed. Fiocruz, 2009); Emily Wakild, *Revolutionary Parks: Conservation, Social Justice, and Mexico's National Parks, 1910–1940* (Tucson: The University of Arizona Press, 2011); Federico Freitas, "A Park for the Borderlands: The Creation of Iguacu National Park in Southern Brazil, 1880–1940," *Revista de Historia Iberoamericana* 7, no. 2 (2014); Claudia Leal, "Conservation Memories: Vicissitudes of a Biodiversity Conservation Project in the Rainforests of Colombia, 1992–1998," *Environmental History* 20 (2015): 368–395.

19. Regina Horta Duarte, *Activist Biology: The National Museum, Politics and Nation Building in Brazil*, trans. Diane Grosklaus Whitty (Tucson: the University of Arizona Press, 2016); Camilo Quintero, *Birds of Empire, Birds of Nation: A History of Science, Economy, and Conservation in United States–Colombia Relations* (Bogotá: Ediciones Uniandes, 2012).

20. Stuart McCook, *States of Nature: Science, Agriculture, and Environment in the Spanish Caribbean, 1760–1940* (Austin: University of Texas Press, 2000).

21. Bogotá and Mexico City have received more attention that other urban centers. Work in Colombia has been largely carried out by students: María Lucía Guerrero Farías, "Pintando de Verde a Bogotá: Visiones de la naturaleza a través de los parques del Centenario y de la Independencia, 1880–1920," *HALAC* 1, no. 2 (2012); María Clara Torres Latorre, "El alcantarillado subterráneo como respuesta al problema sanitario de Bogotá, 1886–1938," and Laura C. Felacio Jiménez, "Los problemas ambientales en torno a la provisión de agua para Bogotá, 1886–1938," in *Semillas de historia ambiental*, ed. Stefania Gallini (Bogotá: Jardín Botánico de Bogotá, Universidad Nacional de Colombia, 2015). On Medellín, see Bibiana Preciado, *Canalizar para industrializar: La domesticación del río Medellín en la primera mitad del siglo XX* (Bogotá: Ediciones Uniandes, 2015). Americans have led the way in the study of Mexico City: Vera Candiani, *Dreaming of Dry Land: Environmental Transformation in Colonial Mexico City* (Palo Alto: Stanford University Press, 2014) and Mathew Vitz, "The Lands with Which We Shall Struggle": Land Reclamation, Revolution, and Development in Mexico's Lake Texcoco Basin, 1910–1950," *Hispanic American Historical Review* 92, no. 1 (2012): 41–71.

22. Some notable exceptions include Mark Carey, *In The Shadow of Melting Glaciers: Climate Change and Andean Society* (New York: Oxford University Press, 2010); Miche-

line Cariño and Mario Monteforte, *El primereEmporio perlero sustentable del mundo: La Compañía Criadora de Concha y Perla de la Baja California S.A., y sus perspectivas para Baja California Sur* (México: UABCS, SEP, FONCA-CONACULTA, 1999); Molly Warsh, *American Baroque: Pearls and the Nature of Empire 1492–1700* (Chapel Hill: UNC Press, forthcoming); and Mikael D. Wolfe, *Watering the Revolution.*

23. Rafael Cabral Cruz and Demétrio Luis Guadagnin, "Uma pequena história ambiental do Pampa: Proposta de uma abordagem baseada na relação entre perturbação e mudança," in *A sustentabilidade da região da Campanha-RS: Práticas e teorias a respeito das relações entre ambiente, sociedade, cultura e políticas públicas,* ed. Benhur Pinós da Costa, João Henrique Coos and Mara Eliana Graeff Dickel (Santa Maria: Ed. UFSM, 2010), 155–179; and Adrián Gustavo Zarrilli, "Capitalism, Ecology and Agrarian Expansion in the Pampean Region, 1890–1950," *Environmental History* 6 (Oct. 2001): 561–583.

24. On arid environments, see Cynthia Radding, *Wandering Peoples: Colonialism, Ethnic Spaces, and Ecological Frontiers in Northwestern Mexico, 1700–1850* (Durham: Duke University Press, 1997) and Micheline Cariño, Aurora Breceda, Antonio Ortega and Lorella Castorena, eds., *Evocando al edén: Conocimiento, valoración y problemática del Oasis de los Comondú* (Mexico City: Conacyt; Barcelona: Icaria, 2013).

25. Andrew Sluyter, "Recentism in Environmental History on Latin America," *Environmental History* 10 (Jan. 2005): 91–93; John R. McNeill and Peter Engelke, *The Great Acceleration: An Environmental History of the Anthropocene since 1945* (Cambridge, MA: Harvard University Press, 2016).

26. Candiani, *Dreaming of Dry Land*; and Barbara Mundy, *The Death of Aztec Tenochtitlán: The Life of Mexico City* (Austin: University of Texas Press, 2015). Previous works include Rosalva Loreto, *Una vista de ojos a una ciudad novohispana: Puebla de los Ángeles en el siglo XVIII.* (Puebla: CONACYT, 2008); Studnicki-Gizbert and Schecter, "The Environmental Dynamics of a Colonial Fuel-Rush." On colonial Peru, see Gregory T. Cushman, "The Environmental Contexts of Guaman Poma: Interethnic Conflict over Forest Resources and Place in Huamanga, 1540–1600," in *Unlocking the Doors to the Worlds of Guaman Poma and His Nueva Crónica,* ed. Rolena Adorno and Ivan Boserup (Copenhagen: Museum Tusculanum Press, 2015). On colonial Brazil, see Diogo Cabral, *Na presença da floresta: Mata Atlântica e história colonial* (Rio de Janeiro: Garamond, 2014).

27. Joana Bezerra, *The Brazilian Amazon: Politics, Science and International Relations in the History of the Forest* (New York: Springer, 2015); William Denevan, "The Pristine Myth: The Landscape of the Americas in 1492," *Annals of the Association of American Geographers* 82 (1992): 369–385; Shawn Miller, *An Environmental History of Latin America.*

28. Paul Sutter, "Reflections: What Can U.S. Environmental Historians Learn from Non-U.S. Environmental Historiography?" *Environmental History* 8 (Jan 2003): 109–129.

29. Alcida Ramos, *O papel político das epidemias: O caso Yanomani,* Série Antropologia, no. 153 (Brasília: Universidade de Brasília, 1993); Mateo Martinic, *Historia de la Región Magallánica, Punta Arenas* (Chile: Universidad de Magallanes, 1992); Libby Robin, "Australia in Global Environmental History," in *A Companion to Global Environmental History,* ed. J. R. McNeill and E. S. Mauldin (Oxford: Wiley-Blackwell, 2012), 182–195;

Libby Robin and Tom Griffiths, "Environmental History in Australasia," *Environment and History* 10, no. 4 (2004): 439–474.

30. Nicholas A. Robbins, *Mercury, Mining, and Empire: The Human and Ecological Cost of Colonial Silver Mining in the Andes* (Bloomington: Indiana University Press, 2011); Carlos Serrano Bravo, *Historia de la minería andina boliviana, siglos XVI–XX* (Potosí, 2004), accessed 6 October 2017, http://www.unesco.org.uy/phi/biblioteca/files/origi nal/370d6afed30afdca14156f9b55e6a15e.pdf.

31. Studnicki-Gizbert and Schecter, "The Environmental Dynamics of a Colonial Fuel-Rush."

32. John McNeill, *Mosquito Empires*; David Watts, *The West Indies: Patterns of Development, Culture and Environmental Change since 1492* (Cambridge: Cambridge University Press, 1987); Shawn Miller, "Fuelwood in Colonial Brazil: The Economic and Social Consequences of Fuel Depletion for the Bahian Recôncavo, 1549–1820," *Forest and Conservation History* 38 (1994): 181–192.

33. Candiani, *Dreaming of Dry Land.*

34. Crosby, *The Columbian Exchange*; Martha Few and Zeb Tortorici, *Centering Animals in Latin American History* (Durham: Duke University Press, 2013).

35. Adrian Franklin, *Animal Nation: The True Story of Animals and Australia* (Syndey: University of New South Wales Press, 2006).

36. Yadvinder Malhi, Toby A. Gardner, Gregory R. Goldsmith, Miles R. Silman, and Przemyslaw Zelazowski, "Tropical Forests in the Anthropocene," *Annual Review of Environment and Resources* 39 (2014): 125–59; Michael Williams, "A New Look at Global Forest Histories of Land Clearing," *Annual Review of Environment and Resources* 33 (2008): 345–367.

37. Estimates based on the Food and Agriculture Organization of the United Nations (FAO), *State of the World's Forests* (Rome, 2011) and Michael Williams, *Deforesting the Earth: From Prehistory to Global Crisis* (Chicago: University of Chicago Press, 2002), 335, 397.

38. Marcy Norton, "The Chicken or the Iegue: Human-Animal Relationships and the Columbian Exchange," *American Historical Review 120, no.* 1 (2015): 28–60; Rebecca Earle, "If You Eat Their Food . . .": Diets and Bodies in Early Colonial Spanish America," *American Historical Review* 115 (2010): 688–713.

39. Eduardo Gudynas, "Buen Vivir: Germinando alternativas al desarrollo," *América Latina en Movimiento* 462 (2011): 1–20; idem, "Buen Vivir: Today's Tomorrow," *Development* 54, no. 4 (2011): 441–447.

40. James E. Sanders, *The Vanguard of the Atlantic World* (Durham: Duke University Press, 2014).

41. Camilo Quintero, *Birds of Empire, Birds of Nation.*

42. Santiago, *The Ecology of Oil.*

43. Mathew Kelly, Claudia Leal, Emily Wakild, and Wilko Graf von Hardenberg, "Introduction," in *The Nature State, Rethinking the History of Conservation,* ed. Wilko Graf von Hardenberg, Matthew Kelly, Claudia Leal, and Emily Wakild (London: Routledge, 2017).

44. Case Watkins, "African Oil Palms: Colonial Socioecological Change and the Making of an AfroBrazilian Landscape in Bahia, Brazil," *Environment and History* 21 (2015):

13–42; Judith A. Carney and Nicolas Rosomoff, *In the Shadow of Slavery: Africa's Botanical Legacy in the Atlantic World* (Berkeley: University of California, 2009).

45. John Soluri, "Something Fishy: Chile's Blue Revolution: Commodity Diseases and the Problem of Sustainability," *Latin American Research Review* 46 (2011): 55–81; Stuart McCook and John Vandermeer, "The Big Rust and the Red Queen: Long-term Perspectives on Coffee Rust Research," *Phytopathology Review* 105, no. 9 (2015): 1164–73.

46. Warren Dean, "The Green Wave of Coffee: Beginnings of Agricultural Research in Brazil (1885–1900)," *Hispanic American Historical Review* 69 (1989): 91–115; Stuart McCook, *States of Nature*; Soluri, *Banana Cultures*.

47. Mauricio Folchi, "Una aproximación a la historia ambiental de las labores de beneficio en la minería del cobre en Chile, siglos XIX y XX" (Ph.D. diss., Universidad Autónoma de Barcelona, Bellaterra, 2003); Christoper F. Jones, *Routes of Power: Energy and Modern America* (Harvard: Harvard University Press, 2014).

48. Nontropical biomes include semiarid deserts (11.3 percent); temperate grasslands (7.9 percent); and temperate forests (2.0 percent). From a historical perspective, these biomes have been extremely important. For example, the vast temperate grasslands, characterized by extremely deep, fertile soils, form the basis of an agropastoral economy that made Argentina one the world's wealthiest nations in the early twentieth century.

49. United Nations Environment Program, *State of Biodiversity in Latin America and the Caribbean,* 2010.

50. Nancy Lee Stepan, *Picturing Tropical Nature* (London: Reaktion Books, 2001), 17–18; see also David Arnold, *The Problem of Nature: Environment, Culture, and European Expansion* (Cambridge: Blackwell, 1996).

51. John McNeill, *Mosquito Empires.*

52. Glenn A. Chambers, *Race, Nation, and West Indian Immigration to Honduras, 1890–1940* (Baton Rouge: Louisiana State University Press, 2010); Aviva Chomsky and Aldo Laurie-Santiago, eds., *Identity and Struggle at the Margins of the Nation-State: The Laboring Peoples of Central America and the Hispanic Caribbean* (Durham: Duke University Press, 1998).

53. Shawn Van Ausdal, "Reimagining the Tropical Beef Frontier and Nation in the Early Twentieth Century Colombia" in ed. Gordon M. Winder and Andreas Dix, *Trading Environments: Frontiers, Commercial Knowledge, and Environmental Transformation, 1750–1990* (New York: Routledge, 2016).

54. Literature on the "Anthropocene" is growing swiftly. Among the most widely cited works in English is Dipesh Chakrabarty, "The Climate of History: Four Theses," *Critical Inquiry* 35 (Winter 2009), 197. For Latin American perspectives, see Astrid Ulloa, "Dinámicas ambientales y extractivas en el siglo XXI: ¿es la época del antropoceno o capitaloceno en Latinoámerica?" *Desacatos* 54 (May–Aug. 2017): 58–73 and José Augusto Pádua, "Brazil in the History of the Anthropocene" in *Brazil in the Anthropocene: Conflicts between Predatory Development and Environmental Policies* (London: Routledge, 2017).

CHAPTER 1

Mexico's Ecological Revolutions

Chris Boyer and Martha Micheline Cariño Olvera

The state has acted as the primary mediator between nature and society in Mexico. This is not because its power and stability have made possible control of the social or economic practices of people, businesses, or bureaucratic entities within its borders. Nor has it been a powerful state characterized by its ability to direct the country's political, economic, and ecological destiny. Rather, the Mexican state's influence is the result of governments and changing political circumstances that have created opportunities for various groups of actors in different historical periods, with profound consequences for the nation's population and territory. The state has also experienced radical changes due to the establishment of militant liberalism in the nineteenth century, the social revolution of 1910, and the resurgence of development liberalism beginning in the mid-twentieth century. Transitions from one period to another nearly always have been sudden and unforeseen. In other words, the country has not only experienced a series of political revolutions, but also various "ecological revolutions," in the sense proposed by Carolyn Merchant: dramatic changes in the way people conceive and make use of their surroundings and the country's so-called natural resources.[1]

These ecological revolutions arose in a context of growing—though discontinuous—commodification of nature and in increasingly precarious environmental conditions. Nevertheless, in many specific cases they have given rise to sustainable uses, and even to *new* sustainable uses, of territory and resources.

Beginning in 1854, when the state began to consolidate, up to the present, Mexican territory went through three stages that led to ecological revolutions: the political-liberal movement that erupted in Ayutla in 1854, the social revolution of 1910, and the so-called Green Revolution that began in 1943 and that presaged the neoliberal period beginning in 1992. None of these revolutions completely broke with prior ecological, social, and political conditions, yet each generated new circumstances in which each social group that used natural resources came to new understandings about their surroundings and were likewise affected by changes in the environment. Each revolution left long-term social and ecological footprints, creating the context that led to the following revolution. But each revolution also created countercurrents, that is, historical dynamics capable of counteracting the effects of the revolution itself which, in the long term, constituted unexpected openings for groups and individuals to value and use nature, creating new forms of social organization.

Nineteenth-century liberalism cemented private property's hegemony, opening new investment possibilities leading to the increasing commodification of natural resources. Thus, it contributed to the neocolonial extractive regime that characterized the regime of president Porfirio Díaz (1876–1911, known as the *Porfiriato*), which was characterized by the sacking of minerals, water, forests, and oil by predominantly foreign interests. The social revolution reorganized landholding and permitted its collective use, though neither private property nor the intensive use of natural resources was eradicated. These were subject to a new period of exploitation with the Green Revolution, whose ostensible goal was to promote small-scale agriculture but ultimately favored private landholders and commercial production. As the years passed and with the advent of neoliberalism, market forces became stronger, putting an end to the accomplishments of the 1910 revolution and producing a new wave of commodification in fields, forests, rivers, seas, mines, and on seashores. The commodification of nature has gone hand in hand with an increase in the dispossession of peasants, fishermen, and indigenous communities, thereby sharpening social inequality. The cities overflow with migrants who are hard put to find work even in the informal economy. Insecurity grows, as does pollution in both urban and rural areas.

This situation explains the increase in popular mobilizations of people fed up with the growing power of the transnational corporations that increasingly have acquired control of the nation's natural resources. In response to the widespread reprivatization of land and aquatic ecosystems experienced in the country, since the year 2000 an unprecedented phenomenon has appeared: the slow but unmistakable strengthening of a rural-urban alliance proposing alternatives to the overexploitation of natural resources and the use of genetically modified organisms, and opposing the dismantling of *campesino* agriculture. These same movements seek new ways to reconstruct the country on the basis of its biocultural wealth and diversity.

Biocultural Sketch

Mexico is the world's eleventh most populated country, with more than 119 million inhabitants as of 2014. It is categorized as one of the world's twelve megadiverse countries according to Conservation International. Thirteen percent of the nation's territory is located within 177 protected areas, including biosphere reserves, national parks, natural monuments, natural resource pro-

Map 1.1. Mexico's main biomes.

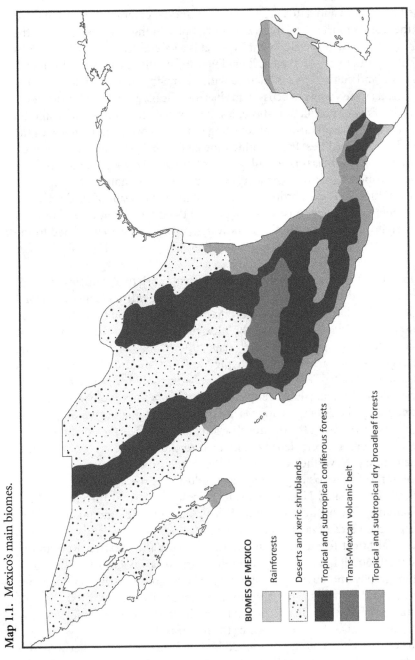

BIOMES OF MEXICO

Rainforests

Deserts and xeric shrublands

Tropical and subtropical coniferous forests

Trans-Mexican volcanic belt

Tropical and subtropical dry broadleaf forests

Source: Anthony Challenger, *Utilización y conservación de los ecosistemas terrestres de México. Pasado, presente y futuro* (Mexico City: UNAM/CN-CUB, 1998), figures 6.2 (p. 278) and 6.3 (p. 280). Simplified version by Camilo Uscátegui.

tected areas, flora and fauna protected areas, and sanctuaries. In terms of GDP, it is the world's fourteenth largest economy, but is in the sixty-first place in the terms of the Human Development Index. It is a federal republic composed of a capital city and thirty-one states. It has been and continues to be a rich country in natural and cultural terms, blessed with five major biomes, as illustrated in map 1.1. Its wide variety of ecosystems has historically translated into an enormous diversity of production strategies. One of the most important examples is the ancient peasant custom of selecting grains of corn from plants with the most desirable qualities. For the nine thousand years since the domestication of *Zea mays* in the Balsas river valley, maize has spread throughout Mesoamerica, and farmers have produced forty-one landraces and more than a thousand local varieties. This extraordinary agrodiversity is the result of seed selection by farmers looking for those best adapted to the microclimatic conditions of their territories. As a result, the agrodiversity of corn is closely related to the diversity of the country's indigenous societies, which currently speak no fewer than sixty-seven autochthonous languages.[2]

Unsurprisingly for a country with such a wide variety of climates and cultures, it is divided into myriad biocultural regions with socioenvironmental characteristics that have marked both their own history and their place in the country's evolution. Beginning at the Mexico–United States border, the great Mexican north is an arid space that opens toward the northwest, toward the long Baja California Peninsula and the Gulf of California—which Jacques Cousteau once called "the world's aquarium"—the only sea owned by a single nation. Most of the north is occupied by the Sonora Desert phytogeographic region, one of the Americas' four largest deserts, but one outstanding for its rich biodiversity. Given the territory's aridity and vast size, the northern states have a low demographic density compared to those of the center and south of the country. Nevertheless, it is also there, and especially near the border, where some of Mexico's largest and most industrial cities are located: Tijuana, Mexicali, Hermosillo, Nogales, Ciudad Juárez, Monterrey, Torreón, Saltillo, and Tampico. The north is a region of vast plains and high mountains. In the former, large herds of cattle once roamed the enormous haciendas that were the special target for agrarian distribution during the Mexican Revolution. Since the 1960s, it has been the Green Revolution's favored territory due to its flat topography and abundant water sources for agroindustrial development. In the latter, logging and mineral mining—especially copper in Cananea in Sonora and El Boleo in Baja California Sur—have driven a dynamic economy and polluted soil and water ever since the nineteenth century.

The center of the country is marked by highlands where the Sierra Madre Occidental and Oriental cordilleras come together. Here the old colonial cities are located, many of them World Heritage sites: Morelia, Guanajuato, San Luis Potosí, Querétaro, Puebla, Tlaxcala, and Mexico City. Toward the south, the cordilleras join in the Mixtec Range and are interrupted by a depression that forms the Tehuantepec Isthmus. The highlands and their valleys have been densely populated since the pre-Hispanic epoch by different indigenous groups, who in the colonial era were often forced to work in the region's gold and silver mines. As population and power centers, these sites have been the stage for important events in the nation's history, especially during the Independence War, the War of the Reform—as the liberal revolution is known in Mexican historiography—and the Mexican Revolution. For example, the independence struggle (1810) was planned in Querétaro, and it was there that Maximilian of Hapsburg faced the firing squad (1867) and the current Constitution of 1917 was written and signed. The Mexico City metropolitan area, with more than twenty million residents, is among the world's largest megacities. In spite of pollution and constant changes in land use, the central mountains and valley still have large conifer forests, which, in addition to providing lumber and cellulose, are partially protected in parks and reserves. The area is also rich in archeological sites and many communities whose residents maintain their indigenous culture.

In the east, under the tropical influence of the Gulf of Mexico, are the Huasteca Mountains with their forest microclimates. In the sierras of Puebla and Veracruz are the coffee-producing regions—organic, for the most part—and the areas that still contain the greatest corn diversity. In the Gulf of Mexico lie the largest oil deposits, and this black gold continues to lead Mexico's exports. In the west, the Pacific coast forms a rich plain, and the coastal area includes an abundance of lagoons, mangroves, beaches with palm stands, and tourist centers, including those with a long history, such as Acapulco, as well as new sites like Huatulco. Also in this area are Mexico's largest ports: Ensenada, Mazatlán, Manzanillo, Lázaro Cárdenas, Salina Cruz, and Puerto de Chiapas.

In the south are the states with the widest biocultural diversity: Guerrero, Oaxaca, Chiapas, Tabasco, Campeche, Yucatán, and Quintana Roo. The last three form the Yucatán Peninsula where the calcareous plain is the site of an intricate complex of underground rivers that produce natural open air wells known as *cenotes*. Here too rises the Petén forest, where Mayan ruins and communities abound. The seven southern states contain exuberant tropical ecosystems whose biodiversity varies from the high mountains to the coast. They also contain invaluable archaeological wealth at numerous large sites of the Mixtec, Zapotec, Olmec, and Maya cultures. There are also beautiful beaches on the Pacific Ocean, the Gulf of Mexico, and the Caribbean Sea, with countless coastal lagoons, mangroves, bays, islands, and tropical reefs, including the Mesoamerican Reef, which ranks as the world's second largest, and nearly sixty protected areas (35 percent of Mexico's total). These are also the states with the largest number of indigenous groups and languages. This wealth has constantly attracted those who seek to exploit the land, the coast, the seas, and the subterranean minerals. Since colonial times, large landed estates have been concentrated in this area, monocultures have been introduced, tons of timber have been extracted, enormous hydroelectric dams have been built, and the coast has been plagued by resorts closed to the majority of Mexicans whose annual income would not pay for a single night's stay.

Mexico's location within world geopolitics has been a key factor in its environmental history. Ever since the colonial era, the country's two ocean fronts have joined Asia to Europe, facilitating colonial Spain's interoceanic communication and making Mexico the most important colonial administrative center; meanwhile, Mexico's natural resource wealth (silver, especially, but also other precious commodities such as cochineal, pearls, and cacao) had an influence on the location and development of human settlements and the institutionalization of an economy based on extractivism. Since the mid-nineteenth century, proximity to the United States has been a decisive factor in the development of another productive wave based on mining, livestock, and large-scale agriculture, especially in the north. Today, the two countries share the world's longest terrestrial border between the global North and South. The border, physically marked by a fence and by the Río Grande has become one of the globe's most dynamic frontier areas. Tons of merchandise—legal and illegal—cross the border between the two trading partners. Millions of people also cross, including both documented and undocumented migrants hailing from Mexico but also from Central and South America.

Mexico is a country of contrasts and contradictions, which have turned into socioenvironmental conflicts whose historical trajectories have culminated in the ecological revolutions analyzed in this article.

The Political-Liberal Revolution: From Mexican Independence to the Fall of the *Porfiriato*

The extractive regime of the colonial economy was destroyed by eleven years of armed movements, beginning with the rebellion led by Miguel Hidalgo in 1810 and ending with Agustín de Iturbide's military uprising in 1821. The major mines (in Zacatecas and San Luis Potosí) flooded, and a half century passed before the mining industry recovered. The sector's decline temporarily ended the environmental damage caused by colonial mining, allowing forests to recover for several decades. Ever since the sixteenth century, exploitation of precious metals had caused deforestation around mineral deposits, the extraction and refining of which required increasing quantities of wood. Mining was also the force behind the development of businesses providing supplies, such as the charcoal haciendas and the small-scale charcoal sellers.[3] As Robert C. West demonstrated more than sixty-five years ago, the variety and scale of inputs that the mines required multiplied their ecological impact, and this influenced the location of human settlements and the use (and overuse) of forest, hydraulic, and agricultural resources, and thus determined the socioenvironmental history of various desert areas in the north of the country.[4] With the outbreak of the independence wars at the beginning of the nineteenth century, insurgent and royal armies sacked the haciendas in Bajío (an area rich in grain production, located in parts of the states of Guanajuato, Querétaro, Jalisco, and Michoacán). As Eric Van Young has shown, the rural population also attacked haciendas during the war.[5] Economic damage, together with the disappearance of the mining market, bankrupted many properties and led to the creation of small agricultural properties known as *ranchos*. These family-run farms came to predominate in regions where grain production diminished and created conditions for the spread of livestock, which in turn had important ecological implications in the form of deforestation, soil compaction, and increased erosion.[6]

The prolonged independence movement undermined the central government's ability to rule, with both social and ecological consequences, especially in the border areas of the new nation-state. In Yucatán, the violent caste war broke out in 1847 as a result of generations of indigenous oppression and competition among local elites. A significant number of Maya communities in the present-day state of Quintana Roo continued fighting until 1883. The rebellion destroyed commercial properties, displaced thousands of persons, and led to the outbreak of epidemics.[7] Lack of security prompted illegal logging of mahogany both on the coast of Tabasco and the border with Guatemala, particularly by British interests.[8] In the north, a series of uprisings by ethnic groups such as the Comanches and the Apaches created constant uncertainty in the states of Chihuahua, Sonora, and Coahuila. Northern indigenous groups settled in the arid, sparsely populated space to establish an economy (and a military way of life) based on raising horses.[9]

Economic catastrophe produced political instability, leaving Mexico vulnerable to two groups of imperialist adventurers: North American (1846–48) and French (1862–67). Not surprisingly, the population grew very slowly, from 6.8 million in 1828 to 8.4 million in 1868, an annual rate of 0.6 percent.[10]

The liberal political revolution began in 1854 at the hands of Ignacio Comonfort and Benito Juárez and others, culminating with the Laws of Reform (1855–57) and the War of Reform (1857–61), which eventually put an end to postindependence instability. The liberal revolution was consolidated through a series of initiatives begun during the Restored Republic (1867–76) and ending with the long regime of Porfirio Díaz (1876–1911), an era known as *Porfiriato* that was characterized by authoritarian political stability and an extended phase of economic growth between 1880 and 1905. A liberal state par excellence, the Porfiriato's development model was based on foreign (especially U.S.) investment in extractive industries, railroads, and other infrastructure, in the agriculture and livestock sector, and in financial institutions. This model generated enormous fortunes, not only for foreign investors but also for Mexican businessmen and owners of large estates who had capital and political ties. But the social and ecological cost of economic growth was extremely high. Thousands of people lost their communal lands to commercial agriculture as well as to the forestry, mining, oil, and transport sectors.

The railroads were the backbone of the extractive model and they grew at an impressive rate: from 650 kilometers of track on the eve of the Porfiriato, the figure rose to 25,000 kilometers around 1910, almost half of which were owned by North American companies. The expansion of railroads threatened lands of indigenous peoples as investors bought up territory where new lines were expected to be opened. This process, which John Coatsworth has called "anticipatory dispossession," led to at least fifty-five local uprisings, beginning in 1877.[11] Forests were also affected by the felling of timber needed to build trestles and ties, water stops, and as fuel for many steam engines prior to the universalization of coal power around 1930.[12] In some cases, railroads established their own lumber companies. The Ferrocarril Noroeste de México, a Canadian-U.S. consortium in Chihuahua, built two enormous sawmills with cutting-edge equipment. The Madera Company received a generous concession and rented additional land for a total of more than 670 thousand hectares of virgin forest. Although only one sawmill ever operated with any regularity, it succeeded in producing a half million board feet daily by 1909. Almost all the timber came from forests easily accessible by train; the rapid devastation of available timber contributed to the consortium's collapse within a decade.[13] In addition to the railroads, Porfirian development affected forests in other ways. Lumber companies were created in various areas in the center of the country to provide lumber to mines and to meet the increased urban demand for construction materials and charcoal.[14] Commercialization of forest resources led to their privatization or delivery to logging interests in the form of concessions, all of which impeded the access communities had traditionally enjoyed to the woods.

In most cases, railroads were built to transport the goods that had structured the Mexican economy for four centuries: minerals. The Porfirian mining renaissance was made possible by foreign investment, new technologies, political stability, and North American demand for industrial metals such as copper. Dozens of mines were opened in the north; the largest were El Boleo in Baja California Sur and Cananea in Sonora. Mining generated new settlements and an increasing demand for lighting for cities and mines, for which sperm whale oil was used; hunting for this species by North American whalers drove it to near-extinction.[15] In the center of Mexico, illumination was provided by turpentine, itself a distillate of pine tree resin, which proved more sustainable.

Population movement toward the border drove the agriculture-livestock industry and the resulting environmental transformation. It also increased water and soil pollution from industrial wastes.[16]

In addition to forests, other collectively owned village resources were privatized, with serious implications for the environment and society. In particular, the commercialization of agriculture led to important changes in water use. In the state of Morelos, more than 100 percent of the water's real disponibility from the Higuerón River was leased to sugar haciendas.[17] Modernized Michoacán haciendas drained wetlands that were the source of reeds and fish for surrounding indigenous communities. The reorientation of water that previously had been, if not communal, at least shared between haciendas and indigenous communities produced social tensions that became evident during the 1910 revolution.[18]

The desire to control water was manifest especially in the country's capital, where engineers hoped to control constant flooding. After a decade of work, they managed to almost completely drain Lake Texcoco by building a huge canal and a tunnel that transferred the water out of the Valley of Mexico and into the Valley of Mezquital. But that civil engineering victory did not solve the flooding problem. Draining the watershed led to shortages and stimulated extraction of groundwater which, in addition to being unsustainable, undermined the capital's subterranean foundation.[19]

The Porfirian economic boom intensified in land use, particularly of irrigated fields, which accounted for approximately 13 percent of all agricultural land in 1907. In central areas, cutting-edge equipment came into use, such as steam-powered tractors, although these technologies were often inappropriate for the Mexican climate.[20] Nevertheless, the great majority of rural dwellers were small-scale farmers, as shown in illustration 1.1.

Many social consequences of the Porfirian agricultural revolution have been studied in detail, especially the privatization of communal lands brought about by the Lerdo Law (1856) and its rigid application to indigenous commons during the final decades of the nineteenth century.[21] The ecological effects are not as well known, though it is clear that monocultures appeared in many areas of the country, including La Laguna (cotton), Yucatan (henequen), and Morelos (sugar). In Yucatan, the demand for twine to feed agricultural machinery in the United States and Canada led to the de facto slavery of the Maya population and the transformation of a small-scale livestock-raising landscape into one dominated by henequen haciendas with the felling of what was left of the Yucatan forest.[22] In the north, immense livestock haciendas were established, like that of Luis Terrazas in Chihuahua, with close to three million hectares. These probably also changed local ecology by favoring certain forage species and by compacting soils, but that subject needs to be studied further to assess these possible effects.

Illustration 1.1. "Corn Patch Fringed with Maquay [*sic*]," Toluca, 1907.

Corn patches
fringed with
Maquay to prevent
washing as well as fencing
On the trail to Toluca.

S⬦M-1-D-179

Note the careful use of maguey to mark the border of the corn patches and to minimize erosion.

Courtesy Milwaukee Public Museum, Sumner W. Matteson Collection, negative number SWMI-D179.

Intellectuals recognized the Porfiriato's ecological impacts. Biologists, engineers, agronomists, and others formed scientific societies to discuss the possible consequences of overexploitation of natural resources. Especially outstanding was Miguel Ángel de Quevedo, a hydraulic engineer who soon became known as the "Apostle of the Tree" for voicing alarm about deforestation and the disappearance of forests around cities, which, in his opinion, were essential to public health. These experts' concerns led to the formation of a forestry service, a school of forestry, and the first national conservation regulations. Although companies that produced the greatest environmental damage ignored the legislation, the nascent intellectual conservation movement was a forerunner of the environmental movement of the postrevolutionary years. But intellectuals were not alone in their concern for nature. In some cases, businessmen themselves warned of the need to protect the resources on which they were economically dependent. In Baja California Sur, Gastón Vives, director of the Compañía Criadora de Concha y Perla de Baja California S.A., received a concession in the Gulf of California and developed the world's first method of cultivating pearl-producing oysters. This stopped the overfishing of the mother-of-pearl species and allowed him to produce pearls in large quantities. Vives's pioneering methods in the pearl industry demonstrated that extractivism is not the only road to economic development.[23]

The Social Revolution and Cardenismo

Some sectors of Mexican society who had paid the price for Porfirian commodification of nature and privatization of the commons (water, soil, forests), as well as with social and political repression, found a way to voice their anger in the early twentieth century. The end of the Porfiriato came in 1910, in the context of presidential succession. Díaz had been elected for five terms and was eighty years old. Two years earlier, he had provoked a wave of political speculation by suggesting that he would leave the presidency to facilitate a democratic opening. In the end, he refused to do so and jailed his opponent, Francisco I. Madero, who declared himself in rebellion on 10 November 1910, sparking a social revolution. The civil war lasted for almost a decade and reduced the population by 6.6 percent, as one million people either perished or fled into exile.[24]

Among the numerous factions involved in the revolutionary struggle, the one with the greatest socioenvironmental impact was led by Emiliano Zapata. Under the popular slogan "Land and Liberty," he headed a mass peasant movement; its major demand was to divide up the large landed estates known as *haciendas* and to return the *pueblos* (peasant communities) their lands, forests, and water. These demands eventually formed the basis of one of the Revolution's principle achievements: land reform codified in the Constitution of 1917. It took decades, however, to become a reality.

Postrevolutionary regimes tried to fulfill the expectations of the popular classes that had been accentuated by the revolutionary experience. Land reform began slowly and intensified during the Lázaro Cárdenas administration (1934–40), which transferred eighteen million hectares to agrarian communities. Postrevolutionary governments also sponsored other changes that had environmental implications, including passage of conservationist legislation, the development of bureaucracies for managing natural resources, and nationalization of some strategic industries related to the exploitation of raw materials such as oil and the primary railroads.

Agrarian reform represented a basic change in land use, though it did not affect all of the country's regions; in Morelos, Yucatan, and La Laguna, the intensive commercial use of land gave way to small-scale peasant production. The few detailed studies of the ecological effects of this productive transformation suggest that peasant uses of the land are less damaging, although more widespread in some cases, than commercial agriculture.[25] But the agrarian bureaucracy also created new power relations between land-reform communities (*ejidos*) and the state, placing the knowledge of experts over that of local residents and establishing a dependency relationship between state officials and peasant communities.[26]

The agrarian reform affected not only agricultural land but forests as well. Many indigenous communities lived in the woods since, as Gonzalo Aguirre Beltrán argued in 1967, the forests (along with jungles and desert areas) represented refuges far from mestizo population centers.[27] Some experts opposed the redistribution of the forestlands, fearing that deforestation would inevitably follow, along with the consequent erosion and changes to the hydrological regime. These concerns inspired the Forestry Law of 1926, an attempt to control use of forests through the establishment of production cooperatives. In practice, the provisions of the new law were not respected until 1934, when Cárdenas established an autonomous forestry department, headed by Quevedo, that was charged with enforcing compliance. There were only six cooperatives in the country before 1935; in 1940, the number rose to 860.[28] In practice, cooperatives experienced a wide variety of problems, including capture by corrupt political bosses and a growing dependence on commercial timber interests. Nevertheless, the experiment introduced by Cárdenas represented one of the world's first attempts at what today we would call community forestry management. Moreover, peasant communities gained capacity and experience in managing their own resources in several parts of the country.

The Constitution of 1917 also foresaw the exploitation of subsoil natural resources for the benefit of Mexican citizens. However, the application of this measure would be delayed for more than twenty years—that is, until the na-tionalization of the petroleum industry in 1938. From the first years of the twentieth century, North American and British companies produced oil in the Huasteca of Veracruz. These companies felled forests to build small in-dustrial cities, where employees, usually foreigners who were paid better than their Mexican counterparts, lived far from the wells and the toxic surround-ings where the locals lived. When Mexican workers complained about health conditions and pay, oil companies refused to negotiate, and Cárdenas nation-alized the industry on 18 March 1938. After six months of administration by the workers themselves, Cárdenas created the state-owned PEMEX company charged with producing, refining, and selling the oil. PEMEX has implemented measures to protect the wellbeing of workers over the years, although less at-tention has been paid to care of the natural environment.[29]

The promises of the revolution of 1910 reached their widest application during the presidency of Lázaro Cárdenas. He was the first chief of state to take broad measures to assure the long-term conservation of natural re-sources. His administration created forty national parks and sponsored the first studies of fishing in Lake Pátzcuaro and the Pacific Ocean in a bid to achieve a sustainable catch and to promote production cooperatives. His ad-ministration also made innumerable improvements in infrastructure (roads, electrification, potable water) in cities and rural areas. The *ethos* of the period aimed at harmonizing the use of nature with society's needs, especially in the agrarian sector. The Cardenistas managed, simultaneously, to modernize and organize the landscape and society.[30] Reforms were not error-free, and some were not sustainable, but they were conceived as a holistic regime that would not alienate human beings from the biosphere. Unfortunately, a few years af-ter its birth, this perspective came up against a third revolution: the so-called Green Revolution.

The Green Revolution and Statist Developmentalism

Since 1945, Mexico has fluctuated between revolutionary-*cardenista* promises of food sovereignty and support for small-scale peasant production versus a modernizing liberal development model centered on industrialization, the commodification of natural resources, and the channeling of economic and administrative assets into the cities. It is no surprise that the latter model grew more powerful given that the nation's population grew from 28.3 million in 1950 to 117.9 million in 2010, a period during which the urban population went from 42.6 percent to 76.8 percent of the country's inhabitants.[31] The demographic shift from countryside to cities produced both social and envi-ronmental changes, as well as a series of debates, beginning in 1970, between a group of so-called *campesinista* academics, such as Rodolfo Stavenhagen, who argued that rural society could survive more or less intact thanks to the economic underdevelopment of the Mexican countryside,[32] and the so-called *de-campesinista* academics, such as Roger Bartra, who predicted the gradual proletarianization of peasant society in the context of capitalist penetration into the countryside and growing inequality between relatively rich peasants and their landless peers.[33]

The viability of peasant society began to be undermined in the early 1940s with the application of new and costly agricultural technologies to increase productivity, a process known as the Green Revolution. In 1942, the Angostura dam began operating in the upper Yaqui Valley, in Sonora. The dam provided water to irrigate sixty thousand hectares of land previously belonging to the Yaqui Indians. The opening of a vast territory to an intensive agricultural regime came to the attention of North American agronomists, who arrived in the region the following year with plans to modernize Mexican agriculture, motivated by humanitarian considerations, as well as by "good neighbor" geopolitics implemented by the Roosevelt administration at the beginning of World War II. The project was sponsored by the Rockefeller Foundation under the leadership of North American agronomists such as Edwin J. Wellhausen and future Nobel laureate Norman Borlaug. North American and Mexican scientists founded the International Maize and Wheat Improvement Center (known by its Spanish acronym CYMMIT), which introduced a variety of Japanese semi-dwarf wheat (Norin 10) in 1952 that was crossed with local varieties to produce a hybrid whose weight did not cause the plants to break (i.e., lodge) when fertilizer was applied. The new wheat variety led to an explosion of biological changes that transformed certain regions of Mexico, and later of Asia and Africa, into agricultural landscapes that were highly productive but also dependent on industrial inputs such as large-scale irrigation, artificial fertilizers, and chemical pesticides.[34]

The Green Revolution spread through the arid Mexican north (where irrigation created favored pockets of production) and brought significant changes to commercial agriculture and peasant production. In spite of the slow demise of farming on communal lands, various state institutions promoted the use of pesticides and chemical fertilizers in peasant maize, strawberry, and coffee production. For example, representatives of the Banco Ejidal promoted the use of agrochemicals in corn and wheat production, while state-owned corporations such as INMECAFE did something similar in communal coffee production lands in the states of Oaxaca, Puebla, and Veracruz. The application of Green Revolution technology produced a number of problems. In many cases, peasants received inadequate training, leading to the overuse of fertilizers and, especially, pesticides. Some small farmers lacked the funds to buy these agricultural technologies and depended on state subsidies, which were gradually reduced until they disappeared with neoliberalism in the 1980s. Thus, the Green Revolution favored commercial farmers with the capital to take advantage of the new technologies. Many members of these wealthier producers grew export crops such as strawberries, tomatoes, vegetables, and (beginning in 1990) avocados. In many cases, agroindustrialists did not supervise workers applying fertilizers and pesticides, with dreadful consequences for the environment and the health of field workers.[35] The development of some commercial crops—such as avocado in Michoacán—also led to large-scale deforestation.

This situation does not imply that there existed—or continues to exist— an innate incompatibility between peasant production and Green Revolution technology. In many cases, peasant producers enthusiastically adopted the fertilizers and pesticides provided by the federal agrarian bureaucracy and used them on their own corn and other small-scale crops, such as coffee and avocado. On the other hand, the new technology's "benefits" did not reach all rural, much less indigenous, communities. The lack of zeal on the part of the federal government and the administration for indigenous development was partially responsible for these failures, but the weight of tradition in agricultural production also played a role. In many communities, techniques for growing the corn crop and even the use of certain varieties of corn are the backbone of the local culture. In almost all regions throughout the country, the planting and harvesting of corn, as well as the preparation of traditional dishes derived from it (tamales, *tlacoyos,* toasted corn, *corundas, sopes,* tacos, and so on), continue to be a major factor in the economic and cultural life of Mexicans.

This does not mean that rural communities have remained unchanging. On the contrary, many rural people have looked for ways to participate in international markets. During the Porfiriato, for example, some "traditional" Huasteca communities opted to privatize their communal lands and form a kind of collective corporation called a *condueñazgo* to sell vanilla in European markets.[36] Elsewhere, rural people have used their natural resources to produce crafts for the tourist market, such as the famous guitars of Paracho, Michoacán, or the lacquered boxes of Olinalá, Guerrero. The strategy of selling in international markets reappeared in the final decades of the twentieth century. One example is the well-known Unión de Comunidades Indígenas de la Región del Istmo (UCIRI) in Oaxaca, which produces crops such as coffee for export under the Fair Trade brand.[37]

The Green Revolution also represented the spearhead of a process leading to the further commodification of nature, which would transform the environment in almost all of Mexico during the second half of the twentieth century. The renaissance of Porfirista-type concessions opened forests, mines, and fishing to private enterprise. In 1952, President Miguel Alemán Valdés granted a three-hundred-thousand-hectare concession to the Bosques de Chihuahua,

S.A., a company in which Alemán himself was a silent partner. In coastal areas, especially in the south, new colonization policies around 1960 led to the transfer of thousands of peasants from the center of the country to the rain forests of regions such as Quintana Roo and Tabasco. The homesteaders felled trees to open land for livestock grazing, a practice supported by the National Land Clearance Program (PRONADE), destroying almost a half-million hectares of forest classified as "useless."[38] As regards oil, the poor administration and lack of investment in equipment that characterized PEMEX (the semi-state oil company) caused many oil spills and industrial accidents, such as the enormous spill caused by an explosion in the Ixtoc I well in 1979, whose effects are felt in coastal communities to this day.[39]

The feverish rate of urban growth also produced new built spaces throughout the country, but especially in Mexico City, where the population reached twenty million by 2010. The capital has the same environmental problems that afflict other megacities in the Global South: transportation bottlenecks, crime, lack of green spaces, and a profusion of informal settlements. But its location in what was the basin of a lagoon implies special challenges, such as thermic inversions (produced by the concentration of hot, polluted air pushed to the surface by cold air) in the winter, which result in extreme atmospheric pollution indices, the perennial problem of flooding during the rainy season (June to October), and the partial sinking of the city due to the volcanic lime and clay composition of the soils in the context of the intensive exploitation of aquifers. These conditions were aggravated by the earthquake of 19 September 1985, measuring 8.1 on the Richter scale, whose epicenter on the Pacific coast was fully 350 kilometers from the capital. High population density produced devastating consequences, with ten thousand deaths, seven hundred thousand persons left homeless in Mexico City and close to three thousand modern buildings destroyed, including part of the General Hospital and two buildings in the Tlatelolco residential complex.[40] The federal government's inadequate response to this "natural" disaster produced popular resentment that has been felt for decades.

Ecological links between the city and its rural surroundings underline the close relation between "the rural" and "the urban." For example, water scarcity in the capital, dating from the end of the nineteenth century, has become worse. Heavy buildings, like the Metropolitan Cathedral and the Fine Arts Palace, are gradually sinking in soils that are undermined by the pumping of water without allowing for aquifer recharge. But since demand for the precious liquid continues to rise, the National Water Commission (CONAGUA) has to bring water from ever more distant rivers to satisfy the great city's thirst. Official neglect of the peasant sector has contributed to a notable drop in productivity in agricultural production in communal lands, and the country has thus become a net importer of basic grains. The neoliberal system, and especially the North American Free Trade Agreement (NAFTA) with Canada and the United States, in effect since 1994, has made problems worse. NAFTA opened the market to such an extent that Mexico imports a third of its corn. While it is true that subsidies have made the country self-sufficient in white corn for human consumption, most of the crop comes from agribusinesses in the state of Sinaloa. Small producers on communal lands lack the resources and the opportunities to participate in domestic production, much less the transnational market. On the other hand, urban residents (with the exception of citizens who participate in movements in support of semi-urban and urban agriculture) tend to eat processed foods rather than consuming crops produced by peasants close to cities.

These contradictions, as well as the ever more widespread recognition of the need to defend the ecological integrity of the national heritage, contributed in the 1980s to the resurgence of an environmental tradition that goes back to the nineteenth century. In the forests, neoliberalism quashed concessions and the paragovernmental enterprises that dominated the sector, while professional forestry engineers and local leaders established community enterprises engaged in the sustainable use of forests in the states of Oaxaca, Quintana Roo, Michoacán, and others. The same official organs, such as SEMARNAT (Environment and Natural Resources Secretariat), supported a local, sustainable production policy.[41] Cooperatives have been created to produce organic coffee (in Chiapas and Puebla), to fish in rivers (in Veracruz and Baja California Sur), and to offer alternative tourism activities (in Oaxaca and Yucatán), among many other sustainable production activities. Urban residents also organized. In March of 1985, poet Homero Aridjis formed the "Group of One Hundred," a collective of urban intellectuals who promoted ecological policy; five years later, the organization led an international movement that paralyzed expansion of the Exportadora de Sal (salt exporter), a business financed by Japanese and Mexican capital on the Pacific coast of Baja California Sur that would have affected the habitat of the sperm whale.[42]

A new wave of ecological activity appeared around 2000 based on the idea of food sovereignty and small-scale, sustainable production. There also appeared a broad opposition to transnational agriculture and transgenic crops. In 2002 and 2003, the El Campo No Aguanta Más (the countryside can take no more) movement appeared and, more recently, the Sin Maíz No Hay País (without corn there is no nation) campaign has proposed to maintain small-scale organic production and to veto transgenic corn in Mexico. However, it seems that the *de-campesinista* argument has predominated, in general terms, though there continue to be rural communities and urban activists who promote sustainable use of the land.

Conclusion: Toward a New Ecological Revolution?

Each of the revolutions studied here has derived from both the successes and failures of its predecessor. The protagonists of these revolutions have included political actors, of course, but they have also involved scientists, agronomists, biologists, and economists, as well as urban and rural producers and consumers. Each revolution promised to increase productivity and improve collective wellbeing; all achieved the first goal but, with the partial exception of the Zapatista/Cardenista revolution, they failed in the second. Two factors explain this phenomenon: the promoters of the revolutions (again, with the partial exception of the Zapatista/Cardenista movement) have proceeded from the neocolonial view of progress that prioritizes economic growth and privileges the dominant classes, yet minimizes the socioenvironmental consequences of unsustainable development. Nevertheless, the series of social, political, economic, and environmental adjustments and readjustments that the Mexican revolutions have produced are a fertile base of historical experience on which Mexican society can build a future.

Developmentalist thinking has demonstrated its profound ability to polarize society and to destroy the environment. The cumulative effects of centuries of overexploitation of natural resources, erosion caused by bad land management—by both large-scale farmers and peasants—and, more recently, global climate change have generated serious problems that affect all Mexicans. Nevertheless, the costs of ecological degradation have been unequally distributed; the poorest members of the population, the most marginalized—indigenous, in many cases—are those who suffer the most dreadful consequences. More than eleven million people migrated to the United States between 1980 and 2010, and it is no exaggeration to maintain that many of them should be classified as environmental refugees.

In spite of their seriousness, however, ecological problems have not managed to overwhelm promises articulated by Zapata, Cárdenas, and generations of their followers. The communally held lands still exist, as do producers' cooperatives and indigenous production methods that thrive in socioenvironmental movements whose participants attempt to use soils, water, and forests in a way that allows them to live with dignity while leaving a legacy to future generations. Popular demands for socioenvironmental justice and the cosmovision of native peoples constitute indispensable components of an other, possible Mexico.[43] The consequences of social struggles have slowly but surely permeated federal environmental legislation and the socially committed scientific and intellectual tradition, and have reinforced social organization. In urban and rural areas, Mexican society is renewing its relationship with its territories, its ecosystems, and its biocultural heritage, deriving a variety of local expressions for sustainability that offer concrete alternatives to overcoming civilization's global crisis. Urban-rural social movements, such as Sin Maíz no hay País and Vía Campesina, have a visible presence in the capital and in provinces, and their members have experienced significant political successes, such as the prohibition on transgenic corn production. In addition, in most states, solidarity movements have sprung up to promote consumption of locally and organically produced items. Indigenous communities have waged legal battles against state abuses (for example, the neo-Zapatistas in Chiapas, who have embraced sustainable production on their lands as a self-sufficiency measure),[44] as well as against organized crime and those who would sack their resources, as in the recent case of Cherán, in Michoacán (where the people rose up in 2013 against illegal timber operations in communal forests and against authorities in collusion with organized crime; today they have undertaken a communal reforestation program and self-government). It is premature to state that a new ecological revolution is taking place, but countercurrents visible today may well gather strength in the years to come.

Chris Boyer is professor of history and Latin American and Latino studies at the University of Illinois at Chicago. He has published widely in English and Spanish on the social and environmental history of Mexico, including a recent a volume of environmental histories of modern Mexico titled *A Land Between Waters,* which represents the first binational reflection on Mexican environmental history. His most recent book is *Political Landscapes: Forests, Conservation, and Community in Mexico.* He coedits the Latin American Landscapes book series published by the University of Arizona Press. His current research deals with agroecology and food sovereignty in Mexico.

Martha Micheline Cariño Olvera is full-time research professor at the Autonomous University of Baja California Sur in Mexico, where she teaches undergraduate history courses and in the postgraduate program on sustainable development and globalization. She has coordinated a number of research projects relating to sustainability, conservation, and environmental history. She is a member of Mexico's Sistema Nacional de Investigadores and served as president of the Society for Latin American and Caribbean Environmental History (SOLCHA). She is the author of thirteen books and more than a hundred articles and book chapters.

Notes

This chapter was translated by Mary Ellen Fieweger.
1. Carolyn Merchant, *Ecological Revolutions: Nature, Gender, and Science in New England,* 2nd ed. (Chapel Hill: University of North Carolina Press, 2010).

2. Rafael Ortega-Paczka, "La diversidad del maíz en México," in *Sin maíz no hay país,* ed. C. Esteva and C. Marielle (Mexico City: Consejo Nacional para la Cultura y las Artes, Dirección General de Culturas Populares e Indígenas, 2003), 123–154.

3. Jerome O. Nriagu, "Mercury Pollution from the Past Mining of Gold and Silver in the Americas," *The Science of the Total Environment* 149 (1994): 167–181; Daviken Studnicki-Gizbert and David Schecter, "The Environmental Dynamics of a Colonial Fuel-Rush: Silver Mining and Deforestation in New Spain, 1522 to 1810," *Environmental History* 15 (2010): 94–119.

4. Robert C. West, *The Mining Community in Northern New Spain: The Parral Mining District.* University of California Publications in Ibero-Americana, vol. 30 (Berkeley: University of California Press, 1949).

5. Eric Van Young, *The Other Rebellion: Popular Violence, Ideology, and the Mexican Struggle for Independence, 1810–1821* (Stanford: Stanford University Press, 2001), 433–444.

6. David Brading, *Haciendas and Ranchos in the Mexican Bajío: León 1700–1860* (Cambridge: Cambridge University Press, 1978); Margaret Chowning, *Wealth and Power in Provincial Mexico: Michoacán from the Late Colony to the Revolution* (Stanford: Stanford University Press, 1999), 286–306; and Luis González y González, *Pueblo en vilo: Microhistoria de San José de Gracia* (Mexico City: El Colegio de México, 1968).

7. Heather L. McCrea, *Diseased Relations: Epidemics, Public Health, and State-Building in Yucatán, Mexico, 1847–1924* (Albuquerque: University of New Mexico Press, 2011).

8. Jan de Vos, *Oro verde: La conquista de la Selva Lacandona por los madereros tabasqueños* (Mexico City: Fondo de Cultura Económica, 1988), 202–227.

9. Brian DeLay, *War of a Thousand Deserts: Indian Raids and the U.S.-Mexican War* (New Haven: Yale University Press, 2009), 86–113.

10. Raúl Benítez Zenteno, *Análisis demográfico de México* (Mexico City: Instituto de Investigaciones Sociales de la UNAM, 1961), appendix A.

11. John H. Coatsworth, *Growth against Development: The Economic Impact of Railroads in Porfirian Mexico* (DeKalb: Northern Illinois University Press, 1981), 174; see 149–174.

12. Luz Carregha Lamadrid, "Tierra y agua para ferrocarriles en los partidos del Oriente potosino, 1878–1901," in *Entretejiendo el mundo rural en el "oriente" de San Luis Potosí, siglos XIX y XX,* ed. Antonio Escobar Ohmstede and Ana María Gutiérrez Rivas (San Luis Potosí and Mexico City: El Colegio de San Luis / CIESAS, 2009), 177–204.

13. Christopher R. Boyer, *Political Landscapes: Forests, Conservation, and Community in Mexico* (Durham, NC: Duke University Press, 2015).

14. José Napoleón Guzmán Ávila, *Michoacán y la inversión extranjera, 1880–1900* (Morelia: Universidad Michoacana de San Nicolás de Hidalgo, 1982), 39–67; José Juan Juárez Flores, "Besieged Forests at Century's End: Industry, Speculation, and Dispossession in Tlaxcala's La Malintzin Woodlands, 1860–1910," in *A Land Between Waters: Environmental Histories of Modern Mexico,* ed. Christopher R. Boyer (Tucson: University of Arizona Press, 2012), 100–123.

15. Cariño, Micheline, *Historia de las relaciones hombre/naturaleza en Baja California Sur 1500–1940* (La Paz, Mexico: UABCS-SEP, 1996).

16. Samuel Truet, *Fugitive Landscapes: The Forgotten History of the U.S.-Mexico Borderlands* (New Haven: Yale University Press, 2008); John Mason Hart, *The Silver of the*

Sierra Madre: John Robinson, Boss Shepherd, and the People of the Canyons (Tucson: University of Arizona Press, 2008).

17. This situation is possible only if we consider that the government oversold the water from the Higuerón River.

18. Alejandro Tortolero, "Water and Revolution in Morelos, 1850–1915," in *A Land Between Waters: Environmental Histories of Modern Mexico,* ed. Christopher R. Boyer (Tucson: University of Arizona Press, 2012), 124–149; Alfredo Pureco Ornelas, *Empresarios lombardos en Michoacán: La familia Cusi entre el Porfiriato y la postrevolución (1884–1938)* (Zamora: El Colegio de Michoacán, 2010).

19. Manuel Perló Cohen, *El paradigma porfiriano: Historia del desagüe del valle de México* (Mexico City: Porrúa, 1999); Matthew Vitz, "'The Lands with Which We Shall Struggle': Land Reclamation, Revolution, and Development in Mexico's Lake Texcoco Basin, 1910–1950," *Hispanic American Historical Review* 92, no. 1 (2012): 41–71.

20. Alejandro Tortolero, *De la coa a la máquina de vapor: Actividad agrícola e innovación tecnológica en las haciendas de la región central de México, 1880–1914* (Mexico City: Siglo XXI, 1995).

21. For a summary of agrarian history, see Antonio Escobar Ohmstede and Matthew Butler, "Transitions and Closures in Nineteenth- and Twentieth-Century Mexican Agrarian History," in *Mexico y sus transiciones: Reconsideraciones sobre la historia agraria mexicana, siglos XIX y XX,* ed. A. Escobar Ohmstede and M. Butler (Mexico City: CIESAS, 2013), 38–76.

22. Sterling Evans, *Bound in Twine: The History and Ecology of the Henequen-Wheat Complex for Mexico and the American and Canadian Plains, 1880–1950* (College Station: Texas A&M University Press, 2007).

23. Micheline Cariño and Mario Monteforte, *El primer emporio perlero sustentable del mundo: La compañía Criadora de Concha y Perla de Baja California, S.A. y perspectivas para Baja California Sur* (La Paz: UABCS, 1999).

24. Robert McCaa, "Missing Millions: The Demographic Costs of the Mexican Revolution," *Mexican Studies/Estudios Mexicanos* 19, no. 2 (2003): 367–400.

25. Gene C. Wiken, *Good Farmers: Traditional Agricultural Resource Management in Mexico and Central America* (Berkeley: University of California Press, 1987).

26. Joseph Cotter, *Troubled Harvest: Agronomy and Revolution in Mexico, 1880–2002* (Westport, CT: Praeger, 2003), 45–123.

27. Gonzalo Aguirre Beltrán, *Regiones de refugio* (Mexico City: Instituto Indígena Interamericano, 1967).

28. Christopher R. Boyer, "Revolución y paternalismo ecológico: Miguel Ángel de Quevedo y la política forestal, 1926–1940," *Historia Mexicana* 57, no. 1 (2007): 91–138.

29. Myrna Santiago, *The Ecology of Oil: Environment, Labor, and the Mexican Revolution, 1900–1938* (Cambridge: Cambridge University Press, 2007).

30. Christopher R. Boyer and Emily Wakild, "Social Landscaping in the Forests of Mexico: An Environmental Interpretation of Cardenismo, 1934–1940," *Hispanic American Historical Review* 92, no. 1 (2012): 73–106.

31. United Nations, *World Population Prospects: The 2012 Revision,* vol. 1, *Comprehensive Tables* (New York: United Nations, 2012), table S2, 58; Instituto Nacional de Estadística y Geografía. Retrieved 6 May 2013 from http://www3.inegi.org.mx/sistemas/temas/default.aspx?s=est&c=17484.

32. Rodolfo Stavenhagen, "Capitalism and the Peasantry in Mexico," *Latin American Perspectives* 5, no. 3 (1978): 27–37.

33. Roger Barta, *Campesinado y poder político en México* (Mexico City: Ediciones Era, 1982).

34. Cynthia Hewitt de Alcántara, *Modernizing Mexican Agriculture: Socioeconomic Implications of Technological Change, 1940–1970* (Geneva: United Nations Research Institute for Social Development, 1976), 148–171.

35. Angus Wright, *The Death of Ramón González: The Modern Agricultural Dilemma*, rev. ed. (Austin: University of Texas Press, 2005).

36. Emilio Kourí, *Un pueblo dividido: Comercio, propiedad y comunidad en Papantla, México* (Mexico City: Fondo de Cultura Económica, 2013).

37. José Antonio Ramírez Guerrero, "La experiencia de UCIRI en México", in *Producción, Comercialización y certificación de la agricultura orgánica en América Latina*, ed. Gómez Cruz et al. (Mexico City: UACH-CIESTAAM, AUNA), 119–132.

38. Arcelia Amaranta Moreno Unda, "Environmental Effects of the National Tree Clearing Program, Mexico, 1972–1982" (Master's thesis, Universidad Autónoma de San Luis Potosí/Cologne University of Applied Sciences, 2011).

39. Lisa Breglia, *Living with Oil: Promises, Peaks, and Declines on Mexico's Gulf Coast* (Austin: University of Texas Press, 2013).

40. The most well-known account of the Mexico City earthquake is Elena Poniatowska's *Nada, nadie. Las voces del temblor* (Mexico City: Ediciones Era, 1988). There is still no environmental (or social) history of the event, but see Louise E. Walker, "Economic Fault Lines and Middle-Class Fears: Tlatelolco, Mexico City, 1985," in *Aftershocks: Earthquakes and Popular Politics in Latin America*, ed. J. Buchenau and L. L. Johnson (Albuquerque: University of New Mexico Press, 2009), 184–220.

41. Leticia Merino Pérez, *Conservación o deterioro. El impacto de las políticas públicas en las instituciones comunitarias y en las prácticas de uso de los recursos forestales* (Mexico City: SEMARNAT/INE/CCMSS, 2004), 194–211.

42. Lane Simonian, *Defending the Land of the Jaguar: A History of Conservation in Mexico* (Austin: University of Texas Press, 1995), 212–215.

43. Víctor Toledo, *Ecología, espiritualidad y conocimiento: De la sociedad del riesgo a la sociedad sustentable* (Mexico City: PNUMA, Universidad Iberoamericana, 2003).

44. There are many other cases of communities—indigenous and mestizo—whose members have confronted the state to avoid confiscation of their territories threatened by the building of dams and aqueducts or by enormous, toxic mines.

 CHAPTER 2

The Greater Caribbean and the Transformation of Tropicality

Reinaldo Funes Monzote

The common sea that bathes the coasts of its many islands and continental territories is one of the factors that makes the Greater Caribbean a unique geographic region marked by the heritage of African slavery and plantations.[1] The region's environmental history has been affected by its location in the tropics, which contributed to defining the roles of its ecosystems and societies when a European-dominated globalization began that gave rise to an international division of labor. Imperial visions from other latitudes fed the emphasis on tropicality along with hegemonic internal powers that reproduced discourses and practices of self-proclaimed Western civilization's domination based on factors such as skin color or climate.

The golden age of Antillean plantations, operating within the framework of an organic, solar-based economy, was often explained by the "comparative advantage" that its ecosystems and enslaved populations offered for obtaining tropical produce. Thus was forged a colonial dependence that produced high dividends for European metropoles and contributed to the beginning of their industrialization. Nevertheless, the advance of an industrial metabolism and its new production and transportation technologies called into question those alleged ecological advantages, due to the appearance of new competitors, such as beet sugar to the detriment of cane sugar. Gradually, the region's role as the source of calories in the diets of distant populations changed to that of a tourist destination for the enjoyment of tropical nature; the change occurred after the eradication, in the early twentieth century, of diseases such as yellow fever, which caused high mortality among immigrants, foreign armies, and occasional visitors.[2]

The Greater Caribbean's ecosystems, including tropical rain forests, tropical dry forests, savannahs, and wetlands, are characterized by their high biodiversity. In addition to being an area where the biotas of North and South America meet, the region possesses a high rate of endemism, especially on islands. Another important regional characteristic is the frequent incidence of natural disasters that cause great ecological and socioeconomic impact. At various points in the past, the devastating effects of volcanic eruptions and earthquakes left their mark on the Caribbean's geography. However, given their frequency and reach, no other phenomena receive as much attention as do tropical cyclones or hurricanes.[3] Droughts also merit mention as their adverse effects can be more lasting than those resulting from hurricanes or earthquakes.

On islands throughout the Caribbean, indigenous populations diminished dramatically after Europeans arrived due to epidemic diseases and severe labor exploitation. But, the indigenous presence remained significant along continental coastlines. The mixture of humans from different places constitutes one of the region's secular characteristics as a place where distinct civilizations and cultures met. The African population brought as labor to replace the indigenous, and the mixing of both groups with people from Europe and Asia, especially China, India, and Indonesia, gave rise to tremendously diverse societies in ethnic, linguistic, and cultural terms.

The rapid proliferation of domesticated mammals from Eurasia and Africa led to extensive livestock production. At first, their ecological impact was limited, but their proliferation eased the way for more intensive forms of agriculture and livestock activities in the future. In contrast to the spread of commercial plantations throughout the Antilles beginning in the seventeenth century, livestock raising, together with subsistence agriculture, remained important along the continental Caribbean through the nineteenth century.

The extraction of minerals and precious woods began in early colonial times. Until the eighteenth century, the metropolis tended to monopolize timber resources for ship construction (for example, in Havana's shipyards), export for metropolitan construction, and use as dyes in the European textile industry. Beginning in the nineteenth century, the selective felling of trees was more commercial in character. Countries like Haiti, the Dominican Republic, and Belize exported great quantities of species, such as mahogany, considered among the world's finest. Other extractive activities included the capture of caimans and crocodiles for their skin; birds, such as the flamingo, for their feathers; and freshwater fish. Ecosystems rich in biodiversity, such as the mangroves, were used to obtain firewood and charcoal; colonial authorities tended to view mangroves as unhealthy places that should be drained.

The ever-present sea gave birth to centuries' old mariner traditions. From early colonial times, the Caribbean offered valuable resources, such as Venezuela's pearls, Cuban and Bahaman sponges, and turtles. There is a widespread perception that in spite of being surrounded by a sea, Caribbean island-dwellers do not include fish in their daily diets. A partial explanation could be that although tropical seas tend to be rich in marine species variety, population densities are generally low. In the greater part of the Antillean arc, island platforms are narrow and nutrient-poor and thus cannot sustain large fish populations, a situation distinct from that of South America's marine environments. In the twentieth century, the use of industrial fishing methods resulted in overfishing, leading to the destruction of marine habitats, the pollution of coastal waters, and the displacement of fishing communities by other economic activities.[4]

Maritime communications represented a difference in interregional and extracontinental ties to American territories distant from major flows of international commerce. This was because they were the stage par excellence from which other European monarchs challenged Spain's imperial hegemony: Great Britain, France, the Low Countries, Denmark, and Sweden occupied territories and established colonies in the region.

Those ties began to break following the slave-led Haitian revolution that led to independence in 1804. Mainland Spanish colonies with coast on the Caribbean followed, in a wave of independence struggles that ended in the 1820s; the Dominican Republic declared itself independent from Haiti in 1844. The Republic of Cuba was not established until 1902; decolonization of British possessions did not occur until much later (Jamaica and Trinidad and Tobago, 1962; Barbados, 1966; Belize, 1981). Nevertheless, various colonial ties remain in the region. Throughout the twentieth century, the influence of the United States grew. The expansion of its economic and political interests occurred primarily through indirect mechanisms of neocolonial domination, although United States forces occupied the Panama Canal Zone from 1903 to 1999, and have remained in Puerto Rico since 1898.

Caribbean Plantations and the First Industrial Revolution

For more than three centuries, the Caribbean was a great center for the export of tropical produce to European metropoles and international markets through the plantation system, an institution considered the main thread running through regional history.[5] Sugar cane was the Antilles' major crop, especially in flat or slightly hilly areas. While it was not the only commercial crop, sugarcane stood out as a result of its large geographical scale and its agroindustrial character: upon being harvested, sugarcane was processed on site. African slaves harvested the cane and made the sugar. In addition to the original reason for using African slaves to replace decimated indigenous populations, one of the pretexts for the brutal trade in men and women from Africa would be their alleged superiority, compared to Europeans, for work in the sun, heat, and humidity of the tropics.

The first sugar plantations worked by slaves were created around 1520 on La Española (Hispaniola), where the plantations enjoyed a boom period before declining during the final third of the sixteenth century. By the mid-

seventeenth century, a "sugar revolution" began on several small islands occupied by Great Britain, tied to the commercial and mercantile preeminence
of the metropole.[6] The emblematic case was Barbados (440 sq. km.), where
planters quickly established sugar monocultures, a model replicated on other
islands. Thanks to their greater amount of cultivable land, Jamaica (English)
and San Domingue, the French portion of La Española, became the leading
producers of sugarcane in the eighteenth century. As the world's number one
sugar and coffee exporter, San Domingue was known as the "pearl of the Antilles"; then, in 1791, the great slave rebellion broke out, leading to the Republic
of Haiti in 1804.

The Haitian Revolution represents a culminating moment in Caribbean
plantations within the framework of a preindustrial organic economy based on
the force of human and animal muscles, aided by wind and water power (primarily used in English and French colonies). In the Lesser Antilles, problems
such as rapid deforestation, scarcity of firewood, erosion, and declining soil fertility led to innovations in production, including more efficient means of boiling cane, the use of cane residue (bagasse) as fuel; the application of manure to
fertilize plantation soils; the introduction of new cane varieties; and irrigation.[7]
Rapid deforestation and its effects on soil, native fauna, and climate explain
why the Lesser Antilles became pioneers in adopting conservation measures,
including the creation of botanical gardens, forest reserves, and protected areas.[8] In socioecological terms, plantation-based colonies symbolize economic
dependence on an outside world due to excessive specialization, creating great
vulnerability in the face of adverse economic conditions.

The Haitian Revolution created a vacuum in the sugar market that was filled
by other Antillean colonies, such as Jamaica, which had its largest harvests
around 1805. But above all, it opened the way for Cuba to become the great
sugarcane exporter during the nineteenth and twentieth centuries. This ascent
by the largest island of the Antilles coincided with new conditions created by
the industrial revolution, including the conversion of sugar into an article of
mass consumption.[9] Though there were never as many slaves in Cuba as in
Haiti when the revolution broke out (452,000), between 1830 and 1870 Cuba
produced ten times the quantity of sugar formerly produced in San Domingue
(76,000 tons). The incorporation of steam engines in the sugar mills was decisive in enabling this giant leap to occur: between 1827 and 1869, the number
of mills powered by steam engines grew from 26 to 949, and Cuban mills further mechanized via the introduction of vacuum vat trains and centrifuges.

At the same time, beginning in 1837 the construction of a dense railway
network (the first and largest in Latin America), together with the growing use
of steamships, revolutionized transport of the final product. Cuba's extensive
plains favored a constant increase in the scale of production and the success of
new technologies. The slave-based agroindustry moved toward forested fron-

Illustration 2.1. Sugar Mill "Tinguaro", established in 1840, in Colon Plains, Matanzas, Cuba.

Eduardo Laplante litography, Source: Justo G. Cantero, *Los ingenios. Colección de vistas de los principales ingenios de azúcar de la isla de Cuba,* Imp. Litográfica Luis Marquier, La Habana, 1857.

tiers in search of fertile land, fuel, and timber for construction, a cycle repeated in new areas when cane production exhausted these resources.

As this routine practice accelerated, influential scientists charged that the slave plantations exploited the soils like an open pit mine, taking advantage of the abundant organic material resulting from the felling and burning of forests.[10] Sugarcane plantations spread throughout Cuba, but their center was in the western half of the island; as a result, deforestation was the most intense in this area, leading to more droughts and a reduction in native biodiversity, especially birds, coupled with water pollution and an invasion of exotic plants.

Cuba's major forested areas remained in the eastern half of the island, dedicated primarily to livestock. A similar situation existed in the Dominican Republic, where the major economic activities were livestock production and the sale of timber. During the same period, other Caribbean islands, such as Puerto Rico and Guiana, established sugar plantations and experienced small economic booms, but in general their sugar industries were limited by the end of forced labor or limited mechanization.

The list of commercial Caribbean crops included ginger, tobacco, cacao, indigo, and cotton. But the second most important one after sugar was coffee. Haiti was the major world exporter until the end of the eighteenth century, a position that was taken over during the first decades of the following century by Cuba and Jamaica. Coffee plantations tend to be seen as gardens in comparison to industrialized sugar plantations, but they also faced difficulties resulting from deforestation, plant pests, and increased erosion due to planters' preference for growing coffee in mountainous areas.

After the Revolution, coffee maintained its presence in Haiti's mountains on small farms, contributing, together with commercial logging, to the deforestation of those areas. Coffee production also continued in areas such as Jamaica's Blue Mountains and Cuba's Sierra Maestra. In the second half of the nineteenth century, the mountainous areas of west-central Puerto Rico experienced a coffee boom. Environmental impacts included changes in the makeup of local forests, alterations in the nesting and feeding habits of birds, such as the green parrot, as well as greater vulnerability to hurricanes. Problems associated with soil exhaustion and erosion soon appeared. The intensive cultivation of coffee, which reached its peak in Puerto Rico in the 1880s and 1890s, led to the neglect of subsistence crops and animal-raising, increased dependence on imported food, and a deterioration in workers' diets.[11]

Plantations and the Conquest of the Caribbean Tropics

The spread of United States interests throughout the Caribbean from the end of the nineteenth century coincided with the idea, shared in Anglo-Saxon and European power and intellectual circles, of the inevitable "conquest of the tropics" by the "white man."[12] A milestone in this respect was the confirmation of the hypothesis, proposed in 1881 by Cuban physician Carlos J. Finlay, that the female *Aedes aegypti* mosquito was the vector transmitting yellow fever during the United States' occupation of Cuba from 1898 to 1902. This led to the eradication of the disease in Havana and prepared the way for public health success during the construction of the Panama Canal. U.S. military physician William Gorgas, chief of public health in first Cuba and later Panamá, wrote in 1909, "I believe that our work in Cuba and Panama should be seen as the first demonstration that the white man can flourish in the tropics and as a starting point for the Caucasian race's effective settlement in these regions."[13]

The inauguration of the Panama Canal in 1914 symbolized the consolidation of United States hegemony in the Caribbean, evidenced by numerous military bases and interventions. The emergence of the idea of an "American Mediterranean" was arguably the origin of the idea of a Greater Caribbean. The new interoceanic route represented a new phase of the region as a focal point of a world whose fate was closely tied to shipping lanes, a condition that would only begin to change with the rise of commercial aviation.

From the last third of the nineteenth century, changes began to take place in the Caribbean sugar industry, especially in Cuba. The spread of mechanization in the manufacturing sector drove concentration in ever greater units (called central sugar mills), which coincided with the end of slavery (between 1880 and 1886). The tendency was to separate the manufacturing and agricultural sectors, favoring greater care in the cultivation of old plantations. It should be remembered that there was already a significant history in other Antillean colonies, especially the French. However, the largest factories were built in new areas on the forest and livestock frontier in mid-eastern Cuba, where land was cheaper. Similar changes occurred in the Dominican Republic, where sugar production took off toward the end of the nineteenth century, and in Puerto Rico from the beginning of the twentieth century. In the three Hispanic Antilles, the new era of dependence on the United States was decisive. Apart from being the principle market, large corporations from that country bought or built the most powerful factories and came to generate the bulk of production in the dominant (North) American sugar kingdom.[14]

In all three cases, the productive leap was dizzying. Cuba—which, after harvesting a million tons in 1894, experienced a marked decline during the independence war (1895–1898)—reached around 2.5 million tons just before World War I and double that amount when the conflict ended. In 1952, Cuban workers harvested more than seven million tons. Puerto Rico's soils produced more than a million tons of cane during the 1930s, and the Dominican Republic, more than four hundred thousand around 1929. The ecological transformations were more visible in Cuba and the Dominican Republic than in Puerto Rico, where space limitations led to more intensive cultivation methods. The new cane lands in both Cuba and the Dominican Republic depended on the massive immigration of workers, especially Antillean, to meet labor needs.

The Greater Caribbean was a kind of first frontier in the materialization of "the conquest of the tropics," a notion that became increasingly popular in the United States in conjunction with the direct investments in plantations made by that country's large corporations. In fact, Conquest of the Tropics was the title of a book dedicated to extolling the United Fruit Company's banana plantations along the Caribbean coast of Central American and Colombia in addition to its industrial sugar mills in Cuba.[15]

The large-scale eruption of banana production was due, to a great extent, to the application of technologies associated with the second industrial revolution, together with the organizational and financial capabilities of U.S. business corporations. Toward the end of the nineteenth century, Jamaica was the leading banana exporter, but U.S. plantations established in Caribbean Central America surpassed it by the early twentieth century. Unlike sugar, coffee, cacao, and tobacco, bananas are exported as fresh fruit; their perishable nature means that they must reach markets before ripening. Thus, the coordination of railroads, steamships, and refrigeration in the tropics was decisive, as well as transportation networks to consumer markets.

The triumph of the banana as an item of mass consumption in the United States and other markets led to a great agroecological transformation in extensive coastal areas in northern Colombia and on the Caribbean coasts of Panama, Costa Rica, Guatemala, and the Honduras. In the Antilles, bananas were produced in Jamaica, the Dominican Republic, Trinidad, and other islands favored by preferential agreements with imperial Britain. Export banana zones were the stage for similar forms of interactions among plantation agriculture, transnational business practices, national governments, social protest, and technological modernization in the forms of railroads, hospitals, electrical grids, and centers for agricultural research. The banana industry fostered labor migrations within the Greater Caribbean, which gave way to foci of conflict as well as contributing to reinforcing multicultural identities.

Two plagues that affected the banana, Panama disease and sigatoka, exemplify the link between socioeconomic and ecological changes. The banana industry combated Panama disease, a soil fungus that appeared at the end of the nineteenth century, by razing diseased plantations and opening new areas for cultivation. The control of sigatoka, an airborne pest, inaugurated the massive application of fungicides as a method of control in the 1930s, changing labor processes and debilitating spray workers' health. The connections between export banana zones, created by frequent steamer traffic, facilitated the regional movement of the pathogens. One factor that limited attempts to eradicate these diseases was the creation of a mass consumer market for the Gros Michel variety, which slowed the adoption of disease-resistant varieties that did not coincide with the popular image and taste of the ideal banana.[16]

Foreign companies participated in the commercial production of other crops, such as tobacco, in which they controlled both cultivation and the manufacture of end products; cacao, with important nuclei in the Dominican Republic and Trinidad; and pineapple and citrus fruits. Some U.S. companies invested in livestock businesses, but this activity remained primarily in Creole hands. In regard to cattle, beginning in the late nineteenth century, ranchers introduced more intensive management techniques through innovations such as barbed wire fences, the introduction and cultivation of exotic, often African, forage species, and the import of new cattle breeds from the United States, Europe, and Asia. Holstein was among the most common dairy breeds, while Cebu, of Indian origins and therefore more adapted to tropical climates, was the preferred meat breed. The intensification of bovine production led to an increase in deforestation for pasture.[17] The rise of motorized vehicles, including tractors, meant that bovine livestock could be raised exclusively for animal protein.

Foreign companies also participated in mining and industrial activities. From the 1880s, United States mining companies established branches in Cuba for extracting minerals such as iron, manganese, and copper. Beginning in 1920, a great oil extraction boom began in the Lake Maracaibo area in Venezuela after the Juan Vicente Gómez administration granted concessions to Standard Oil and other United States firms. The companies refined the crude oil in the neighboring Dutch colonies of Aruba and Curaçao. The oil wells around Venezuela and Trinidad assumed strategic significance for the geopolitical interests of the United States in Latin America because U.S. companies largely controlled oil consumption in South American cities.[18]

The Decline of the Plantations

Caribbean agroexport economies maintained a presence in the tropical fruit trades during the second half of the twentieth century, but their weight tended to be secondary to that of continental producers with access to new land. This pattern began in the mid-nineteenth century with coffee, followed by other crops, such as cacao and cotton, and finally bananas and sugar.

The world economic crisis of the 1930s dealt a hard blow to the sugar industry in Cuba and the Dominican Republic, whose economies would not recover until World War II. In Puerto Rico, the impact of the Great Depression was lessened by preferential access to the United States market. However, Puerto Rican agroindustry began to come undone in the 1950s due to land reform and the creation of incentives for accelerating industrialization. That also seemed to be the path chosen by Cuba after 1959, when the revolutionary government passed two agrarian reform laws, nationalized major sugar mills, and reoriented the economy toward agricultural and industrial diversification.

Nevertheless, the growing conflict with the United States government led to the cancelation of the Cuban sugar quota in 1960, ending a trade agreement that had been in force since 1934. The Cuban government responded to the U.S. economic blockade of the island (1960) and severing of diplomatic relations (1961) by establishing tight economic and political ties with the Soviet Union, China, and the Communist bloc. New trade agreements with the Soviet Union in 1964 led to the return of sugar as the economic engine: the Cuban government set an ambitious goal of a ten-million-ton sugar harvest in 1970. Cuba did not reach its lofty goal, but produced a record 8.5 million tons.

In order to increase sugar production, the Cuban government devised a strategy to expand plantations and increase output based on three pillars: the large-scale application of agrochemicals, mechanization, and an increase in irrigation. During the 1970s and 1980s, Cuba's sugar harvest remained between seven and eight million tons. At the same time, promotion of more specialized livestock production led to progress in the diversification of animal protein sources. In the bovine category, efforts were concentrated on producing dairy products, and great progress was achieved in hog and poultry production.

Illustration 2.2. Sugar Central España Republicana (former España), in Colón municipality, Matanzas, Cuba.

Photo by Reinaldo Funes Monzote, June 2007.

However, these food production practices were highly dependent on the import of animal feeds and other inputs; when the Soviet Union dissolved, Cuban industrial agriculture suffered a drastic decline, contributing to a broader crisis.

The Cuban Revolution benefitted other Caribbean sugar producers due to the redistribution of sugar quotas in the North American market. For example, during the 1960s and 1970s, the Dominican Republic enjoyed greater harvests (over one million tons annually) that accounted for more than 90 percent of its export earnings. Unlike Cuba, the level of mechanization of the agricultural sector remained low due in part to the availability of low-wage Haitian workers. Beginning in the 1980s, sugar production declined in the Dominican Republic, a process that was accentuated in 1988 following a reduction in the United States import quota.

Cane sugar remained the major agricultural export item until around 1980 in Guiana, Jamaica, Belize, Barbados, Trinidad, St. Kitts, and Surinam. Most of the harvest went to the United States, where a tariff increase in 1982 hit Caribbean producers hard. Ten years later, new obstacles appeared with the loss of preferential treatment granted to former European colonies due to measures linked to the creation of the European Union. As many mills closed, agroindustry underwent a high degree of entrepreneurial concentration.

Banana plantations experienced similar processes of boom and bust in the mid-twentieth century. Jamaican exports reached their peak on the eve of World War II and recovered to a degree between the 1950s and the 1970s, but production subsequently diminished significantly. However, a number of the Lesser Antilles—Dominica, Santa Lucía, Saint Vincent, and Granada—increased their exports to the United Kingdom, a preferential market, between the 1960s and the 1990s. France extended banana producers in Martinique and Guadalupe similar protections.[19]

Banana plantations along the Central and South American Caribbean coast also experienced change after World War II. Beginning in the late 1950s, fruit companies increased production of high-yielding Cavendish varieties shipped in cardboard boxes. Other innovations included the use of cables to transport harvested bananas from fields to packing plants, and the rise of national producer associations. The United States remained the principle market for nearly the entire period.

The enclave character of banana producing areas led to their decline. For example, the Atlantic coast of Costa Rica, around the port of Limón, entered into a deep crisis as a result of the 1929 depression and the transfer of banana plantations to the country's Pacific coast.[20] Another example was the decline of banana plantations in Colombia's Magdalena region beginning in the 1960s. The boom and bust phases of the banana trade were a constant in the environmental history of the Greater Caribbean.

One of the region's defining characteristics has been foreign dependence, particularly in the Antilles as a result of its history and geographical fragmentation.[21] In spite of the huge quantity of calories exported in the form of sugar and bananas, the high-volume of imports of food and basic products has been a constant. To diminish that dependence following World War II, efforts were made to increase animal protein via livestock production. Various Caribbean governments devoted themselves to programs to encourage the dairy industry, based on breeds with great milk production potential, the creation of scientific research centers, and improvement in pasture or the use of imported feed. In spite of obstacles, such as droughts or difficulties in acclimatizing certain breeds of dairy cows, countries such as Puerto Rico, Jamaica, Trinidad, the Dominican Republic, and Cuba achieved successes that would later be affected by free trade and economic crises. The case of Puerto Rico stands out. Its achievements in the dairy industry around 1960 were characterized as a revolution in tropical agriculture.[22]

With a view to increasing bovine meat production, Cebu and Santa Gertrudis breeds were introduced. These breeds, adapted to tropical climates, had great possibilities for success in places with expansive pasturelands, including Cuba, the Dominican Republic, and Jamaica. Raising livestock for meat was also promoted in the continental Caribbean, including Colombia, Venezuela, Guiana, and Central America, regions from which livestock had been shipped to island markets since colonial times. In a number of these countries, mid-twentieth-century livestock production was oriented toward producing beef for fast food businesses in the United States, creating a "hamburger connection."[23] At the same time, huge strides were made in obtaining animal protein from hog and poultry livestock, including egg production.

Industrial Metabolism and Urbanization

Beginning in the mid-twentieth century, the role of plantations in Caribbean societies gradually began to give way to other forms of natural resource exploitation and to the search for new economic alternatives. This process took place in the context of the socioenergy transition tied to oil, urbanization, and efforts to achieve development through agricultural and industrial modernization. One manifestation of the shift to an industrial metabolism was mining activity.

During World War II, Cuba became the world's second leading nickel producer with heavy investment from North American companies responding to wartime demand. Following the 1959 revolution, the Cuban government nationalized mining companies and forged agreements with the Soviet Union. The USSR's disintegration in 1991 paved the way for Canadian capital to enter into nickel mining. Other important nickel deposits are found in the Dominican Republic. In both countries, the metal is extracted through open-pit processes, leading to serious environmental impacts. Important deposits of nickel are located in areas of high ecological value, including Cuba's largest national park, Alejandro Humboldt.

Bauxite is also widely mined in open pits in the region. During the first half of the twentieth century, deposits in Guiana and Surinam were vital to U.S. industry. Bauxite mining boomed in the 1950s and 1960s. Jamaica became the world's leading exporter in 1970, when 18 percent of its territory was held in mining concessions controlled by six United States companies and one Canadian firm; most value-added processing took place outside of the Caribbean. In this context, the governments of Jamaica and Guiana began to fashion nationalization proposals and imposed new redistributive taxes, policies that generated hostile responses from foreign companies. Caribbean bauxite production, which represented two-thirds of the world total mined in the 1960s, fell to one-sixth of total output two decades later. The legacy of mining has been antiquated equipment left behind on polluted lands. The ecological degradation sparked protests in Jamaica, leading to the creation of environmental protection laws and the restoration of abandoned mining areas.[24]

A new wave of mining, beginning in the 1990s, is taking place in countries such as the Dominican Republic (gold and nickel) and Cuba (nickel), and, above all, in Central American nations such as Honduras, Nicaragua, and

Guatemala, and is generating multiple social conflicts. Canadian and U.S. companies predominate, and demand from emerging powers, such as China, is also a factor.

Venezuela and Trinidad remain the Caribbean's most important centers of fossil fuel extraction. The Caribbean is present in oil geopolitics with local producers, as well as those of the Middle East and Africa, selling the oil in the United States market. A portion of production is processed in refineries in Aruba, Curaçao, Trinidad, the United States' Virgin Islands, and Puerto Rico. The petroleum policies of the Venezuelan revolution, led by Hugo Chávez, challenged U.S. hegemony in the region by supplying Caribbean nations with subsidized oil through the Petrocaribe project.

Industrialization was the goal of economic development policies implemented since the mid-twentieth century. The most emblematic example was Puerto Rico's "Operation Bootstrap" (*Operación Manos a la Obra*), beginning in the late 1940s. Through a policy of "industrialization by invitation," authorities intended to substitute the sugar monoculture with private investment in the industrial sector, providing fiscal incentives to United States companies. That process produced massive emigration, rapid urbanization, and the gradual abandonment of agriculture so that more than 90 percent of the island's food had to be imported. At the beginning of the twenty-first century, 94 percent of Puerto Rico's population lived in cities, marking the island with one of the highest human densities per square kilometer in the world.[25]

From an energy supply based on hydroelectric dams around the mid-twentieth century, Puerto Rico quickly moved to thermoelectric plants that supplied 99 percent of the island's energy by the century's end. The island has one of the highest levels of fossil fuel consumption in the Caribbean due primarily to the transportation sector (0.75 vehicles per inhabitant, the highest of any United States jurisdiction).[26] Transnational corporations consume a significant portion of Puerto Rico's electricity supply.

At first, industry in Puerto Rico was concentrated in the textile and food sectors before being replaced during the 1960s by petrochemicals, metallurgy, and factories producing machinery. These in turn gave way to high tech, pharmaceutical, and electronics industries. Many factories benefitted from comparatively weak environmental protection regulations, leading to serious damage to ecosystems and natural resources.[27] Numerous local communities and the environmental movement formed organizations, such as the Puerto Rican Industrial Mission (1969), to defend the environment and support poor communities with pollution problems.

Regardless of the degree of actual industrialization, Caribbean cities registered high indices of urban concentration. In 1920, Santo Domingo had fewer than thirty-one thousand inhabitants (4 percent of the nation's total); by the beginning of the twenty-first century that figure had climbed to around three million (40 percent). In that same period, rural residents went from 84 percent to less than 30 percent of the total population. In Puerto Príncipe, Haiti's capital, the population rose from 150,000 in 1950 to more than 2,500,000 fifty years later. On some small islands, more than 90 percent of the population lives in urban areas.[28] The Caribbean coast of South America is dotted with several urban centers with populations of at least one million inhabitants including Caracas and Maracaibo in Venezuela, and Barranquilla and Cartagena de Indias in Colombia. In contrast, there is little urbanization on the Caribbean coast of Central America in part because of its marginal status in countries like Costa Rica, Nicaragua, and Panama. However, in Honduras, San Pedro Sula (700,000 inhabitants) was one of the fastest-growing urban areas in late-twentieth-century Central America.

The New Tropicality of Mass Tourism

The boom in mass tourism since the mid-twentieth century marks a new period in the Greater Caribbean's environmental history, where the so-called industry without chimneys and the service economy come to the fore. Among factors explaining this evolution is the change in the image of the tropics from places of sickness and death for the "white man" to earthly paradises filled with crystalline beaches, exotic landscapes, and hedonistic pleasures, places to escape from freezing winters and the tensions of modern life. At the same time, in industrial countries, tourism gradually stopped being a privilege of the elites, a change coinciding with the revolution in passenger air transportation and cruise ships.

The plantations established in the region by United States companies from the beginning of the twentieth century facilitated the arrival of tourists. For example, the United Fruit Company and its fleet offered pleasure trips on the same ships that transported bananas and other tropical fruits. The major destination until 1959 was Cuba, served by regular airline from the United States as early as the 1920s. The number of tourists increased from 33,000 in 1914 to 272,256 in 1957, and 85 percent were from the United States. At the time, Cuba received around 25 percent of all visitors; most stayed in Havana, famous not only as an attractive city but also as the Caribbean's gambling and prostitution center. After the 1959 Cuban revolution, the activity collapsed to the point that, in 1962, only 361 tourists arrived.[29] The second most popular destination during the first half of the century was Jamaica, where the colonial government provided incentives to investors and where, by the 1920s, tourism was already the fourth most important source of income.[30]

The 1950s marked the definitive take-off of mass tourism in the Caribbean, thanks to changes in destination preferences from mountains to sunny beaches. During the 1960s, amid independence processes in various British islands, international tourism was seen as a lever for economic development. Established destinations, such as Jamaica and the Bahamas, faced new competition from Barbados, Antigua, the Virgin Islands, Puerto Rico, San Martín, and the Dominican Republic, which became especially popular beginning in the 1980s. After the deep economic crisis of the 1990s, Cuba was also reincorporated into Caribbean-based tourism in order to take advantage of the wave of tourists coming into the region.

Under these circumstances, the number of vacationers went from 3,500,000 in 1970 to 12,800,000 in 1980 and to 20,400,000 in 2000.[31] Most came from the United States, Canada, and Western Europe, which contributed to identifying the activity as a new hedonistic phase of colonialism, with the creation of a "pleasure periphery," according to the logic of a new specialization in the globalized neoliberal order.[32] Thus, while the image of earthly paradises for foreign visitors was strengthened, local community members saw themselves reduced to employees in the service economy or separated from the enjoyment of beaches and other goods that were part of their way of life, as a consequence of the commodification of nature and the culture of these countries.[33]

The mass tourist era in the Caribbean had, as a counterpart, an increase in massive migration from the region to former colonial metropoles and the so-called developed countries in general. This phenomenon led to the formation of transnational Caribbean cultural communities, whose members returned later as visitors to their countries of origin. A lesser-known subject is internal tourism by wealthier sectors or trips by urban residents in continental countries to Caribbean beaches, in many cases based on stereotypes similar to those that drive foreign tourists, together with enjoyment of a wide repertory of Caribbean dance rhythms. An example is the rise of tourist development in Caribbean Colombia, including places like Santa Marta and San Andrés Island. A similar relationship formed between affluent Venezuelans and Margarita Island.

Enjoyment of sun and beach in coastal enclaves is the dominant factor in regional tourism, which implies a degree of isolation from local societies, reinforced in many cases by "package deals." In addition, the boom in cruises (which grew from 35 percent of visitors in 1980 to 44 percent in 2001) means growing competition with land-based hotel infrastructure. Thus, income per tourist is lower, reinforcing subordination to outside operators, and cruise ship waste tends to end up on the sea floor.[34]

The relationship between tourism and environment is complex. Among the positive aspects is a revaluing of the natural environment that provides a source of income for protected areas, thus improving management. The provision of jobs to residents near tourist poles could diminish pressure on natural resources within a subsistence economy. Something similar occurs when tourism acts as an incentive for diversifying local food production.[35] However, the direct negative impacts are many, including erosion of beaches, loss of native biodiversity, introduction of exotic and invasive species of flora and fauna, changes in natural drainage, pollution of soil and water, changes in the visual landscape, and competition with communities for potable water. Sun and beach tourism is a constant threat to fragile ecosystems, such as mangroves that provide many environmental services.

In the early decades of the Caribbean tourism boom, the environmental implications tended to be overlooked. More recently, tourist infrastructure tends to be smaller-scale, situated more distant from sand dunes, and better integrated into local ecosystems. Problems such as land and water pollution are addressed, as are unnecessary use of water, lack of effective solid and liquid waste treatment, mangrove destruction and damage to coral reefs, and reduction of high energy and electricity demand. The uncertainty that climate change represents is also recognized as a serious threat to the low coastal areas that are today's preferred tourist destinations.

As regards sustainability, mass tourism touches in many ways on the history of plantations, with cycles of boom and bust related to the demands of the global economy. But tourism's vulnerability may be even greater than that associated with export agriculture, as it is not a matter of taking cheap calories beyond the sea but of attracting visitors to consume local nature and culture (or folklore). And for that to be possible, there needs to be an unprecedented deployment of energy inputs into the Caribbean region. The move to a tourism economy is not the result of an endogenous impulse to come into contact with pristine landscapes after having achieved development, but is a question of mere survival.[36] It is simply a matter of occupying the place reserved by capitalist powers within a new stage of globalization that reinforces the former structure of historical dependence.[37]

One type of tourism with tremendous potential is ecotourism, though it is not free of serious contradictions due to dispossession or limitations on access to resources by local communities. Specialists believe that this is closer to an ideal tourism with low inputs, privileging quality over quantity, and with greater sensitivity toward local ecosystems and cultures. In Costa Rica, ecotourism and adventure tourism have become the predominant types, and this corresponds to policies that favor protected areas.[38] An outstanding feature of ecotourism is that it does not require massive investment and uses designs and materials in harmony with the surroundings. It attracts tourist who are better educated and respectful of local customs, allows small-scale entrepreneurs to participate in the activity, and contributes money for conservation, encouraging respect for all life forms. At the same time, it represents an incentive for learning about and better appreciating a country's natural environment and, thus, conserving it.[39]

Conservation and Sustainable Development

The Caribbean tropical tourism boom coincides with international progress regarding conservation ideas and practices, as well as with the implementation of environmental policies. For the Lesser Antilles, as we have seen, the creation of protected areas began very early, in response to the impact of plantations on territories with limited space, especially due to adverse influences of deforestation on local climatic conditions and soil fertility.

In the Greater Antilles, formal protection of natural sites came much later, though there were earlier measures, like those designed to preserve trees for shipbuilding in Cuba (seventeenth to eighteenth centuries), the Montes Ordinances for Puerto Rico and Cuba, promulgated in 1876, and the 1884 law protecting forests in the Dominican Republic. In 1902, Puerto Rico had its first forest reserve, the Luquillo Forest (designated a National Forest in 1907). In the Dominican Republic, the Yaque Aquatic Reserve and Model Forest were created in 1926; in Cuba, the Sierra de Cristal National Park was established in 1930; in Haiti, the San Rafael National Forest was created in 1936. During the first half of the twentieth century, Caribbean governments also adopted measures for the creation of flora and fauna refuges to protect threatened species, such as Cuba's flamingos and Guatemala's quetzals.

The most recent boom in conservation policies coincides with the plantation's diminished importance, which has led to the relative abandonment of rural spaces. In Puerto Rico, recovery of forest cover rose from 5 percent in 1940 to 30 percent in 1990, though only 8 percent of that island land is protected. On the other hand, the rise in constructed areas, constituting 15 percent, is worrisome. The Dominican Republic increased its forest cover from 14 percent in 1967 to 30 percent, nearly one third of which is protected. The situation is far from idyllic but is certainly much better than Haiti's, with its territory almost completely deforested and with high levels of erosion. In Cuba, recovery of the forest cover was slow between 1959 and the 1980s, from 14 percent to 18 percent of the national territory. But, during the economic crisis of the 1990s and early 2000s, forests expanded to cover 25 percent of the nation. The increase in forests is not always equivalent to protected areas, but may also refer to forestry plantations or secondary vegetation. In Cuba, 22 percent of the national territory is covered by various conservation categories; most of these areas (14.5 percent) are marine ecosystems.

Continental Caribbean territories appear to be better preserved. That is due in large part to low population densities. This panorama characterizes most of the coastline, from Yucatan to the Gulf of Maracaibo. Even an active port city like Colon appears abandoned in comparison with Panama City and its flashy skyscrapers. On the other hand, close to 80 percent of Venezuela's population is concentrated in a narrow strip between the cordillera and the Caribbean coast, where economic activities are more oriented toward the sea.[40]

The relative isolation of Central America's Caribbean coasts contributes to the existence of areas for conservation. One example is the Path of the Panther Project, proposed in 1990, within the framework of peace processes and integrationist efforts, in order to protect biodiversity and contribute to economic recovery through activities such as ecotourism and agroforestry. The principal axis is a route along the entire Caribbean side, with around half of the area protected, in order to create a Central American Biological Corridor through which wild fauna can move.[41]

One of the challenges for the service and tourism economy's current phase is that people are abandoning agriculture and, with that, losing the ability and the local knowledge needed to work the land and produce food. Cuba's experience during the acute crisis of the 1990s demonstrated industrialized agriculture's vulnerability when faced with lack of inputs. The country was forced to move to a less dependent agricultural model based on organic production, which included a return to animal traction, urban gardens, and the distribution of lands belonging to state enterprises.[42] But the largest of the Antilles has by far the best correlation of all the Caribbean islands between agricultural lands and population. So, what would happen if a crisis of similar proportions hit one of the islands with a density completely incompatible with generating a life that does not depend on the constant flow of fossil fuels?

A Brief Balance

Weighing the interaction in the Greater Caribbean between its societies and the natural world of which they are a part is a subject with multiple variables that cannot be reduced to declarations of environmental degradation. That does not mean that we should be satisfied with a linear history of progress and economic growth, colored by social consequences such as slavery, class, race, and general inequality, or colonial dependence and imperialism. Their location in the tropics and the changing way in which these countries have related to hegemonic powers are interwoven with concrete material transformations in the region's ecosystems and the ways these are perceived.

Caribbean plantations depended initially on tropical potential to take advantage of sunlight, high temperatures, and annual precipitation, as well as forests and organic matter in the soil. But these conditions were not perpetual and it was necessary to turn to fertilizers, irrigation, mechanization, and scientific knowledge. The boom in the Caribbean's commercial crops was much more than an expression of the "comparative advantage" bestowed by nature that multiplied harvests. Adopting an intensive agricultural model helped to overcome ecological realities that acted as a brake on plantation and livestock productivity, but in the long term the model created new problems due to increased technological and energy dependence as industrial agriculture become consolidated.

The idea of the conquest of the tropics was part of the imaginary that came with economic and geopolitical expansion of the United States into the Greater Caribbean. Of course, it was not only a matter of discourse but a way of intervening in concrete landscapes that, above all, was intended to incorporate the region into the flows of that country's industrial metabolism as a source of raw materials and as markets for its products. In the context of the Cold War and development policies promoted by new international organisms or the contending blocs, that metaphor of the conquest of the tropics was replaced in large part by the conquest of, or domination over, nature. Nevertheless, no matter how hard one tries to apply industrialization models or struggles for food sovereignty, tropicality comes back with unexpected force in the form of longing for an earthly paradise to escape cold winters and modern life in the former colonial metropoles. After the fear of tropical diseases was overcome, the images and perceptions of the Caribbean reinforced the idea of an Eden over that of a hell, without abandoning deeply rooted stereotypes regarding the inhabitants and local ecosystems.

But international tourism depends on more than beaches and exotic landscapes. It requires transportation possibilities and the movement of materials and energy by industrial society, as well as making real the odd idea of a paradise that combines sunbathing with a resort's air conditioning. Caribbean societies, and above all the powerful groups that benefited the most from colonial and postcolonial relations, took advantage of opportunities that were available in the hegemonic centers of an ever more interconnected world. In any case, we cannot ignore the fact that historical resource use in Caribbean territories led to great transformations in ecosystems and agroecosystems on which human beings and other species depend. Nor can we forget the paradoxical fact that the construction of independent nations and popular sovereignty is based, in contemporary society, on unsustainable material flows beginning with the limited resources of most of the Greater Caribbean.

Reinaldo Funes Monzote studied at the University of Havana and obtained his Ph.D. in 2002 from the University Jaume I in Spain. He directs the geohistorical research program at the Núñez Jiménez Foundation in Cuba and is a professor of history at the University of Havana. His book *De bosque a sabana. Azúcar, deforestación y medio ambiente en Cuba, 1492–1926* (Siglo XXI, 2004), revised and translated into English as *From Rainforest to Cane Field: A Cuban Environmental History since 1492* (UNC Press, 2008), won multiple awards in Mexico and the United States. He is currently visiting professor in the MacMillan Center at Yale University (2015–2019).

Notes

This chapter was translated by Mary Ellen Fieweger and John Soluri.

1. Broader definitions include the area of the Gulf of Mexico or start from concepts such as Afro-America. This text refers only to those territories in the Caribbean Sea basin with very similar economic, social, and environmental processes; the Mexican Caribbean is not included. See Antonio Gaztambide-Géigel, "La invención del Caribe a partir de 1898 (Las definiciones del Caribe como problema histórico, geopolítico y metodológico)," in his book *Tan lejos de Dios . . . Ensayos sobre las relaciones del Caribe con Estados Unidos* (San Juan: Ediciones Callejón, 2006), 29–58.

2. John McNeill, *Mosquito Empires: Ecology and War in the Greater Caribbean, 1620–1914* (Cambridge: Cambridge University Press, 2010).

3. Louis A. Pérez Jr., *Winds of Change: Hurricanes and the Transformation of Nineteenth-Century Cuba* (Chapel Hill: North Carolina University Press, 2001); Stuart B. Schwartz, *Sea of Storms: A History of Hurricanes in the Greater Caribbean from Columbus to Katrina* (Princeton: Princeton University Press, 2015).

4. Manuel Valdés-Pizzini, *Una mirada al mundo de los pescadores en Puerto Rico: Una perspectiva global* (Puerto Rico: Centro Interdisciplinario de Estudios del Litoral, 2011).

5. Frank Moya, *Historia del Caribe. Azúcar y plantaciones en el mundo atlántico* (Santo Domingo: Ediciones Ferilibro, 2008).

6. Barry Higman, "The Sugar Revolution," *The Economic History Review* 2 (2000): 213–236.

7. David Watts, *The West Indies: Patterns of Development, Culture and Environmental Change since 1492* (Cambridge: Cambridge University Press, 1987).

8. Richard Grove, *Green Imperialism: Colonial Expansion, Tropical Island Edens and the Origins of Environmentalism, 1600–1800* (Cambridge: Cambridge University Press, 1995)

9. Sidney Mintz, *Sweetness and Power: The Place of Sugar in Modern History* (New York: Viking-Penguin, 1985).

10. Reinaldo Funes Monzote, *From Rainforest to Cane Field in Cuba: An Environmental History since 1492*, transl. Alex Martin (Chapel Hill: University of North Carolina Press, 2008), ch. 3.

11. Fernando Picó, "Deshumanización del trabajo y cosificación de la naturaleza: los comienzos de café en Utuado," *Cuadernos de la Facultad de Humanidades* 2 (1979): 55–70.

12. Richard P. Tucker, *Insatiable Appetite: The United States and the Ecological Degradation of the Tropical World* (Berkeley: University of California Press, 2000).

13. William C. Gorgas, "The Conquest of the Tropics for the White Man," *The Journal of the American Medical Association* LII, no. 25 (June 1909): 1967–1969.

14. César Ayala, *American Sugar Kingdom: The Plantation Economy of the Spanish Caribbean, 1898–1934* (Chapel Hill: University of North Carolina Press, 1999).

15. Frederick U. Adams, *Conquest of the Tropics: The Story of the Creative Enterprises Conducted by the United Fruit Company* (New York: Garden City, 1914).

16. John Soluri, *Banana Cultures: Agriculture, Consumption, and Environmental Change in Honduras and United States* (Austin: University of Texas Press, 2005).

17. Shawn Van Ausdal, "Labores ganaderas en el Caribe colombiano, 1850–1950," in *Historia social del Caribe colombiano:Territorios, indígenas, trabajadores, cultura, memoria e historia,* ed. José Acuña and Sergio Solano (Medellín: La Carreta Editores, Universidad de Cartagena, 2011), 123–161.

18. Jonathan Brown and Peter Linder, "Oil," in *The Second Conquest of Latin America: Coffee, Henequen, and Oil during the Export Boom, 1850–1930,* ed. Steven Topik and Allen Wells (Austin: University of Texas Press, 1998), 125–187.

19. Peter Clegg, *The Caribbean Banana Trade: From Colonialism to Globalization* (New York: Palgrave MacMillan, 2002).

20. Ronny Viales, *Después del enclave: Un estudio de la región atlántica costarricense, 1927–1950* (San José: Editorial de la Universidad de Costa Rica, 1998).

21. Bonham Richardson, *The Caribbean in the Wider World, 1492–1992: A Regional Geography* (New York: Cambridge University Press, 1992).

22. Donald MacPhail, "Puerto Rican Dairying: A Revolution in Tropical Agriculture," *Geographical Review* 53, no. 2 (April, 1936): 224–246.

23. Tucker, *Insatiable Appetite,* 323–332.

24. Richardson, *The Caribbean in the Wider World,* 118–120.

25. Tania López-Marrero and Nancy Villanueva, *Atlas ambiental de Puerto Rico* (San Juan: Universidad de Puerto Rico, 2006).

26. Edwin Irizarry, *Fuentes energéticas: Luchas comunitarias y medioambiente en Puerto Rico* (San Juan: Universidad de Puerto Rico, 2012).

27. Julio Muriente, *Ambiente y desarrollo en el Puerto Rico contemporáneo: Impacto ambiental de la Operación Manos a la Obra en la región Norte de Puerto Rico: Análisis geográfico-histórico* (Rio Piedras: Publicaciones Gaviota, 2007).

28. Héctor Pérez-Brignoli, *Breve historia de Centroamérica* (Madrid: Alianza, 1990), 112.

29. Eros Salinas, *Geografía y turismo: Aspectos territoriales del manejo y gestión del turismo* (La Habana: Editorial Si-Mar, 2003): 224–225.

30. Robert Goddard, "Tourism, Drugs, Offshore Finance, and the Perils of Neoliberal Development," in *The Caribbean: A History of the Region and Its People,* ed. Stephan Palmié and Francisco Scarano (Chicago: University of Chicago Press, 2011), 571–582.

31. David Duval, "Trends and Circumstances in Caribbean Tourism," in *Tourism in the Caribbean: Trends, Development, Prospects,* ed. David T. Duval (London: Routledge, 2004), 3–22.

32. Mimi Sheller, "Natural Hedonism: The Invention of Caribbean Islands as Tropical Playgrounds," in *Tourism in the Caribbean: Trends, Development, Prospects,* ed. David T. Duval (London: Routledge, 2004), 23–38.

33. Marian A. Miller, "Paradise Sold, Paradise Lost: Jamaica's Environment and Culture in the Tourism Marketplace," in *Beyond Sun and Sand: Caribbean Environmentalism,* ed. Sherrie L. Baver and Barbara Deutsch Lynch (New Brunswick: Rutgers University Press, 2006), 35–43.

34. Robert Wood, "Global Currents: Cruise Ships in the Caribbean Sea," in *Tourism in the Caribbean: Trends, Development, Prospects,* ed. David T. Duval (London: Routledge, 2004), 152–171.

35. Duncan McGregor, "Contemporary Caribbean Ecologies: The Weight of History," in *The Caribbean: A History of the Region and Its People,* 39–51.

36. Roderick Nash, "The Exporting and Importing of Nature: Nature Appreciation as a Commodity, 1850–1980," *Perspectives in American History* 12 (1979): 517–60.

37. Marian A. Miller, "Paradise Sold, Paradise Lost: Jamaica's Environment and Culture in the Tourism Marketplace," 36–38.

38. Sterling Evans, *The Green Republic: A Conservation History of Costa Rica* (Austin: University of Texas Press, 1999).

39. Tamara Budowski, "Ecoturismo a la tica," in *Hacia una Centroamérica verde: Seis casos de conservación,* ed. Stanley Heackdon et al.(San José, Costa Rica: DEI, 1999), 73–92.

40. Monique Bégot, P. Buléon, and P. Roth, *Caribe emergente. Una geografía política* (Paris: Harmattan, 2013), 39.

41. In this territory are located Guatemala's Tigre National Park, the Honduran Río Plátano Biosphere Reserve, the Nicaraguan Cayos Miskitos Marine and Coastal Reserve, the Costa Rican Tortuguero National Park, and Panama's Darien National Park. Anthony G. Coates, ed., *Paseo pantera. Una historia de la naturaleza y cultura de Centroamérica* (Washington: Smithsonian Books, 2003).

42. Fernando Funes-Aguilar, *Sustainable Agriculture and Resistance: Transforming Food Production in Cuba* (Oakland: Food First Books, 2002).

 CHAPTER 3

Indigenous Imprints and Remnants in the Tropical Andes

Nicolás Cuvi

In the recent past, Ecuador (2008) and Bolivia (2009) reformed their constitutions, incorporating the concepts of *sumak kawsay* and *sumaq qamaña* (living well, beautiful life, or good life). For the first time in the world, Ecuador also recognized the rights of nature. The reforms aroused great interest within and beyond the Andes because they show an intention to enter into a dialogue with indigenous thought. They are statements that acknowledge cultural diversity and its expressions within those countries, valuing them as key elements for rethinking society in the twenty-first century.

The formalization of such discourses has been possible largely due to dense indigenous populations that live in Ecuador and Bolivia. With a heightened intensity since the 1990s, indigenous actors established links with both the formal powers of national politics and sectors of the population that are convinced that it is necessary to rethink development and society by looking at certain indigenous beliefs, knowledges, and practices that survive to the present day.

Although some observers argue that *sumak kawsay* and *sumaq qamaña* are "many-sided, ambivalent and nebulous concepts which are difficult to specify" and "convergent or divergent depending on the ideological and political use which is made of them,"[1] their enshrinement in contemporary constitutions is a sign of something singular. In the countries of the tropical Andes there have historically existed—and still exist—ways of existing/being/thinking/acting that are different from the ones derived from a Eurocentric cosmovision and praxis that have been hegemonic in the projects of development, progress, and modernity in Latin America.

In contrast with other regions where the diseases and wars related to the European conquest annihilated the native populations, dense populations of indigenous peoples survived in the tropical Andes. In 2009, the indigenous population of the Andes (whose main languages are Quechua and Aymara) surpassed nine million (table 3.1), although one should take into account the difficulties national censuses face in defining who is (or is not) indigenous.

Table 3.1. Indigenous population of the Andean highlands, circa 2009–2015.

Country	Total indigenous population*	Indigenous population of the Andean area*	% of the indigenous population who live in the Andean area	National population**	% of the indigenous sector of the total population in the Andean area***
Bolivia	5,002,646	4,535,066	90.65	10,027,254	45.23
Peru	3,920,450	3,696,509	94.28	27,412,157	13.48
Ecuador	582,542	415,061	71.24	14,483,499	2.87
Colombia	1,392,623	372,538	26.75	42,888,592	0.87
Argentina	600,329	143,757	23.94	40,117,096	0.36
Chile	692,192	78,889	11.39	16,572,475	0.48
Venezuela	534,816	9,722	1.81	28,946,101	0.03
Total	12,725,598	9,251,542		180,447,174	

Sources: (*) Inge Sichra, "Andes," in *Atlas sociolingüístico de pueblos indígenas en América Latina,* ed. Inge Sichra, 513–644 (Cochabamba: UNICEF and FUNPROEIB Andes, 2009), 516; (**) latest census of the population in each country; (***) the percentages are approximations because some censuses took place after Sichra's study.

The indigenous population has played different roles in the history of the Andean republics. For example, armies made up of indigenous people were crucial in the nineteenth-century revolutions in Bolivia, which they joined out of the conviction that they would thus prevent the loss of their communal lands. They have also organized themselves to fight against republican interventions in their territories and culture, at times through armed groups.

The coexistence of indigenous peoples and the rest of society has been sometimes negotiated, sometimes imposed, sometimes peaceful, sometimes violent. The tensions are symbolized, for example, in the *Yawar fiesta,* which is very popular in Peru and features a fight between a condor and a bull, a re-creation of the battle between the Andean and the Spanish cultures. The condor, tied to the back of the bull, has an opportunity to fight for a few minutes, causing rivers of blood to be shed, after which it is freed.

From those relationships there have also emerged hundreds of *mestizajes,* hybrid identities and syncretisms, which are usually, but not always, identified with the *criollo* culture. Typical expressions of this fusion are the *chagras* of Ecuador, cowboys of the highlands who mix the indigenous poncho with the Spanish *zamarro* (leather chaps), admire both cultures, and are at ease with their combination. It is also seen in the indigenist writers who expressed their views on the relations between indigenous people, *criollos,* and *mestizos* in novels such as *Raza de Bronce* (Bronze race, Alcides Arguedas, 1919), *Tempestad en los Andes* (Tempest in the Andes, Luis Eduardo Valcárcel, 1927),

Huasipungo (Jorge Icaza, 1934) and *Los ríos profundos* (Deep rivers, José María Arguedas, 1958), among others.

Indo-European, Mediterranean, and Andean technologies have been blended to grow potatoes along with wheat and barley, or to rear llamas, alpacas, and *cuyes* (guinea pigs) along with pigs, chickens, and cows. The same is seen in an architecture that joins earth to ceramic tiles, glass to stone, and straw to metal. Many indigenous persons have included tinned fish, rice, and noodles in their diets, or have replaced *chicha* with beer. They have assimilated Western communication technologies, machinery, seeds, modern agricultural inputs, and consumptive practices that produce pollution. And many have joined the republican powers and their forms of government.

Despite these mixtures, not all of the indigenous (or *criollo*) people are the same. And while there is "an elastic ethnic frontier" in which one has to take into account the "heterogeneous universe which is hidden behind the category 'indigenous,'"[2] this chapter proposes an environmental history that regards the differences between indigenous people and *criollos* as crucial, since they have led to singular and distinctive landscapes and artifacts. I share the idea that indigenous people have "other modalities of relating to nature which, originating several thousand years ago, still exist at the start of the new millennium in the rural areas of those nations which, due to resistance or marginalization, have managed to avoid the cultural and technological expansion of the industrial world."[3]

Long recognized as agents of social, political, and cultural histories of the tropical Andes, indigenous people have also been key actors in its environmental history. And they have left traces and remnants in the realm of ideas and material practices: animals, seeds, the management of soils and forests, uses of water, slopes, and altitudinal gradients, tools and agricultural practices, among other elements associated with particular forms of social relations that in some cases survive to the present day. In short, this not a history of what it has meant to be an indigenous person, but one which calls attention to the geographical imprints left by indigenous people as a key approach for explaining the environmental history of the tropical Andes.

That said, I do not mean to argue that the most visible remnants of the past two hundred years can be attributed to indigenous people: the project of transforming the territories of the republics into primary exporting economies has above all been led by the *criollo* sphere. Much has been written about export economies, some of which I synthesize in this chapter. Following the formation of the tropical Andean republics, there was an aspiration to organize their territories in accordance with the French model that inspired Simón Bolívar, regardless of whether the government was liberal or conservative, dictatorial, *caudillista* (run by a *caudillo*), or of the right or left. Government institutions and *criollo* society have constantly pursued modernity as defined by North Atlantic societies.

But the socioenvironmental changes that occurred within the framework of state or private-sector projects did not happen without stresses, since many territories within these nations had forms of government, settlement patterns, and resource uses that were not wholly or even partly transformed by colonial institutions.

Finally, I do not set out to defend or highlight indigenous practices as pure and sustainable, nor portray the *criollo* society as the source of all the negative socioenvironmental externalities in the Andean countries. My premise is that semiotic and material imprints and remnants of the indigenous peoples have existed and still exist, and that they are relevant for the environmental history of the tropical Andes and elsewhere in Latin America where dense clusters of ancestral populations live (for example in Mexico).

The chapter consists of four parts. The first explains the biogeophysical context of the tropical Andes and some crucial indigenous adaptations to the highlands, such as the vertical use of space. The second part goes into some of the elite and state-led transformations that occurred after ca. 1820, especially the efforts to obtain raw materials and mobilize labor for production for both export and internal markets. The third part highlights how, in the midst of these dizzying transformations—and sometimes as part of them—the indigenous inhabitants of the Andes, as they had done during the Colony, kept alive adaptive practices that have had influence on the agrodiversity, water-management technologies, soils, and forests, and even in urban and peri-urban settlements. I conclude by calling for heightening the visibility and recovery of Andean indigenous modes of living; for these purposes, environmental history may serve not only as a timely narrative but as a tool.

Adaptation to a Mountainous Environment

The tropical Andes, also known as the Northern and Central Andes, run from the north of Chile and Argentina to western Venezuela. They include large stretches of land in Bolivia, Peru, Ecuador, and Colombia. Their highest point is the *nevado* (snow-capped peak) Huascarán, at 6,768 meters above sea level, and their lowest elevations lie around 600 meters above sea level on both sides of the cordillera, with variations due to slope and latitude.

They cover more than a million and a half square kilometers, of which 78 percent correspond to natural ecosystems and the rest to lands that have been modified by human activity. They are one of the regions with the greatest microclimatic and biological diversity found on Earth: they harbor 45,000 species of vascular plants (of which 20,000 are endemic) and 3,400 species of vertebrates (including 1,567 endemics) on 1 percent of the planet's continental mass.[4]

Four countries of the tropical Andes (Bolivia, Peru, Ecuador, and Colombia) head world lists of endemism and number of species per unit of area.[5] This diversity exists overall thanks to the evolutionary processes brought about by the barriers and gradients created by the mountains. For example, the tapirs or *dantas* (genus *Tapirus*) include an Amazonian species, a coastal species, and an Andean species.

In contrast with the Himalayas and the mountains in east and central Africa, the tropical Andes run from south to north, and their coasts on the Pacific Ocean are bathed by two sea currents—one cold and the other warm—which converge at the Equator. And in contrast with the southern portion of the Andes (primarily in Argentina and Chile), they are not subject to severe seasonal changes in temperature or luminosity: instead, microclimates predominate in which every day of the year has almost the same number of hours of darkness and light. They are subject to drastic daily oscillations of temperature, above all at the highest altitudes and during the dry season. Most tropical Andean territories have a dry and a rainy season, but they do not coincide throughout the tropical Andes, neither in the time of the year nor in the amount of rainfall. Intermittent El Niño events add to the variability of annual cycles of precipitation.

Coping with that diversity, complexity, and uncertainty was a challenge for the different populations of the region: the indigenous inhabitants who dynamically adapted to the region over thousands of years; the Spanish colonizers; and later those of the republics, who were determined to link the mountains with the nation, the world, and European visions of modernity.

One extreme challenge was to use places at altitudes over 4,000 meters above sea level, marked by low concentrations of oxygen, a harsh cold, and high daily doses of ultraviolet radiation, similar to the conditions in Tibet. The archaeological sites of Pucuncho, at 4,355 meters above sea level, founded at least 12,800 years ago, and Cuncaicha, a 12,400-years-old transitory hunting encampment situated 4,480 meters above sea level, are testimonies to the long-term human occupation of the roof of the Andean world.[6] In the city of El Alto (Bolivia), more than 850,000 people (mostly indigenous) live at an altitude of 4,070 meters above sea level.

But, over the past few millennia, most Andean people have lived in areas situated in a range between 2,000 and 3,500 meters above sea level. Centers of indigenous, colonial, and republican political and administrative power formed in this altitudinal range, including the cities of Arequipa, Bogotá, Cochabamba, Cuenca, Cusco, La Paz (*Chuquiagomarka* in Aymara), Pasto, and Quito.

Life in the high Andes has been possible largely due to the connections with the surrounding foothills and hot plains, which have allowed for the vertical integration of the territory. The settling and maximum leveraging of the different altitudinal ecological floors by societies and families have led to the exchange of highland produce, such as potatoes, maize, and vegetables, with those of the intermediate zones (for example, chili peppers, coca, and coffee) and those of the lowlands, including cotton, salt, and river and sea fish. By banking on several microclimates for resources, the inhabitants of the tropical Andes have overcome restrictions, uncertainties, and risks.

Different names have been given to this practice of integrating ecological floors.[7] Anthropologist John Murra coined the term "vertical archipelago," but he acknowledged that the concept was not a novel one for the region.[8] However, the notion of "islands" separated by altitudinal gradients in a continuous or broken manner requires greater precision, as the vertical archipelago functioned differently across the tropical Andes. For example, it worked differently in the expansive, relatively arid highland region that housed the Andean empires studied by Murra than it did in the wetter *páramos* of the northern Andes (modern-day Colombia and Ecuador), marked by compact ecological gradients and indigenous societies organized into *señoríos étnicos* (chiefdoms).[9]

This exchange of products between vertical zones of varying proximities has been driven and regulated (though not necessarily determined) by networks of extended families and communal affiliations that have operated at different intensities in the course of the region's history. In the northern Andes, people known as Yumbos were also responsible for many of the exchanges between altitudinal zones. In the southern Andes, the *chasquis* carried seafood from the beaches to Cusco. In the highland and lowland marketplaces of the tropical Andes, one can still see the mosaic of foods, fibers, woods, medicines, dyes, and perfumes from the adjacent altitudinal zones, which now includes hundreds of plants and animals introduced since the beginning of the Columbian Exchange.

The Spanish colonizers adapted to the vertical landscape in their own manner, extracting silver from the high peaks at Potosí, raising sheep in Quito and its surroundings, and establishing their vice regal capitals and ports in Bogotá, Lima, Guayaquil, and Cartagena. The Spanish used the highest lands to produce foodstuffs and, since many of their original indigenous inhabitants had survived, the highlands served as "reserves of manpower" for mines and large *haciendas* (estates) situated at any altitude.

After the end of the colonial rule, some *haciendas* in the northern Andes continued to exploit multiple ecological floors:[10] these were multi-*haciendas* holdings, more like discontinuous islands that constituted "a notable but not unusual effort to assemble a vertical gradient into agricultural and stock-raising zones at different altitudes (and thus with different microclimates), which complemented each other in an overall scheme of production."[11]

Part of the republican longing to link the highlands to the coastal plains (and to the world) was also evidenced by the building of railways that began in the mid-nineteenth century. This iron embrace united internal markets and enabled the flow of certain merchandise toward export markets. The Bolivian railway arose to serve the needs of the silver mines and later of tin extraction. In Peru, the railways similarly expanded in response to the production of copper, zinc, lead, and silver. The railroads advanced alongside telegraph lines and the dreams of forging unified, modern nations. Workers built two hundred kilometers of railways on the savanna of Bogotá and more than three thousand kilometers in the whole of Colombia, linking the mountains and lowlands. The trains strengthened cities including Barranquilla, Guayaquil, Lima, Bogotá, Quito, Medellín, Arica, and La Paz, among others.

The railways gave importance to the places through which they passed,[12] contributing to their spatial expansion and population growth. Railroads remained the most important link through the mid-twentieth century, when highways, which also linked different zones, displaced them in conjunction with the rise of a global economy of petroleum and automobiles.

Vertical integration has also been strategic in the face of earthquakes and volcanic eruptions, which have destroyed cities and caused massive migrations. A great tectonic plate in the Pacific Ocean is subducting under the western edge of the South American plate, occasioning earthquakes like the one in 1906 that caused a tsunami to crash on the coasts of Colombia and Ecuador and kill approximately one thousand people.

The dozens of active volcanoes in the north represent latent risks. One high-risk site is the city of Latacunga and its surroundings, devastated several times in the past, the last after the Cotopaxi volcano erupted in 1877. There is less volcanic activity toward the south, where the mountains are older and higher and there is the presence of the singular *altiplano* (plateau), with ecosystems such as the humid and dry *punas* and the Uyuni and Coipasa salt flats, witnesses to the fact that those lands once lay under the sea.

Some mountains (with active or extinct volcanoes) harbor glaciers that begin to appear around 4,500 meters above sea level but, in recent times, have been constantly receding.[13] If the masses of melted ice were to burst, they could cause landslides, like the ones from the Nevado del Ruiz that destroyed the town of Armero and its surroundings in 1985, killing more than twenty thousand persons. Earthquakes can also loosen masses of ice and rocks, as happened in 1970, when a landslide hurled down from Huascarán peak and buried the city of Yungay, killing more than twenty thousand persons.

Another fundamental feature of the Andes has been water. The mountains collect and distribute water to the Amazon basin, the Pacific Ocean, and the Río de la Plata basin, leaving a singular imprint in the form of deep streams and riverbeds. The high Andes are home to the lakes of Titicaca and Poopó; the former lies at 3,800 meters above sea level and has a surface area of 3,800 square kilometers, while the latter is gradually drying up.

Andean ecosystems, like the humid *punas, páramos,* and cloud forests, store water for human consumption, irrigation, and hydroelectricity, although heavy rainfalls may cause landslides and floods. Some *punas* suffer occasional droughts, which is why the indigenous inhabitants have maintained and restored systems for harvesting water in addition to drainage and irrigation systems, including raised-bed agriculture and *amunas* (see below).

The Great Criollo Transformations

Most of the changes in the landscapes of the tropical Andes that have occurred since ca. 1820 have resulted from the visions and actions of the *criollos,* aimed at strengthening an export-oriented national economy and promoting the growth of internal markets and population centers. The governments, elites, and a broad sector of the population of the Andean republics have fomented the incorporation of territories in order to extract or produce raw materials, causing severe deforestation and pollution while creating landscapes dominated by monocultures and extractive enclaves.

This model, which intensified in the mid-twentieth century, created a dynamic marked by the advance of agricultural, ranching, logging, urban, petroleum, and mining frontiers, through boom and bust cycles. The unequal exchange of materials with the rest of the world is evidenced in negative trade balances in terms of matter flows; between 1970 and 2010 the Andean nations exported far more materials to the industrial world than they imported.[14]

By 1800, many of the highlands had been deforested, but the rhythm of destruction has increased since then. In Colombia, for example, the greatest destruction took place in Andean forests situated one thousand meters above sea level. Between 1800 and 1850, some twenty thousand hectares were deforested annually; during the following seventy years the ecosystems between one thousand and two thousand meters above sea level were deforested at a rate of twelve thousand hectares annually, primarily due to ranching. By 1920, the 350 thousand head of cattle that existed in 1850 had multiplied by a factor of five. Between 1920 and 1970, more than forty thousand hectares were deforested annually in all of the ecosystems of the Colombian Andes, primarily for coffee and cattle production. Afterward, the annual rate of deforestation rose to more than 170 thousand hectares in the Andean forests of Colombia including fifty thousand hectares in forests situated over two thousand meters above sea level. By 2000, more than eleven million cattle ranged throughout the Colombian Andes.[15] Many Andean forests were also felled to fuel the boilers of locomotives and for sleepers on railway lines.[16]

By the end of the twentieth century, 30 percent of Andean *páramos* situated at altitudes between three thousand and thirty-five hundred meters above sea level (including those dominated by shrubs, scrubs, or trees) had been completely transformed or degraded. Some 40 percent had been modified, and only 30 percent (all in inaccessible areas) remained in a natural state.[17]

The consequences of deforestation did not go unnoticed by local intellectuals. In his 1895 novel, *Capítulos que se le olvidaron a Cervantes* (chapters that Cervantes omitted), Juan Montalvo, who witnessed the impact of land clearings in the inter-Andean valleys of Ecuador, wrote that when don Quijote saw a man cutting down two cypresses, he asked why he was felling the trees, "destroying in a moment a work which Nature needed so many years to create," and added, "Wouldn't there be a way . . . to avoid this slaughter. If you are worried about the values of those cypresses, I will pay you and they will remain standing. . . . Cut down, they aren't worth anything . . . alive and beautiful as they are, they are worth more than the pyramids of Egypt. And thus I implore and beg you to consider that it is better to change your mind and make a gift to Mother Nature, who enjoys the shade of her children."[18]

At the end of the nineteenth century, efforts were made to ameliorate the negative consequences of highland deforestation by planting eucalyptuses above all, which made the yellowish landscape a common feature. They were also planted to dry out marshes. Other exotic tree species that gained favor at the time included conifers, cypresses, and planes.

Among the export products responsible for deforestation, two plants endemic to the tropical Andes stand out: coca and cinchona, the former cultivated and the latter wild. Coca links formal and informal markets; it has sacred meanings, is analgesic, and chewing its leaves helps one to adapt to high altitudes. Coca cultivation, which usually takes place at altitudes between eight hundred and two thousand meters above sea level (with exceptions as in Colombia, where since 1980 the great majority of the coca grown to produce cocaine has taken place at less than six hundred meters above sea level), intensified during the colonization of the territories that are now Peru and Bolivia (illustration 3.1).

Despite a fall in the production of coca due to the violence that occurred during the early nineteenth century, it soon returned to the external markets, partly because it was an essential ingredient of a popular U.S. soft drink: around 1900 that country imported up to one thousand metric tons of coca leaves annually, mainly from Peru. The Andean-wide surge in coca production during the late twentieth century resulted from a boom in cocaine production and as an unforeseen consequence of the colonization of lands that offered few viable alternatives for commercial agriculture in comparison with coca.[19] In Bolivia, an official monitoring program recorded nearly thirty thousand hectares of coca in the early 2000s. In Colombia, the growing of coca has provoked deforestation, while efforts to eradicate its production by fumigating with a controversial herbicide, Glyphosate, have contaminated

Illustration 3.1. Gathering the coca plant in Bolivia.

GATHERING THE COCA PLANT (*Erythroxylon coca*) IN BOLIVIA.

Wood engraving, 1867, Henry Walter Bates, *Illustrated Travels*, London, 1867–1875. Wellcome Library, London (V0043210).

Andean ecosystems. Both *criollo* and indigenous peasants participate in the coca enterprise.

For their part, the group of cinchona plants (*Cinchona* spp.) that are spread around the tropical Andean region (map 3.1) have also been the subject of regional booms and busts. Due to its content of alkaloids, especially quinine, which serves to prevent and cure malaria, the cinchona bark was extracted from different Andean regions for centuries, at varying intensities. Beginning in the forests of Loja in the seventeenth century, cinchona groves were destroyed, along with the ecosystems where they grow; since cinchona trees are scattered in small stands found at different altitudes, gaining access to the places where the trees grew resulted in collateral destruction. Later, the species *C. calisaya*, which has the greatest concentrations of quinine, was discovered and exploited in Bolivia. Finally, it was the turn of the stands in Colombia, which were intensively exploited during a *fiebre de tierra caliente* (hot land fever) during the nineteenth century.[20] South American production of cinchona bark lasted until the 1860s, when thousands of seeds were successively smuggled out of the tropical Andes to British and Dutch colonies in Southeast Asia. In a matter of decades, production from British plantations sufficed to supply

Map 3.1. The Cinchona regions of South America.

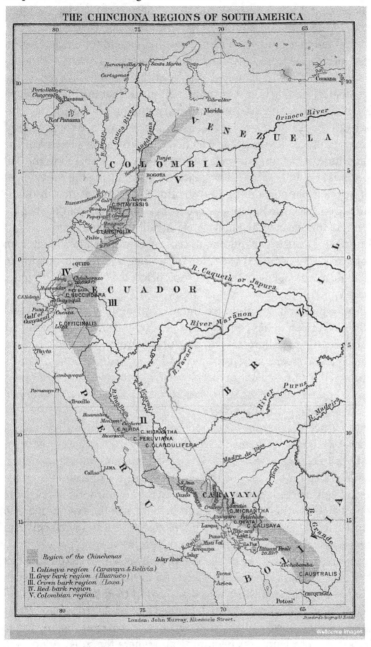

Clements Robert Markham, *Peruvian Bark: A Popular Account of the Introduction of Chinchona Cultivation into British India, 1860–1880* (London: John Murray, 1880), Wellcome Library, London (L0025458).

imperial demand, and the Dutch monopolized world production: by 1910 Java produced 90 percent, India 8 percent, and the Andean republics barely 2 percent of the twenty-five million pounds of bark produced on the planet.

World War II led to a short-lived resurgence in demand for cinchona bark from forest stands in the tropical Andes, supplemented by plantation production from Mexico to Bolivia. Although the boom lasted only five years, the degree of ecological destruction was significant given that during that short period the United States imported around forty million pounds of dry cinchona bark from the Andes. Even leaving aside the volume of bark processed in Latin American factories and that which was destroyed at the end of the war, the magnitude of U.S. imports was enormous when compared to the 350 thousand pounds sent to the Royal Pharmacy of Spain during the thirty-eight years that it held a monopoly on the Andean cinchona barks during the eighteenth century. In other words, exports rose from a rate of approximately nine thousand pounds per year to as many as eight million pounds per year, another testimony to how the exploitation of natural resources intensified from the 1940's onward. That occurred as a result of the biggest scientific mission in search of a single medicinal plant that ever took place, the Cinchona Program, during World War II.[21]

The Andean republics also promoted intensive mining. In the highest elevations of Bolivia, the earth has been opened in search of silver, tin, and other minerals. In Peru, the metallurgical complex of La Oroya is among the most conspicuous. There, copper, zinc, silver, and lead, along with indium, bismuth, gold, selenium, tellurium, and antimony, have been extracted from mountains that reach 3,700 meters above sea level. As a result, La Oroya counts as one of the most contaminated places in the world.

Many of the processes of resource extraction required the mobilization of a workforce. Labor migrations intensified in the 1950s, driven in part by government incentives to colonize areas that were called *baldíos* (wastelands), even though they had been inhabited for thousands of years. For example, in the postwar period, government policies encouraged the colonization of the Santa Cruz region in Bolivia, the Amazon region of Peru, and the region of Santo Domingo in Ecuador. This process was intensified by agrarian reforms and other processes associated with the long Green Revolution.[22]

People migrated to regions like el Chapare in Bolivia, where, despite the fact that less than 30 percent of the soils are suited for agriculture, new crops such as pineapple were introduced without any analysis of their suitability to local ecological conditions or market potential. In need of a viable crop, many farmers shifted to growing coca. Others migrated to the Amazon region, where the exploitation of petroleum was beginning and where the soils are not well suited for intensive agriculture. There were also significant rural-to-urban migrations, along with rural migrations to mining areas, and also toward highlands with *páramos* and *punas*, where the presence of *haciendas* was limited and where intensive systems of farming and ranching were introduced following agrarian reform measures.

Indigenous people participated in many of the *criollo*-led transformations of space, growing coca or pineapple, extracting the bark of cinchona trees, or toiling in the mines. They also played a part in the building of railways and highways and the founding of new towns. When some of them inserted themselves into the dreams of nation and modernity, they lost their traditional practices, but others retained them because they lived in marginal territories, were politically marginalized, or chose to resist. Those processes left imprints and traceable remains that can be found in the realm of both ideas and landscapes.

Indigenous Imprints and Remnants

The *comunas* (communes) have been fundamental for the maintenance of certain indigenous practices, since they promote (and oblige) people to participate in the organization and cohesion that are required for living in the territory, and are based on shared local interests. Throughout the nineteenth century, republican governments tried to do away with *comunas*. In Bolivia, there were repeated attempts to privatize the land and divide it into individual properties, beginning with decrees issued by Simón Bolívar in the 1820s, which abolished communal property, and continuing to the 1874 Law of Disentailment, which attempted to abolish the *ayllu* (a form of local organization, see below), privatize landowning, and create a land market that would lead to the establishment of large *haciendas*.[23] The Andean republics encouraged the creation of *latifundios* (large *haciendas*) and its principal variant, the *hacienda*, which were governed by a regime based on sharp ethnic differences in social and economic relations and matters of power.

In spite of these measures, throughout the tropical Andes and particularly in Bolivia, some *comunas* maintained (or restored) forms of organization such as the *ayllu* in which endured collective property, reciprocity, shared work, a symbolism tied to the territory, a rhythm of work related with the annual and interannual cycles of the earth, the vertical use of the space, the regeneration of the wild, the use of medicinal plants, family networks, and a stress on the collective rather than the individual.

In the *ayllus* and other forms of communal organization, work is done in a collective way in a territory that is made dynamic—not separated—by mountains and rivers. Many of the goods, services, and labor in the *comunas* are distributed by means of reciprocal relationships, not only the payment of wages or other monetary exchanges. The Aymaras have maintained the *ayni* (mutual services among families, based on reciprocity), the *chuqhus* (work parties), and *minkas* (work gangs whose members are paid with a share of the harvest). In other places, the *minkas* relates to the work on community projects, like the construction of a road, an irrigation ditch, or a communal food plot. In late-twentieth-century Bolivia, the *ayllu* has reemerged as an important component of land tenure, but that does not apply to the tropical Andes in general. For example, in some parts of Ecuador, community-level regulation of resources or labor no longer exists on any ecological floor.[24] But in the places where they persist, communal structures are fundamental for the existence of praxis that conserve forest fragments, manage waters and soils, and preserve the diversity of domesticated plants, an area in which the ancient inhabitants of the Andes advanced "probably more than any other society."[25]

There are *ayllus* in the *altiplano* that, due to repeated subdivisions, only span two ecological floors, yet this has not diminished the symbolic meanings of the importance of benefitting from diverse zones. In Huachacalla, Bolivia, the four corners of the town square represent the four *ayllus* of the region, two of the *urinsaya,* or high-altitude holdings, and two of the *anansaya,* or lower-altitude parcels.[26] In Paucartambo, Peru, as well, some families have maintained a vertical specialization in which the highest-elevation fields are planted in potatoes, and the lowest are reserved for maize. They also maintain strategies of devoting different *chacras* (food plots) to diverse crops, while also incorporating some aspects of twentieth-century agricultural modernization.[27]

The communal approach and nonmonetary relations are evident in agricultural fairs of little commercial importance, which "are primarily mechanisms for the collection and exchange of small amounts of products. These fairs coincide with community festivals, such as the *prestes,* that are an expression of reciprocity and the redistribution of the gifts."[28] In the valley of Lares (Cusco), barter markets known as *chalayplasa* are based on principles of reciprocity, redistribution, and self-sufficiency. They function to supply vital nutrients and vitamins while promoting the conservation of biodiversity, the quality of the soils, and the control of pests.[29]

Indigenous cultivators have maintained agrodiversity via constant experimentation, a requisite for the viability of agriculture in a changing environment. These experiments include establishing *chacras* (plots) at different altitudes and planting a diversity of crops—as many as forty species—in each one. In addition to the diversity within each *chacra,* a family may have thirty or more *chacras* dispersed along an ecological gradient as insurance against crop failure, a strategy known as "mass parallelism"—that is, undertaking many different activities in the widest range of ecological conditions possible to deal with unforeseen events. The Incas practiced this kind of adaptation of crops to different altitudes, traces of which can still be seen in places like Moray (near Cusco), where they maintained a form of an agricultural experiment station with its own irrigation system.

The basic idea was to take many variables into account before making decisions, through a repetitive process of trial and error—for example, experimenting with how the yield of a crop would respond to cold weather by planting it at a high elevation. Still today, "simulation and experimentation are routine practices for the native farmer of the Andes."[30] Constant experimentation also implies the staggered sowing of crops throughout the year to deal with frosts, variable rainfall, or pests. Indigenous practices like these, which created unique landscapes and environments, were maintained even under the harsh *huasipungo* system, in which landowners ceded marginal lands for subsistence planting to landless laborers.

The survival of these dynamic environments helps to explain why repeated efforts on the part of governments to promote agricultural modernization in the second half of the twentieth century were not always accepted in full: they were not well suited to maintain and reproduce traditional Andean life. The scientific approach buried or ignored indigenous experiences and knowledges. It is a paradox that Andean countries, among others in Latin America, did the opposite of what their import substitution policies were meant to achieve in the case of agriculture and diet: government policies drove the producers of food to poverty and despair in order to clear the way for agroexport businesses based on monocultures, while importing foodstuffs that could be produced in the Andes, including cow's milk or wheat.

Even so, there are several places where indigenous cultivators have not used scientifically modified seeds or other industrial inputs. Historically, they have preferred to use local seeds selected by cultivators (mostly women) and to practice polycultures. Thanks to this management of seeds and foodstuffs, nearly four thousand varieties of potatoes are grown in the Andes,[31] in addition to other native crops. Moreover, indigenous communities have maintained traditional methods for storing and processing food, fighting pests, and dealing with droughts or frosts.[32] In many places, Andean people continue to prepare cooked foods using ingredients not associated with exports or market demand. For example, the indigenous communities of the Andes conserve quinoa, a plant whose high nutritional value is increasingly recognized.

The practices of indigenous Quechua-speakers in highland Peru experienced complex dynamics with the advent of the long Green Revolution in the 1960s, but the introduced systems did not cause a total loss of traditional practices. The role women play in choosing seeds based on their size and quality has been a fundamental factor in the conservation of agrodiversity. Cultivators maintained such customs as the exchange of seeds among farmers, crop rotation, and food plots at different altitudes, even as they periodically participated in market exchanges and used highly toxic agrochemicals. In the midst of tensions between traditional and modern methods, some indigenous people have tried to take the best of both systems (to the extent possible) with changing fortunes.[33]

If true that many technologies associated with systems of communal organization fell apart following the conquest, others have been conserved to the present day,[34] an indication that it has not been possible to improve upon them. According to anthropologist John Earls, "Although the details vary a great deal . . . the same basic principles are seen everywhere . . . the spatial heterogeneity is reduced by sophisticated procedures of zonification and technologies, most of which also serve to reduce climatic uncertainties. Experimentation, selection of crops, and acclimatization are intrinsic characteristics of agricultural routines, from the family *chacra* to the built fields of the Inca State in Moray."[35]

Indigenous people have also continued to raise native animals, including varieties of edible guinea pigs (*cuyes*), the black duck, the Peruvian Hairless Dog (*perro viringo*) and its hairy cousin, and llamas and alpacas (which are related to wild vicuñas and guanacos). The Andean communities use these species of camelids as sources of protein and fibers and as beasts of burden. In colonial times, the populations of llamas and alpacas rose, but they declined in republican times. Today, there are more than three million alpacas and more than three million llamas in South America—most in Peru and Bolivia.[36] While some breeds of llamas and alpacas did not survive to the present day (including a chubby, short-legged variety that may have been a source of meat), the animals have been reintroduced in recent decades to places like the Cordillera Negra in northern Peru. As a consequence of the reintroduction of alpacas, artificial bogs comprised of succulent plants have been created, a landscape only possible due to the animals maintained by indigenous people, not by criollos or state institutions.

Indigenous imprints can also be seen in water management strategies. The raised fields (*camellones* in Spanish, *suka kollu* in Aymara, *waru-waru* in Quechua) that are prevalent in the highlands have served, among other things, for drainage and irrigation, heat conservation, and the breeding of fishes and water birds, as well as to fertilize the soil. In the tropical Andes, traces of this technology exist from the savanna of Bogotá to the *altiplano* of Bolivia. In the northern highlands of Ecuador, there are two thousand hectares of raised fields, and in the basin of Lake Titicaca there are 120 thousand hectares, situated at 3,800 to 3,900 meters above sea level. Although most of these were abandoned prior to the fifteenth century, extended families and *comunas* continued to use them for subsistence agriculture. Programs to revive raised-field agriculture, initiated in the 1980s, produced high yields, but approximately 90 percent of the 420 recuperated hectares were abandoned again, due to "the serious conceptual and communicative gaps between agronomists, technical personnel, anthropologists and archaeologists, and the local communities."[37]

The *amunas*, ancient systems to collect rainwater and runoff in the highlands in order to feed springs in lowland farming areas (among other uses), are another important form of water management technology. In contrast with the raised fields, they have remained in use since precolonial times.[38]

The same continuous use applies to the use of field terraces, platforms, or embankments (*taqanas*) that are spread over millions of hectares. At the end of the twentieth century, there were 256,950 hectares of *taqanas* in Peru, 4.4 percent of the productive land in the country, of which more than half was used for production.[39]

In addition, systems of "communally regulated fallow"—still found in Peruvian communities—have been used to conserve soils.[40] In high-altitude soils used to grow potatoes, this fallow system serves, among other things, to control a nematode that infests the crops but dies when the soil is given a chance to rest. The *aynuqa* or *aynoka*—"a very complex Andean technology for agricultural production with a rather long history"—is also worth noting.[41] That system is used to manage and maintain surveillance of communal food plots that lay distant from settlements and whose harvests are shared between families and markets. *Aynuqa* also refers to a community banquet in which one fundamental objective is to create a feeling of satisfaction among all.[42]

Many communities have conserved the practice of observing the constellation of the Pleiades (*Qullqa* in Quechua, which means a granary) in order to predict the rainfall in October and November and thereby calculate planting, production, and storage needs. In 1976, a researcher in the Misminay community observed that all of the members of the community attentively watched the *Qullqa* night after night. The start of the rainy season had been very slow, and only a few community members had planted potatoes. If planting did not commence soon, they might face a desperate situation. When the researcher asked his young guide why everyone was watching the stars so intently, the guide answered "because we want to live."[43] Even today, some communities observe the *Qullqa* and other stars in order to make decisions about agriculture.[44]

As for the native forests, indigenous people established (and still establish) forestry plantations "in a deliberate or strategic manner that contrasts with the unthinking use or destruction of the forests during and after the colonial period."[45] It has been suggested that the retreat of the high Andean forests owes less to the extractive activities of the colonial era or to the introduction of pines and eucalyptuses in the republican period, than to the decay of communal forms of land tenure and forest management. In the Quichua language, the word for some planted trees (*mallki*) is different from the word for some related wild ones (*sacha*). The fact that different Andean peoples have planted trees partly explains why the earliest European chroniclers wrote about the dense forest cover in the high Andean valleys that sharply decreased after a few decades of colonial management. The conservation of remnants of *Polylepis* forests (the *quenua* or paper tree) throughout the Andes is probably owed to their management by indigenous people long before the creation of protected areas. Tara gum, which is extracted from a thorny shrub whose pods are also rich in tannins, is an example of the benefits that the conservation of such forests provides.

There are also indigenous imprints on the cities. Urban areas in the high Andes grew slowly until the start of the twentieth century, before exploding in the second half of that century, partly because of migrations triggered by reform processes in the countryside. Cities grew in a disorderly way, shaped more by informality than by urban planning. As a consequence, many urban populations are vulnerable to earthquakes, volcanism, landslides, and floods and lack adequate access to energy, water, and other resources. The structure of the Andean population reversed, from one that was overwhelmingly rural to one that is mostly urban, with Bogotá representing the largest urban center in the high Andes, with more than seven million inhabitants in 2010.

The cities have appropriated the water of the countryside, polluted rivers and soils, and have extended into agricultural and forest lands. The severe air pollution found in cities such as Bogotá, La Paz, and Quito is due in part to the use of bad fuels in vehicles whose engines have an incomplete combustion in the altitude. The mining of construction materials and razing of infrastructure release toxic fine particulate matter into the air. These entropic[46] urban centers have relied on resources drawn from more or less distant places, located in rural low and highlands that supply them with copper, gold, tin, sugarcane, coffee, fish, flowers, petroleum, flowers, rice, shrimp, palm oil, and bananas, among other products.

These processes of urban expansion have not been the same in all places, partly due to the presence of communal lands with singular dynamics. One example is found in the peri-urban area east of Quito. Until a short time ago, the region consisted of indigenous and peasant communities, agricultural fields, *haciendas,* and weekend retreats. Today, the sector is being dominated by the affluent permanent homes of the local elite. The wealthy have purchased land, creating an ostentatious architectural mosaic of metal, glass, and concrete that contrasts sharply with the rural housing, often made of adobe, inhabited by working people. In the midst of this real estate buying frenzy, there exist *comunas* such as Lumbisí, which covers more than six hundred hectares and where there are entrance signs that read, "Welcome to the *comuna* of Lumbisí. We do not sell land here." The *comuna* has resisted successive attempts to appropriate its lands since colonial times, with consequences for the peri-urban landscape.

There are also spatial expressions of indigenous identities in unexpected places, like the Muisca indigenous group in expanding residential neighborhoods on the outskirts of Bogotá.[47] In Bolivia, 51 percent of all urban residents call themselves Quechuas, and 60 percent of the Aymara-speaking population inhabits cities where traditional mechanisms of reciprocity persist.[48] In cities such as El Alto, as in the countryside or the forests, singular dynamics operate that are associated with practices derived from the indigenous world.

Toward the Recovery of Tropical Andean Knowledges and Practices

The ideas and especially the practices described here seem to reveal that certain environmental histories of the tropical Andes may be told through the imprints and remnants left by two manners of thinking and acting, with historical particularities. The history of *criollos* and indigenous inhabitants of that region has not only been one of assimilations and mixtures, but also of populations that have constructed territories inside of nation-states, nations inside of republics, and communities inside of nations, with different manners of being/existing/thinking/acting.

Cultures and people different from those of the tropical Andes have also set out to think and act on the basis of different ontologies. Deep ecology, systemic and complex thought, ontological anthropology, perspectivism, and romanticism are some of the names given to these ideas. But it would appear that, at least in Latin America, it is only where indigenous populations survive that these beliefs and forms of thought are clearly shaped into biocultural landscapes, the result of a praxis that has created and re-created spaces, plants, and animals, even after the harsh interventions caused by the conquest, the colonial era, and republican rule.

In Bolivia, plurinationality has been recently constitutionally recognized, apparently creating the possibility of giving decision-making power about the territory to populations that have been marginalized, denigrated, and colonized for five centuries, yet have nevertheless maintained valid and resilient ideas about biodiversity and land, the territories and its cycles. Indigenous people in Bolivia have sought to act on the institutions responsible for state policy in order to recover their territories and rights for more than a century. As the Aymara leader Manuel Chachawayna declared in 1927, "We can not only be electors, but the elected—good for us; let us start with the *diputación* [legislature] and then win the Presidency of the Republic, since we are the majority."[49] That desire was achieved with the election of Evo Morales as president of Bolivia in 2006. In Ecuador as well, the indigenous movement has won formal representation in politics since the 1990s, yet the political victories have not led the people to abandon their traditional acts of resistance to the state in matters of water, seeds, agriculture, lands, and territories.

The proof of the historical persistence of certain modes of being/existing/thinking/acting is important for several reasons. If it is true that the modern world "is leading the whole species to the edge of a general state of amnesia, to a metaphorical brain death,"[50] it is possible that paths toward a utopian sustainability, which includes socioenvironmental justice, lie in certain beliefs, knowledges, attitudes, and practices of the indigenous people we are speaking of, which have been adaptive over thousands of years. For example, when more people become aware of the consequences of feeding themselves with products that lack variety and are grown with toxic agrochemicals that come from distant places, or of the importance of maintaining sovereignty over the seeds of crop plants, some of the alternatives for a sovereign and healthy production of food may be found in indigenous knowledges and practices. Just as potatoes, once regarded as the "food of the devil," have become a basic foodstuff for the world population, other indigenous ideas and practices are increasingly seen as valuable and wise. I believe that there is a need to study and strengthen these wisdoms, which include reformulating our linear understanding of time that associates modernity with the present and future, and the past with the uncivilized. We should pay more attention to that and other concepts, but without swallowing them up or co-opting their meaning in order to fit them into the preexisting models and myths of modernity.

It would be useful to reflect, for example, on the technological, environmental, and cultural incompatibilities between some of the strategies of the traditional agriculture of the tropical Andes and those of development models aimed at economic growth and profit. We have seen some good intentions to work along those lines, above all since the end of the twentieth century, but their application faces difficulties. The failure of many programs of technical assistance aimed at the recovery of indigenous technologies seems to owe more to a cultural problem, since the academic analyses have limited themselves to recording the presence of techniques and tools "without an in-depth reflection on the ecological significance of the practices of using them and how they are inserted into the construction of living realities. Andean knowledges . . . are inseparable from the traditional forms of social organization and their conception of space and time."[51] That seems to be a crucial lesson not only for the current transformations of the landscape but for the new environmental histories as well.

To care for the thousands of varieties of potatoes (and other agricultural species and varieties), the promotion of modern biotechnologies, or an International Potato Center charged with their *ex situ* conservation, does not seem to be sufficient. Adaptions have existed both in culture and on the material plane, in the *chacras* and the ways of cultivating them, in the exchange of seeds and the constant battle with the vagaries of the climate and the threat of pests. An initiative that is closer to a dialogue of knowledges, even though it has not been free of conflicts, is the *Parque de la Papa* in Cusco (Potato Park).[52] It seems necessary to strengthen the recovery of vertical complementarity (not only global complementarity) and mass parallelism with adaptive aims. In a context of instability and climatic uncertainty, the millenary wisdom of the Andean cultures seems to be adaptive because it does not occur in a laboratory but on the land, in reality, and traceable via enduring imprints and remnants.

But above all, we need to acknowledge that not all of the indigenous practices have been sustainable, nor all of the *criollo* ones disastrous from a socioenvironmental standpoint. There is a need for situated analyses, without concealing the fact that the Eurocentric and *criollo* cosmovision has been hegemonic—a view that has been more imposed than negotiated over five hundred years—and that it has led to crucial imprints in the tropical Andes.

Nicolás Cuvi is a research professor in the Department of Development, Environment and Territory of the Latin American Faculty of Social Sciences (FLACSO) in Ecuador. He researches on the edges and common territories of various socioenvironmental epistemologies, such as environmental history, urban ecology, political ecology, environmental ethics, agroecology, ethnobiology, and ecological economics, together with the history and sociology of science, within the framework of the transitions to sustainability. Since 2010, he has directed the journal *Letras Verdes. Revista Latinoamericana de Estudios Socioambientales*. His latest publication is "Las ciudades como mosaicos bioculturales: el caso del Centro Histórico de Quito" in *Etnobiología* (May 2017).

Notes

I would like to thank Claudia Leal, John Soluri, and José Augusto Pádua for their detailed comments on several versions of this article. And also Chris Boyer, Víctor Bretón, Micheline Cariño, Francisco Cuesta, Reinaldo Funes, Sebastián Granda, Shawn Kenneth Van Ausdal, Stuart McCook, Ramiro Rojas, Myrna Santiago, Lise Sedrez, Robert Wilcox, Adrián Zarrilli, and the readers of Berghahn Books for their important contributions and critiques of these ideas, as well as the Rachel Carson Center for having supported this initiative of reflection.

 1. Víctor Bretón, David Cortez, and Fernando García, "En busca del sumak Kawsay," *Íconos* 48 (2014): 9–24, 17–18.
 2. Víctor Bretón, *Toacazo: En los Andes equinocciales tras la reforma agraria* (Quito: FLACSO Ecuador, Abya Yala and Universitat de Lleida, 2012), 384.
 3. Víctor M. Toledo, "¿Por qué los pueblos indígenas son la memoria de la especie?," *Papeles* 107 (2009): 27–38, 28.
 4. Carmen Josse et al., *Ecosistemas de los Andes del Norte y Centro: Bolivia, Colombia, Ecuador, Perú y Venezuela* (Lima: Secretaría General de la Comunidad Andina, 2009). For the figure of 78 percent of ecosystems are natural areas, one must take into account that (1) the cartographic information dates back to 2008 and does not include the deforestaton of the piedmont areas during the past few years, and (2) it includes areas in the *puna* of Peru and Bolivia where vicuñas are pastured, and lands in the northern *páramos* where the vegetation cover is regularly burned. There are no reliable figures

on the vegetated area of those zones (Francisco Cuesta, personal comunication, 15 September 2015).

5. Russell A. Mittermeier, Cristina Goettsch, and Patricio Robles, *Megadiversidad: Los países biológicamente más ricos del mundo* (México: CEMEX, 1997).

6. Kurt Rademaker et al., "Paleoindian Settlement of the High-Altitude Peruvian Andes," *Science* 346, no. 6208 (2014): 466–469.

7. Mass paralleism, multicyclic agriculture or production, zones of production.

8. John V. Murra, "El control vertical de un máximo de pisos ecológicos en la economía de las sociedades andinas," in *El mundo andino: Población, medio ambiente y economía,* ed. John V. Murra (Lima: Pontificia Universidad Católica del Perú e Instituto de Estudios Peruanos, 2002/1972), 85–125. While searching for reports of this system in the past, it is necessary to come across the drawings Francisco José de Caldas made of the arrangement of plants on the gradient; see Mauricio Nieto, *La obra cartográfica de Francisco José de Caldas* (Bogotá: Ediciones Uniandes, 2006), an evidence of the knowledge which the inhabitants of the high Andes had and which, among other things, might have been a crucial inspiration for Humboldt´s geography of plants, Jorge Cañizares-Esguerra, *Nature, Empire, and Nation. Explorations of the History of Science in the Iberian World* (Stanford: Stanford University Press, 2006).

9. Frank Salomon, *Los señores étnicos de Quito en la época de los incas,* 2nd ed. (Quito: Instituto Metropolitano de Patrimonio and Universidad Andina Simón Bolívar, 2011).

10. Guerrero (1977), quoted in Bretón, *Toacazo,* 49.

11. Mark Thurner, "Políticas campesinas y haciendas andinas en la transición hacia el capitalismo: una historia etnográfica," in *Etnicidades,* ed. Andrés Guerrero (Quito: FLACSO Ecuador, 2000), 349.

12. Olivier Dollfus, *El reto del espacio andino* (Lima: Instituto de Estudios Peruanos, 1981), 114.

13. For example, the Chacaltaya and Antisana volcanoes; see Bernard Francou et al., "Glacier Evolution in the Tropical Andes during the Last Decades of the 20th Century: Chacaltaya, Bolivia, and Antizana, Ecuador," *Ambio: A Journal of the Human Environment* 29 (2000): 416–422.

14. Fander Falconí and María Cristina Vallejo, "Transiciones socioecológicas en la región andina," *Revista Iberoamericana de Economía Ecológica* 18 (2012): 53–71, 65, 69.

15. Andrés Etter, Clive McAlpine, and Hugh Possingham, "Historical Patterns and Drivers of Landscape Change in Colombia since 1500: A Regionalized Spatial Approach," *Annals of the Association of American Geographers* 98: 2–23.

16. Alexander Herrera and Maurizio Ali, "Paisajes del desarrollo: La ecología de las tecnologías andinas," *Antípoda. Revista de Antropología y Arqueología* 8 (2009): 169–194.

17. Robert Hofstede, Pool Segarra, and Patricio Mena Vásconez, "Presentación," in *Los páramos del mundo,* ed. Robert Hofstede et al. (Quito: Global Peatland Initiative, NC-IUCN and EcoCiencia, 2003), 11–13.

18. Juan Montalvo, "Capítulo XVI—De la casi aventura que casi tuvo don Quijote ocasionada por un viejo de los ramplones de su tiempo," in *Capítulos que se le olvidaron a Cervantes* (Ambato: Letras de Tungurahua, 1895/1999).

19. Andreu Viola, "Crónica de un fracaso anunciado: Coca y desarrollo alternativo en Bolivia," in *Los límites del desarrollo: modelos "rotos" y modelos "por construir" en América Latina y África,* ed. Víctor Bretón et al. (Barcelona: Icaria, 1999), 161–203.

20. Germán A. Palacio, *Fiebre de tierra caliente: Una historia ambiental de Colombia, 1850–1930* (Bogotá: ILSA, 2006).

21. Nicolás Cuvi, "The Cinchona Program (1940–1945): Science and Imperialism in the Exploitation of a Medicinal Plant," *Dynamis* 31 (2011): 183–206.

22. Raj Patel, "The Long Green Revolution," *The Journal of Peasant Studies* 40, no. 1 (2012): 1–63.

23. Tristan Platt, *Estado boliviano y ayllu andino: Tierra y tributo en el norte de Potosí* (Lima: Instituto de Estudios Peruanos, 1982), 14–15.

24. Luciano Martínez, *Economía política de las comunidades indígenas* (Quito: ILDIS, Abya-Yala, OXFAM and FLACSO Ecuador, 2002), 9.

25. Jürgen Golte, *Cultura, racionalidad y migración andina* (Lima: Instituto de Estudios Peruanos, 2001), 16.

26. Ramiro Rojas, *Estado, territorialidad y etnias andinas: Lucha y pacto en la construcción de la nación boliviana* (La Paz: Universidad Mayor San Andrés, 2009), 232.

27. Karl S. Zimmerer, *Changing Fortunes: Biodiversity and Peasant Livelihood in the Peruvian Andes* (Berkeley: University of California Press, 1996).

28. Peigne (1994) quoted in Rojas, *Estado, territorialidad y etnias andinas,* 187–188.

29. Neus Martí, "La multidimensionalidad de los sistemas locales de alimentación en los Andes peruanos: los chalayplasa del Valle de Lares (Cusco)" (Ph.D. dissertation, Universitat Autònoma de Barcelona, 2005).

30. John Earls, *La agricultura andina ante una globalización en desplome* (Lima: Pontificia Universidad Católica del Perú, 2006), 153.

31. "Potato," International Potato Center, accessed 10 March 2014, http://cipotato.org/potato/.

32. Óscar Blanco G., "Tecnología agrícola andina," in *Evolución y tecnología de la agricultura andina,* ed. A. M. Fries (Cuzco: Instituto Indigenista Interamericano 1983), 17–29.

33. Zimmerer, *Changing Fortunes.*

34. Golte, *Cultura, racionalidad y migración andina.*

35. Earls, *La agricultura andina,* 157.

36. Alexander Herrera W., *La recuperación de tecnologías indígenas: Arqueología, tecnología y desarrollo en los Andes* (Bogotá: Uniandes, CLACSO, Instituto de Estudios Peruanos, and PUNKU, 2011).

37. Herrera and Ali, "Paisajes del desarrollo."

38. Earls, *La agricultura andina.*

39. Efraín Gonzáles and Carolina Trivelli, *Andenes y desarrollo sustentable* (Lima: IEP and CONDESAN, 1999).

40. Daniel Cotlear, *Desarrollo campesino en los Andes: Cambio tecnológico y transformación social en las comunidades de la sierra del Perú* (Lima: Instituto de Estudios Peruanos, 1989).

41. Herrera, *La recuperación de tecnologías,* 58–60.

42. Herrera, *La recuperación de tecnologías,* 58–60.

43. Gary Urton, *En el cruce de rumbos de la Tierra y el Cielo* (Cusco: Centro Bartolomé de las Casas, 2006).

44. Erick Pajares and Jaime Llosa, "Relational Knowledge Systems and Their Impact on Management of Mountain Ecosystems: Approaches to Understanding the Motivations and Expectations of Traditional Farmers in the Maintenance of Biodiversity Zones in

the Andes," *Management of Environmental Quality: An International Journal* 22, no. 2 (2010): 213–232.

45. Herrera and Ali, *Paisajes del desarrollo.*

46. In the sense that they draw energy and matter from all over the ecosphere and return all of it in degraded form back to the ecosphere; see William Rees and Mathis Wackernagel, "Urban Ecological Footprints: Why Cities Cannot Be Sustainable—and Why They Are a Key to Sustainability," *Environmental Impact Assessment Review* 16, no. 4–6 (1996): 223–248.

47. Christian Gros, *Políticas de la etnicidad: Identidad, estado y modernidad* (Bogotá: Instituto Colombiano de Antropología e Historia, 2000).

48. Sichra, *Andes.*

49. *La Razón* (1927), quoted in Esteban Ticona A., "Pueblos indígenas y Estado boliviano: La larga historia de conflictos," *Gazeta de antropología* 19 (2003).

50. Toledo, "¿Por qué los pueblos indígenas…," 38.

51. Herrera and Ali, *Paisajes del desarrollo.*

52. Raúl H. Asensio and Martín Cavero Castillo, *El parque de la papa de Cusco: Claves y dilemas para el escalamiento de innovaciones rurales en los Andes (1998–2011)* (Lima: Instituto de Estudios Peruanos, Centro Internacional de Investigaciones para el Desarrollo and Fondo Internacional de Desarrollo Agrícola, 2013).

 CHAPTER 4

The Dilemma of the "Splendid Cradle"
Nature and Territory in the Construction of Brazil

José Augusto Pádua

> "To forget about space in Brazil is to be condemned
> to understand nothing, whether in the present or in the past."
> —Fernand Braudel (1943)[1]

Territory, Tropicality, and Nation-Building

Milton Santos, one of the most influential Brazilian geographers of the twentieth century, insisted upon the necessity of developing "a territory-based theory of Brazil."[2] Recognizing the importance of this intellectual challenge—in addition to opening up new channels of fertile dialogue between environmental history, historical geography, and other disciplines—brings us to one of the central aspects of Brazil's presence on the contemporary international scene.

In the historical context of the diffusion of the concept of emerging countries and the invention, in 2001, of the idea of the "BRICs" (Brazil, Russia, India, and China), territorial dimension has become increasingly important in the global political image of the countries. It is not a coincidence that the BRIC countries each occupy large territories and, on different scales, have large populations. In the cases of China and India, the size of their populations has been a decisive factor in their economies in the short term. For Russia and Brazil, despite the considerable size of their respective populations, each countries' most significant geopolitical feature is the magnitude of its territorial expanse. These countries' national space encompasses vast areas, with their respective populations occupying a relatively restricted portion of their territory, despite significant levels of national integration through communications and transportation systems. The population density of a country such as Brazil (23.8 inhabitants per square kilometer) reflects a standard that starkly contrasts with a country such as India (376.1 inhabitants per square kilometer).[3]

An analysis of these countries, at any rate, should not be limited to territorial size. Of greater importance is the question of their ecological contents. One of the founding principles of environmental history is the necessity to go beyond both "political maps" and the conception of territories as "empty," abstract spaces, which are filled in exclusively by human actions. Territories are never empty; rather they are always full and adorned by a variety of ecosystems. Social dynamics interact with these full spaces, producing places where natural and sociocultural diversity mix in a highly complex manner.[4]

The continental space on which Brazil has been constructed as a country is marked by its notable opulence and ecological diversity. It is crucial to note that among the five largest national territories in the world, Brazil is the only one entirely situated in the tropical and subtropical latitudes, encompassing around 60 percent of the enormous water-forest complex of the South American Amazon region. Brazil is thus not only a large territory (covering around 8.5 million square kilometers) but also one that is very rich in natural resources that have taken on new meaning in the contexts of both the present planetary environmental crisis, and of the new frontiers that technological research has been reaching. Brazil stands out for its concentration of tropical forests (approximately 30 percent of the world's remaining cover), biodiversity (between 10 and 20 percent of the world's stock), and water (around 12 percent of the world's availability of fresh water resources). This is also a space marked by the strong presence of solar rays and by the large reproductive capacity of the biomass and capacity for carbon storage, essential factors for controlling global warming. It is, furthermore, marked by the circulation of an immense network of rivers grouped in eight major drainage basins. In other words, the national territory features an enormous natural power source drawing on renewable energy and also, by way of recent discoveries, an important store of petroleum reserves.[5] It is thus unsurprising that contemporary Brazil's international image is in large part identified with its territory, whether in terms of its ecological wealth or, on the negative side, by the destruction of that wealth.

It is important to understand the general pattern of the kaleidoscopic array of ecosystems that exist in Brazil. To create a more intelligible, synthetic overview, this multiplicity of ecosystems has been grouped into six large biomes. Clearly, this categorization scheme cannot be understood in a rigid and absolute way, as the biophysical world exists in permanent dynamism. Each biome is a complex of different ecosystems that bear a considerable degree of similarity. There are also many areas of transition among the different biomes, with mosaics of different types of vegetation. To carry out an analysis on a regional scale, it is necessary to focus in a more detailed, concrete way on local ecosystems and their interactions. From an analytical point of view that takes the country's history as a whole, however, the classification in six biomes is quite revealing. At the moment when the European colonizers arrived in the sixteenth century, what is today part of Brazil's national territory in the northern region and along the northeastern-southern axis of the Atlantic coast was characterized by two magnificent, continuous tropical forest complexes: the Amazon Forest (originally spanning around 4 million square kilometers, if we consider only the areas that are part of Brazil today) and the Atlantic Forest (originally measuring approximately 1.3 million square kilometers). Great expanses of different types of savannas stood between these two forest complexes, especially the Cerrado with its twisted trees and acidic soils (measuring around 2 million square kilometers), the semiarid Caatinga, including significant areas that experienced periodic droughts (covering around 840 thousand square kilometers), and the Pantanal, a region with vast wetlands and abundant animal life (covering around 150 thousand square kilometers). Finally, in the extreme south, the area covering approximately 176 thousand square kilometers that encompasses the Brazilian portion of the open grasslands of the Pampa that are shared with Argentina and Uruguay. It is worth noting, too, the continuous coastline that stretches for over eight thousand kilometers of sandbanks (*restingas*), mangrove swamps, and other coastal formations.[6]

It is necessary to highlight the critical importance of this territorial reality in the formation of Brazil as a country, both literally as physical space and figuratively as an idea. The national anthem, whose lyrics were composed in 1906, exalts the fact that Brazil "eternally lays in a splendid cradle" and is "a giant by thine own nature." It is clear that the idea of the country's immenseness "by thine own nature" is just an ideological tool. Brazil, like all countries, is a social construction. But it is important to characterize the continental space with which this construction interacted. One should also note some of the singular historical characteristics of this process of construction as it unfolded in the context of the formation of new countries out of the breakdown of the European colonial empires in the Americas: (1) in contrast with the Spanish American countries, from the nineteenth century on, Brazil managed to encompass into one single political unit the various regions of Portuguese America; and (2) unlike the United States, Brazil did not need to expand by way of treaty negotiations or military conquests to obtain an enormous expanse of territory. The country received as its political inheritance, at least technically speaking, all of Portuguese America, a territory that already encompassed a space that was nearly the country's current size. Postcolonial Brazil was thus configured, one might say, as a case of precocious territorial gigantism. It is also true, however, that in large part we may characterize Brazil's national sovereignty in the nineteenth century as what we would today call "virtual," as the country's national territory existed for the most part only as a formal dominion on imperfect maps and in previous diplomatic negotiations between the Spanish and Portuguese empires. The effective occupation of this territory by the state and by the dominant European-descended society was very limited.[7] Thus, the process of the construction of an independent country in the nineteenth century was not guided by a political anxiety for external territorial expansion, but rather through the effort to prevent the country's fragmentation and to promote the gradual occupation of the enormous space under its jurisdiction.

Today, Brazil has achieved dominion over its national territory, which although extremely unequal in terms of population and economic density, has reached a rather advanced level of institutional and geoeconomic integration. Other concerns now confront the country. What are the levels of the quality of human life and ecological sustainability of the social and economic structures that the country has been constructing in its interaction with the diversity of biomes that exist in its national territory? What should the political identity be of a Latin American country of continental dimensions? What will the destiny be of the vast ecological resources found within the sphere of Brazil's political responsibility? These questions need to be examined beyond the superficial level in order to transcend stereotypes and simplistic answers. As the composer Antônio Carlos Jobim liked to say, "Brazil is not for beginners." A deep understanding of the country requires a dense contextualization, in both space and in time, of the complicated socioenvironmental processes that are part of Brazil's nation-building trajectory and of the dilemmas that present themselves for the country's future.

The Construction of a Territory and Its Environmental Histories

In the period that followed the beginning of the construction of Brazil as an independent country, from 1822 on, the neighboring countries that emerged out of the ruptures in Spanish America in general accepted the borders that the Portuguese and Spanish colonial empires had negotiated in the eighteenth century. This was the case even though these borders, for the most part, were tenuous and poorly defined and almost completely lacking in any military presence. Brazil's new "national territory" was little occupied by the society produced under Euro-descendent dominance. It is necessary to note that, relative to the magnitude of the country's formal spatial expanse, the population included in "Brazilian" society was quite small, numbering around 4.6 million persons in 1822. In 1900, the population had reached around 17.4 million. But these numbers do not include the various indigenous societies that lived in the vast spaces that were nearly untouched by the dominion of European-descended society. These indigenous societies inhabited a stretch of territory that the modern world considered Brazil, yet which was not really part of the country as a political entity. This was a different situation than the one experienced by indigenous societies that, over time, were forced to insert themselves into Brazilian society, usually in a subaltern condition. But even today, in the confines of the Amazon region, there are dozens of indigenous groups who do not know of the existence of "Brazil."[8]

The settlements that emerged from European colonialism—including individuals of European, African, and Amerindian origin—were distributed throughout national space in an unequal and fragmentary manner, forming a sort of archipelago of spots of territorial occupation. These areas were concentrated in the region of the Atlantic Forest, closer to the coast, with less dense occupations in the various savannahs and along the Amazon River. The control of these areas was in the hands of local elites and was founded on the economic exploitation of various natural resources. However, one should not think of these spots that make up the "archipelago" as isolated islands. On the one hand, they were linked by several kinds of common cultural traits, like the dominance of Catholicism and the Portuguese language. On the other hand, there were currents of exchange, whether of products, people, or cultural practices, that occurred at different levels of intensity and along different geographical expanses. Mule trains and cattle herds, for example, generated important flows that linked occupied areas in the interior and on the coastline. The country also experienced the constant movement of ships along its coasts.

These occupied spots, often based on slave labor, were encircled by vast spaces that were barely under the dominion of Euro-descended people. Since the colonial times, these areas—which some geographers today call "territorial funds," a term that refers to the lands and natural resources on them that lie inside the limits of national sovereignty but on which a nation had not yet drawn[9]—have been labeled as *sertões,* a word that refers to backland areas and was derived from "great deserts" or *desertões.* They were spaces with a higher level of density of wild flora and fauna, where indigenous populations still existed rather autonomously, sometimes interacting with *quilombos* (communities of Africans who escaped slavery) and with rural inhabitants of European origin who sought free lands for informal occupation. In addition to indigenous societies, therefore, relatively autonomous, mestizo populations were formed in the *sertões,* intimately coexisting with the diversity of the natural world. Several words, such as *caipiras* (derived from the Tupi-Guarani word *caá,* meaning forests, and denoting "inhabitants of the forests"), came to be used to denominate these populations, who were frequently denigrated by the urban elites and by the political authorities.[10] However, the *sertão*/civilization duality cannot be understood in a rigid manner. The concept of *sertões* was neither uniform nor did it refer to places that were clearly defined on the map of Brazil. The *sertões,* as areas that are barely occupied by the "civilization" of Euro-descended dominion, could have been represented in a variety of different places, even in the environs of cities. The socioeconomic density of spaces considered "civilized," in contrast, could in fact vary a great deal: from the largest coastal cities where European-derived cultural forms were more visible, to the large cattle ranches and small country towns that, from the coastal point of view, were also often called *sertanejas* (of the *sertão*). The different inhabitants of the *sertões*, furthermore, were not completely isolated, establishing contacts and exchange practices among themselves and with "civilized" areas of greater population density.[11]

For the political elites and the intellectuals who carried out the process of the construction of the national state—initiated with a monarchial regime that ruled between 1822 and 1889, with Rio de Janeiro as the capital city—holding on to formal domination of these *sertões* was imperative. They were seen as culturally "empty" territories, occupied in a rarified manner by populations that they saw as backward and ignorant. But the magnitude of the spaces and the opulence of Brazil's untamed nature generated an evident potential resource for future economic "progress."[12] In fact, the concept of "territorial funds" suggests this double meaning: backland spaces on the one hand, and, on the other, sources of wealth to be gradually consumed. For local elites, in the context of the areas that were already occupied, the adjacent wilderness areas provided space for the horizontal expansion of the economy, which attenuated conflicts over land and over the inheritances of the families of rural property owners. For an economy that almost always maintained the rudimentary and ecologically destructive technical standards of the colonial period, with little intensification of production by way of technological innovation, horizontal growth through the incorporation of new lands was essential. As soils became exhausted as a result of such imprudent practices as frequent burnings, new areas of native vegetation could be opened up with iron and fire, thus extending the devastation.[13]

The political unity of the national territory significantly facilitated this process. The Brazilian empire was threatened by regional conflicts, especially in the 1830s, but nonetheless continued. Even when there were violent regional wars, these were won by way of alliances between the monarchical state and majority sectors within the local elites that preferred to be part of a unified nation. In other words, these elites believed that a Brazilian empire would better guarantee the defense of their interests, including the maintenance of slavery, of the social order, and of the possibility of the accumulative conquest of the *sertões*. The monarchy, for its part, took pains to produce political confidence among the conservative base, presenting itself as the guarantor of territorial unity and stimulating by such means as the arts and sciences the symbolic invention of a national identity.

While the nation-state underwent the process of construction, at the same time a people was also in formation, which in general the social and political elites disdained. In the most densely populated places in the country, as well as in the borderlands of the wilderness, new populations arose by way of multiple interactions and both cultural and physical mixing. Nature's diversity was a fundamental component in the types of material life and social imagination that sprouted in different regions of the country. Tropical biodiversity was the raw material from which populations developed their instruments of both work and leisure. In cities and in the countryside, a vibrant, hybrid cluster of popular cultural practices was beginning to develop, despite the oppression, inequality, and elitism that characterized the exercise of power in a country characterized by slavery. The feeling of nationhood, however, was only quite tenuously present in that network of social interactions.

The idea of the nation was being invented from the top down, drawing heavily on the image of the greatness of Brazil's unified territory. Already in 1824, in the diplomatic documents seeking Portuguese and English recognition of Brazil's independence, the leadership of the new nation argued that the country needed to be an independent power because of its extensive territory and its natural riches.[14] It is important to note that the Brazilian state had a very limited fiscal base over the course of the nineteenth century, particularly due to the lack of direct taxation on the rural properties that made up the axis of the wealth production.[15] The state had a limited capacity to intervene concretely in the socioeconomic life at the regional level. The maintenance of social order, which to a certain degree reproduced a model that already existed in the colonial period, was carried by way of informal agreements between the central state and regional elites. In exchange for a degree of autonomy and certain privileges, regional elites accepted the institutional unity of the country and maintained order in the spaces of local dominion.

The central state—in addition to covering the basic costs of the judiciary, the armed services, and so on—was particularly decisive with regard to what one might call "political culture." It sponsored, for example, the Brazilian Historical and Geographic Institute—IHGB, founded in 1838, which developed the narrative of Brazil's historical continuity and of the legitimacy of its territorial domain. In the second half of the nineteenth century, with the growth in its stability, the monarchical regime supported the production of more precise and detailed cartographic representations of the nation, whether for diplomatic, didactic, or practical uses. At the same time, it supported artistic and literary works that might "educate" the regional elites in a feeling of nationhood. These cultural movements had a powerful environmental dimension. Brazil's natural wealth appeared as a central theme or motif in literature, in the visual arts, in the sciences, and in the official iconography of power. The production of images of the empire adorned with plants and tropical products was common, whether in print or in the production of objects. Museums sought to emphasize the diversity of the country's fauna and flora, disseminating the figure of spectacular, picturesque nature that held universal scientific value. The Brazilian displays at the international expositions focused on the riches of the country's forests, together with indigenous artisanal production, seeking to leverage a certain exoticism to inspire cultural admiration and the economic interest of foreigners.[16]

Most of the social processes and the dynamics of regional territorialization that unfolded in Brazil's national space, though, were neither under state control nor even within the capacity of the national state to oversee and intervene in them. It is, however, by understanding these processes and regional dynamics as geographically based, as having developed out of movements already anticipated in colonial-era Portuguese America, that we can visualize postindependence Brazil's environmental history—or, more accurately, environmental histories. It is thus necessary to consider the archipelago of human settlements as a complex collection of interactions among societies and biophysical conditions.

One social actor whose presence should be analyzed beyond stereotypes and facile generalizations are the indigenous peoples. It is important to analyze these populations' concrete existence, considering their diversity of languages, cultures, and territorial circumstances. In the precolonial world, there were no state structures or deep social stratification, but rather a large variety of autonomous agrarian villages or groups of hunter-gatherers. Knowledge of the fauna, flora, and physical space was very intense in these societies. One indicator of this fact is the current quantity of words in the contemporary Brazilian Portuguese related to biodiversity and geography derived from indigenous languages.[17]

Continental spatiality, however, was important in determining the outcome of European conquests of these societies. In general, indigenous populations were significantly reduced with the advance of European presence, whether by violence or by epidemiological shock brought about by the introduction of microorganisms previously unknown to the immune system of native populations. In the wake of advancing colonialism, the indigenous populations had to decide between direct confrontation and coexistence, almost always in a subordinate position, with the dominant, European-descended societies. A third option was to migrate to the interior of the continent. In the areas that were the farthest from the coast, particularly in the Cerrado and in the Amazon Forest, several peoples survived with different levels of autonomy until the effects of new waves of occupation in the second half of the twentieth century finally reached them.[18] Yet the construction of territory in Brazil was uneven and segmented in nearly all the country's regions, and situations that brought renewed friction or open conflict between the expansion of the areas under Euro-descended dominance and indigenous groups, however debilitated, continued to occur throughout the nineteenth and twentieth centuries.

Prior to the mid-twentieth century, the expansion of economic activities was concentrated in the Atlantic Forest, with less intensive penetration into Brazil's other biomes. Settled areas were established, renewed, and expanded by way of sugar cane, cotton, tobacco, cacao, and coffee plantations, as well as the extraction of minerals (after the peak of gold and diamond extraction in the eighteenth century). In addition, the production of meat, milk, manioc flour, distilled sugar cane (*cachaça*), and other foodstuffs was essential to the dynamics of territorialization in different regions. Brazilian historiography in recent decades has called attention to the relevance of production for internal consumption, transcending the simplistic image of an economy based entirely on the export of tropical products.

Despite the diversity of economic activities and processes of territorial occupation, some environmental practices were used almost everywhere. The use of fire, for example, could be considered a common denominator. It is important to understand that the forest was both a barrier and constituent part of agricultural production. The burning of trees was the principal method of clearing the land and fertilizing the soil.[19] However, one can cite some exceptions, including the cultivation of tobacco in the northeastern state of Bahia—where cultivators fertilized the soil with manure produced by the nocturnal grazing of cattle herds (the so-called *malhadas*); and the production of cacao in southern Bahia, which used the method known as *cabruca*: the partial opening of the lowest-lying strata of the forest, with the planting of cacao trees in the shade of taller trees.[20]

The intense use of burning can be understood as being based in a combination of environmental and social factors. The abundance of the Atlantic Forest—a green ocean, in the eyes of economic actors—provided no incentive for the planting of trees or the adoption of conservation measures. The ample availability of lands that were free for occupation created the notion of a frontier that was always open for economic advancement. When the soils became exhausted, new forest masses were available to be burned. It was easier and more lucrative to destroy forests than to invest in methods of careful management and conservation of the soil. Another agricultural model would have demanded more skilled workers, whereas the availability of cheap—basically slave—labor supplied the hard work of deforestation and low-quality cultivation. This type of calculation, in a certain form, remained dominant until the first decades of the twentieth century. Slavery ended in 1888, but the availability of cheap labor remained by way of the large presence of people from the rural poor, both Brazilians and immigrants.

The most emblematic and politically influential example of slash-and-burn agriculture occurred in the nineteenth century with the expansion of the cultivation of coffee along the Paraíba do Sul River valley, between the cities of Rio de Janeiro and São Paulo, as well as its shift in the twentieth century toward the western part of the state of São Paulo and the north of the state of Paraná. Nineteenth-century Brazil flooded the international market with low-quality coffee, facilitating its transformation into a massive stimulant for the workers of the urban-industrial world. The chains of small mountains in the valley had their slopes stripped bare by fire, together with the system called *picaria* (the cutting and rolling of large trees downhill, taking with them the tangle of the tropical forest). The notable level of soil erosion and the unsustainability of these plantations resulted in the constant abandonment of old productive areas in favor of new waves of deforestation. The planting of coffee in vertical lines along the hillsides to facilitate the oversight of the exhausting slave labor used there was an important factor in the loss of soils. At the end of the nineteenth century, generalized environmental degradation contributed to the economic debacle that coffee estates in the middle of the Paraíba valley in the state of Rio de Janeiro suffered, in turn having an impact on the end of slavery, the proclamation of the republic, and the transfer of the axis of coffee production to the state of São Paulo.[21]

It is relevant to mention here that, since the end of the eighteenth century, dozens of men of science denounced the negative results of agricultural production based on deforestation and slash-and-burn techniques, defending the introduction of more rational and careful land use methods. Despite the near absence of any change in the methods of agricultural production, these authors managed to establish intellectually a consistent and profound critique of the destruction of natural resources in the period.[22] In 1858, for example, at the height of coffee production in monarchical Brazil, Guilherme Capanema criticized the "agricultural system" that always sought new "virgin lands" and left behind "exhausted and unproductive terrains." Without changes to this system, in his words, the railroads would end up becoming "instruments of devastation."[23] Indeed, the railways that ran through the middle of the Paraíba valley were practically abandoned with the subsequent downfall of coffee production in the region. New railways, however, catalyzed the opening up of new farming estates and cities in the tropical forests and savannas of the western part of the state of São Paulo. Thus, the practice of burning as a means of agricultural production remained dominant, and coffee exports remained central to the Brazilian economy into the first decades of the twentieth century.[24]

Between 1884 and 1940, the arrival in Brazil of approximately 4.7 million immigrants, especially Italians, Portuguese, Spanish, Germans, and Japanese,[25] held considerable importance in the opening of new fronts of deforestation in the Atlantic Forest. In addition to serving as laborers in the new coffee estates in São Paulo, many immigrants became landholders of small properties in the forested mountains in the states of southern Brazil. Local oligarchs, who dominated the lower-lying areas by way of livestock breeding, above all on the Pampa biome, steered the immigrant farmers to the dense and less valued highland forests, where indigenous and mestizo populations made their livelihoods in subsistence economies. Cultural frictions, or even direct confrontations, were constant. European farmers accustomed to the age-old techniques of soil conservation, because of guidelines received from local authorities or pragmatism, adopted slash-and-burn agriculture as the basis of their economy, in addition to the timber industry, especially using araucaria trees. Nonetheless, governments and private colonization companies continued to place immigrant settlers in the forests of the southern Brazilian states, whose descendants migrated later in the direction of the savannas of these states' western regions.[26]

Beyond the Atlantic Forest, the processes of territorialization that unfolded elsewhere in Brazil were less demographically dense, despite producing considerable environmental changes. They allow us to visualize the place of savannas in the construction of Brazilian territory. The main catalyst was cattle raising—and on a smaller scale, more concentrated in the northeast, goat herding—for the production of fresh meat, dried meat, dairy products, and leather goods. The Atlantic Forest and the Amazon Forest were difficult environments for livestock herding on a large scale. Their total conversion into grazing fields, in turn, did not tend to compensate for the costs of deforestation. Thus, animal husbandry came to be pragmatically redirected to the *sertões* further from the coast, areas that were dominated by more open types of biomes like the Cerrado, Caatinga, and Pampa, as well as some natural open fields that existed in the Amazon. The occupied areas in the central-western area of Brazil, which had been opened up to development in the colonial period by way of gold mining, were later maintained and slowly expanded by way of livestock rising (see the chapter by Van Ausdal and Wilcox).

A particularly striking environmental consequence of this concentration of animal husbandry in the backland savannas was the tendency toward the spatial separation of agriculture and livestock, preventing the diffusion of polyculture and of the fertilization of the soil with animal manure. Furthermore, extensive animal husbandry generated numerous specific environmental problems, such as the degradation of natural pastures by the recurrent use of burning and the compacting of the soil as a result of overgrazing.[27]

Another macropattern of land occupation—generating processes of territorial construction characterized by low population density, although taking place in geographically extensive areas—was the direct extraction of components of the flora and fauna. This type of activity took place in various different biomes. Hunting for commercial rather than subsistence purposes, an understudied historical phenomenon, was significant in some parts of the Atlantic Forest at the turn of the nineteenth to the twentieth century. In the first decades of the twentieth century, for example, hundreds of thousands of killed birds were exported to serve as ornaments on hats and clothing. An international network of scientists was established to apply pressure to control this type of activity.[28] The extraction of plant resources, too, came to stimulate more permanent processes of territorialization. One important example was the commercial extraction of stocks of *maté* in the forests of the south and southwest from 1880 on, using indigenous and mestizo labor and allowing a few large companies to assert control over large territories.[29]

The most extraordinary case of plant extraction, however, which reached singular proportions, was the rubber boom in the Amazon region between 1850 and 1915. A new system of socioenvironmental interactions emerged in Brazil's northern region when latex extracted from rubber trees (*Hevea brasiliensis*) began to be used in industrialization processes overseas, especially in the production of tires for the emerging automobile industry. The economy that had existed until then, concentrated on the margins of the great rivers, had been based on the extraction of flora and fauna, on fishing, and on the commercial cultivation of a few agricultural products, particularly cacao. The new system connected extensive areas in the interior of the forest, divided into private domains called *seringais* (from the word for rubber tree, *seringueira*), with sectors at the forefront of global capitalism (see chapter by Leal). This global flow took place by way of a chain of exchanges—in fact, a chain of debts—that involved international shipping companies, businesses in northern Brazilian capital cities that dealt with foreign trade, local commercial agents who penetrated the forest by way of rivers and traded rubber for overpriced consumer goods, the owners of the *seringais* (called *seringalistas*), their staff, and, at the end of the chain, the rubber tappers, or *seringueiros*, spread out throughout the forest without any legal protections and powerfully exploited by their employers, who paid very little for the rubber that they extracted.[30]

In the enormous space of the Amazon region, the population incorporated into the dominion of the Euro-descended elites since the colonial period was traditionally small. During the cycle of rubber extraction, however, the region received a large wave of poor migrant workers, particularly those coming from regions that were subjected to periodic droughts in the Caatinga of Brazil's northeast. Professionals and adventurers from various countries also flowed into the Amazon. In 1900, the official population in the states where the exploitation of rubber was concentrated—Pará, Amazonas, and Acre—was around 1.2 million. However, rubber extraction did not demand the removal of trees. On the contrary, in order to sustain the daily extraction of latex, rubber tappers had to maintain not just the rubber trees but also the forest landscapes that gave them ecological support. Conservation was a practical requirement of the economy. Moreover, when the boom ended in the second decade of the twentieth century—as a result of the growing global hegemony of the rubber produced from the trees planted in Southeast Asia—the environmental consequences became even more attenuated. At the beginning of the 1970s, when a cycle of intense deforestation began, the Brazilian Amazon Forest still had approximately 99 percent of its original cover.[31]

The Great Acceleration and the Dilemma of the "Splendid Cradle"

The Brazilian population grew tenfold between 1900 and 2000, when it reached 170 million people. But the great acceleration happened after 1950. In that year, the population was 51.9 million, with a life expectancy of 43 years, a 50.6 percent illiteracy rate, and urbanization measured at 36.2 percent of the country. If we compare these figures with the same indicators in 2010, one can understand the dimensions of the changes that have occurred over those sixty years: 199 million inhabitants, a life expectancy of 73.4 years, a 9.02 percent rate of illiteracy, and 84.3 percent urbanization.[32]

Until the middle of the twentieth century, Brazil was in large part an extension of the model established in the nineteenth century. This is the case even when we consider that history is a constant play of permanence and change. It is true that important transformations occurred. The monarchy ended in 1889, and in its place Brazil became a federalist republic that strengthened the power of local oligarchies. But rural production continued to serve as the economic basis for those oligarchies, with the coffee export sector as its central axis. Enslaved workers were substituted on the large estates by extremely poor laborers who had been practically excluded from formal education. Nonetheless, a certain level of urban growth took place, bringing to the fore public health problems generated by the expansion of cities without adequate sanitation (see chapter by Sedrez and Duarte). Pollution became more evident with the growth of tanneries, slaughterhouses, and small-scale food production and clothing industries.[33] Industrialization advanced slowly in part due to the scarcity of available coal. National production of coal only gained some relevance in the 1920s following the discovery of coal reserves in southern Brazil. Hydroelectric energy also advanced very slowly.[34]

At any rate, the great wave of immigration and the beginning of urban-industrial growth catalyzed important political processes. The desire for modernization became pronounced among the new urban middle classes. Intellectual and scientific life gained greater intensity, with vibrant artistic innovations and political debates that exposed the lethargy of the oligarchical republic. A revolution that took place in 1930, with a strong military presence, pushed the nation-state to take firmer action in favor of urban-industrial advancement. However, policy reforms avoided posing any direct challenge to the concentration of landholding and the power of the rural elites. The Brazilian state expanded its presence in many ways including the promotion of public health measures and research; the establishment, starting in 1934, of legal codes concerning the exploitation of the forests, waters, and other natural resources; the creation of the Brazilian Institute of Geography and Statistics (IBGE) in 1938; the formation of regional development agencies and some initiatives of basic industrialization in the 1940s; the encouragement of internal colonization by way of private colonizing companies or of "national agricultural colonies"; and the construction of roads and cities in the interior of the country.[35]

Yet if we observe the map showing the anthropization of Brazil's biomes in 1960, it is not difficult to recognize the limited reach of this set of policies in the national territory.

While the concept of "anthropization" is debatable, because it potentially obscures the presence of indigenous and rustic populations who managed large territories and landscapes much more lightly, the map demonstrates the areas that experienced the greatest presence of the modern market economy and the conventional standards of progress. The biome that was the most modified was the Atlantic Forest, whose occupation significantly increased in the first half of the twentieth century. One estimate indicates that deforestation

Map 4.1. Anthropization of Brazilian Biomes, 1960.

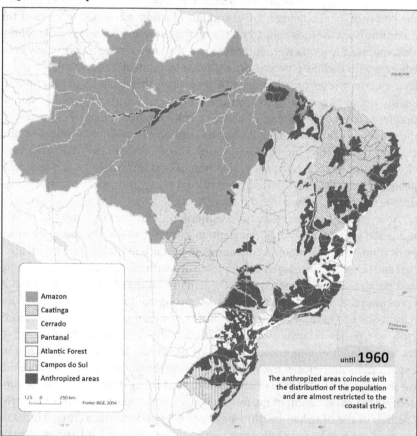

Amazon
Caatinga
Cerrado
Pantanal
Atlantic Forest
Campos do Sul
Anthropized areas

125   0   250 km
⎣⎯⎯⎣⎯⎯⎯⎣   Fonte: IBGE, 2004

until **1960**

The anthropized areas coincide with the distribution of the population and are almost restricted to the coastal strip.

Designed by William Torre, based on maps from the Instituto Brasileiro de Geografia e Estatística.

of the Atlantic Forest reached approximately 336 thousand square kilometers between 1910 and 1947.[36] Settled areas were expanding and becoming more connected to one another. New spaces of occupation, furthermore, were being established. In the Caatinga and in the Pampa, older settlements were slowly advancing as a result of the animal husbandry practiced there. But the Cerrado and the Amazon Forest were very little "anthropized." Despite the abundant presence of native ecosystems depicted on the map from 1960, in the growing spots of effective occupation the economy tended to be highly destructive with respect to the natural world, bringing into the modern world the predatory model inherited from the colonial past. In agrarian areas—as in those urban areas and industrial frontier zones centered on the railroads and mining industries, especially the iron production in southeastern Brazil—the practice of burning and the unbridled cutting down of trees for lumber were the rule, whether for construction, firewood, or the production of charcoal.

From the 1950s on, in turn, a great transformation was underway that shaped contemporary Brazil. This process profoundly modified innumerable rural and urban landscapes, producing a veritable explosion of environmental problems. To understand the meanings of this process, one must consider a series of factors. The reforms and investments carried out from the 1930s on provided the foundations of this process, like the perfecting of the state machine, the advancement of basic industrialization, and the growth of urban infrastructure. The international postwar context—which marked the beginning of a "great acceleration" on a planetary scale[37]—was an essential factor. The abundance of cheap and easily importable petroleum, as well as the new waves of consumption that came with technological innovations (automobiles, telephones, household appliances, and so on)—encouraged the growth of aggressive transnational companies in search of new markets. The increased availability of international credit attracted governments to the dream of developmentalism, even if at the cost of massive public debt. The slogan of President Juscelino Kubitscheck, Brazil's leader between 1955 and 1960, perfectly summarizes the ideological seduction of the era: to advance fifty years in five! Brazil possessed natural resources, a growing consumer market (especially in cities), an abundance of cheap labor, and a continental space to be exploited. It was not difficult to attract capital and transnational businesses to join in state initiatives and national companies. The "march to the west," which had already been the object of state action since the 1930s,[38] received a powerful boost with the inauguration of Brasília in 1960, the nation's new capital city in the heart of the Cerrado. But it was in the context of the military regime that took over in 1964, with technocratic authoritarianism stifling the country's previously existing political conflicts, that this process reached its peak. Between 1967 and 1973, the annual growth rate of Brazil's GDP was 11.2 percent. Even with the occasional stagnation and inflation crises in the following decades, the change in patterns of consumption was notable. The number of automobiles, for instance, went from 650 thousand in 1960 to 51.2 million in 2014![39]

It is essential to remember, however, that the growth process was powerfully elitist, both in geographical terms (with the process of urban industrialization concentrated in the southeast region) and with respect to the increasing concentration of wealth. A more pronounced spatial and income decentralization only began to happen in recent decades, a process embodied by the two presidential administrations of Lula da Silva between 2003 and 2011.

When considering environmental history, it is necessary to be aware of the ecological dimensions of the growth model being discussed. The energy regime that held force until the middle of the twentieth century, based on biomass (especially on firewood), was transformed by the advancement of hydroelectricity and fossil fuels.[40] Hydroelectric plants, on the one hand, provided 68 percent of the total electricity supply in 2014. In that same year, the National Registry of Dams (Cadastro Nacional de Barragens) listed fourteen hundred in Brazil, including several gigantic ones such as Itaipu and Tucuruí. The impact of these constructions on the rural populations and landscapes generated a vigorous popular resistance movement, the Movement of People Affected by Dams (Movimento de Atingidos por Barragens). In the entire Brazilian energy grid in 2014, on the other hand, fossil fuel accounted for 58.6 percent of the total, as opposed to 15.7 percent from sugarcane products and 11.5 percent from hydroelectricity.[41] The construction of highways and wheeled vehicles were clearly a priority in the last decades of developmentalism, taking advantage of the importation of petroleum and the deactivation of railroads.

Agriculture underwent a profound shift in the direction of large agribusiness units based on the use of machines and agrochemicals. The absence of agrarian reform in the second half of the twentieth century reconciled technical modernization with the permanence of a severe concentration of landholding. In 2012, the 13.7 percent of properties with more than one hundred hectares held 82.8 percent of the total area (just 225 properties greater than a hundred thousand hectares held 13.4 percent of the total area!). All of this enjoyed abundant state support in the form of credit and technical assistance. In contrast, family agriculture in small landholdings received little support, but managed to persist in different regions of the country. According to data from 2006, family farms make up 24.3 percent of the total area of agriculture and ranching. Yet they are a large employer of agricultural labor (74.4 percent of the national total) and a large producer of foodstuffs (including 70 percent of black beans and 46 percent of corn production).[42]

In territorial terms, the expansion of agribusiness began in the region of the Atlantic Forest, stimulated in part by government policies in the 1970s that provided support to agrocombustibles, most notably ethanol. But its great frontier of expansion, from 1980 on, has been the Cerrado, based on the agronomic research that was able to modify the natural acidity of the region's soils and to permit a large increase in the production of soybeans, corn, and so on.[43] National grain production grew from 17.2 million tons in 1960 to 207 million in 2014.[44] Although this agriculture is far more sophisticated than it was in the past, employing innovative soil conservation practices such as "direct planting" in straw, its environmental impact is immense. For example, Brazil became the world's largest consumer of pesticides, applying 936 thousand tons during the harvest of 2010/2011.[45]

An enormous rural exodus took place in the last decades of the twentieth century. A large part of the poor rural population lived on old, landed estates, without formal ownership titles, exchanging their labor for permission to live in small lots. Agricultural mechanization expelled most of these families from these lands. Cities received a sizeable portion of this exodus, which also flowed to the frontiers of colonization in the Amazon. The absolute population of cities increased by around 150 million people between 1950 and 2010. Without the support of urban authorities, a large part of rural-to-urban migrants had to occupy the urban spaces with little market value, such as hillsides and mangrove swamps. The multiplying shantytowns known as favelas were a logical consequence of this process. Cities also suffered from real estate speculation and the destruction of green areas and old architectural structures. The pollution produced by vehicles, industries, and warehouses storing dangerous substances considerably increased, as did violence and criminality.[46]

In the remote parts of the country, still covered by tropical forests and savannas, violence also became intense. With the advance of capitalist agriculture in the Cerrado and the geopolitical obsession of the military regime with the occupation of the Amazon—by way of colonization projects, as well as fiscal subsidies and exemptions for corporations with no regard for the social and environmental consequences—the struggle over land became explosive. Expanding farms frequently came into violent friction with indigenous and mestizo populations that traditionally lived in those regions.[47] The same occurred with the construction of roads, mining, and large hydroelectric projects.

The impact of all of these dynamics on Brazilian territory, as well as some others that cannot be discussed here, appear clearly on the map showing the anthropization of the biomes between 1960 and 2000:

The Atlantic Forest was profoundly destroyed, and today there remains only fragments that together represent approximately 12.5 percent of its original cover. The Amazon Forest also suffered immense deforestation, losing around 760 thousand square kilometers, or nearly 19 percent of its original cover. The Cerrado became an even worse case, having lost around half of its cover in just a few decades.[48] Indeed, these processes are not disconnected, since the principal economic actors responsible for the devastation of the Cerrado and

Map 4.2. Anthropization of Brazilian Biomes, 2000.

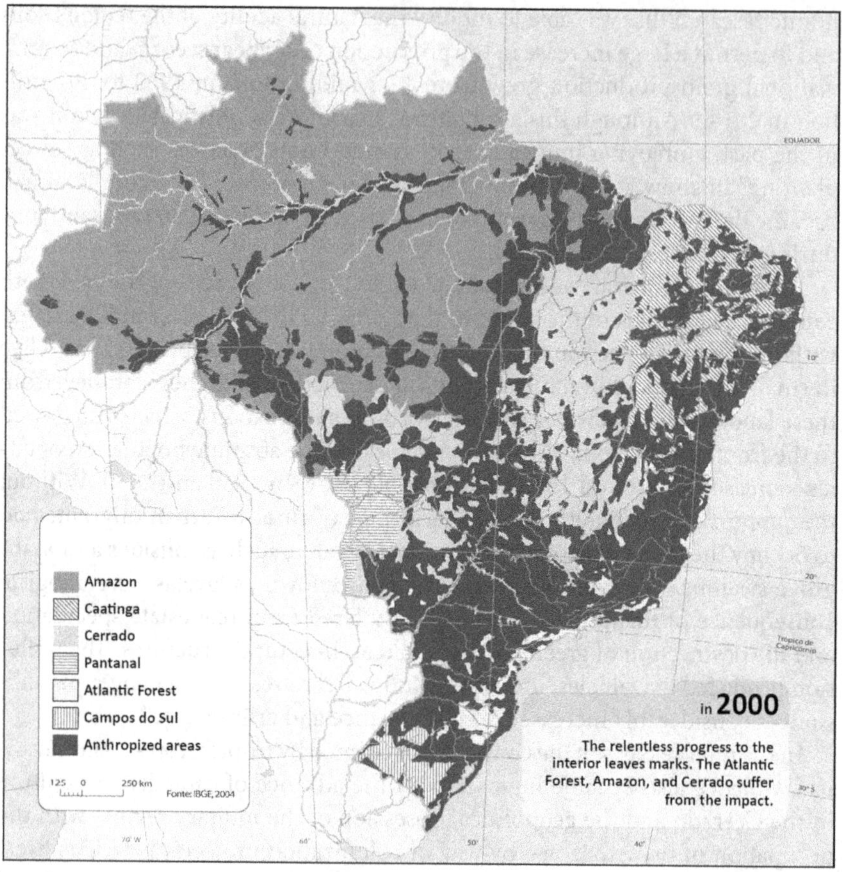

Legend:
- Amazon
- Caatinga
- Cerrado
- Pantanal
- Atlantic Forest
- Campos do Sul
- Anthropized areas

125   0   250 km
Fonte: IBGE, 2004

in **2000**

The relentless progress to the interior leaves marks. The Atlantic Forest, Amazon, and Cerrado suffer from the impact.

Designed by William Torre, based on maps from the Instituto Brasileiro de Geografia e Estatística.

the Amazon Forest came from previous frontiers of devastation of the Atlantic Forest.

From the mid-1970s on, Brazilian society became especially politically vibrant, with the struggle against the military dictatorship serving as a common denominator for a large number of human rights, social, and environmental movements. With respect to the environment, the set of processes summarized above, especially the destruction of the Amazon Forest, made Brazil into one of the central focal points of international environmental debates, a phenomenon that intensified when the United Nations Conference on Environment and Development took place in Rio de Janeiro in 1992.[49] The great expansion of local socioenvironmental conflicts included some that garnered international attention, such as the 1988 murder of the leader of the rubber tappers, Chico Mendes, which became symbolic of the global emergency of the so-called "environmentalism of the poor."[50]

The historical consequences of all of these debates and social struggles have become evident as the subject of the environment has taken on a prominent, although ambiguous, role in Brazil's postdictatorship governments and in the coalition of political forces from the center-left that governed the country between 2003 and 2015 under the leadership of the Workers' Party (Partido dos Trabalhadores). Some tendencies seen in recent decades, emanating from both historical-structural changes and from political decisions, might signal a shift in the growth model discussed above.

The most compelling fact was the reduction in historical rates of deforestation in the Amazon Forest due to a combination of political will and institutional and technological innovations. These changes manifested themselves in a sharp (84 percent) decline in the annual rates of deforestation between 2004 and 2012. This phenomenon is also evident in the fact that Brazil had been responsible for 73 percent of the volume of protected areas created in the world between 2003 and 2009.[51] In late 2010, approximately 43.9 percent of the Amazon Forest fell under different types of protected areas.[52] In general, the existence of the neighboring Cerrado region, an immense space into which agribusiness has advanced, helped make possible the improvements in the conservation of the Amazon Forest; the Cerrado provided an economic escape valve. Yet, the annual rate of deforestation of the Amazon in recent years—hovering around five thousand square kilometers—remains very high and may increase again in the future.

The annual rate of population growth, which reached a rate of 2.99 percent between 1950 and 1960, plummeted to 1.12 percent between 2000 and 2010. In other words, Brazil has experienced a true demographic transition in just a few decades. This reduction strengthens the country's ability to confront old environmental problems. The availability of basic sanitation, for example, has been gradually advancing, with 83.3 percent of the population now having access to potable water. What remains to be done, however, becomes clear in the fact that only 50.3 percent of the population has access to sewage removal and just 42.67 percent of the sewage systems in the country are treated.[53]

The territorial question remains at the center of Brazil's historical dilemmas. The development of Brazil's national economy was not restricted to its export sector economy, but the country's insertion in the world economy has traditionally been as a provider of raw materials for other economies. One of the possible scenarios is the radicalization of this role. The advancement in the production of grains has already been mentioned, particularly in the Cerrado. The presence of large mineral reserves also points in this direction. The country is already among the three largest world exporters of niobium, iron, bauxite, manganese, tantalum, and graphite.[54] The discovery of large petroleum reserves in the pre-salt layer of the ocean platform is transforming the country into an exporter of fossil fuels, notwithstanding the risks of pollution and of accidents associated with deep-sea extraction.

Taking the path that emphasizes the exportation of natural resources is the easiest for the traditionally dominant sectors of the Brazilian economy. It could take on a new international dimension in the wake of the growing demands of an international market that will increase by at least two billion persons by 2050. This is even more true if we consider that various national economies have already suffered the consequences of environmental degradation on a local and a global level, as we can observe in the loss of fertile soils, in the reduction of available water, in climatic imbalances, and so on. Today, Brazil's exports are once again increasingly dominated by primary products. What are the environmental and social consequences, at the regional level, of a powerful intensification of the exportation of grains, biofuels, lumber, minerals, petroleum, and so on?

It is true that the primary sector today has a more advanced science and technology component. Its general logic, however, brings with it the old pattern of overexploitation of the biophysical world in a manner that only considers the present. Could this possibly be the economic model that an increasingly complex society in pursuit of rights, quality of life, and new political values (including greater environmental diligence) seeks? The same international stage that demands ever more primary products from Brazil also demands— although often these demands are contradictory or formulated by different actors—the conservation of its forests and tropical ecosystems in the name of global environmental equilibrium. How can Brazilian society reconcile these different tendencies and historical imperatives?

This chapter has sought to argue that the fate of Brazil's immense territory represents a central theme in the country's history. Dominion over this "splendid cradle" brings dilemmas and responsibilities that have become increasingly evident. How can we make sure that this ecological wealth is used to promote truly democratic and sustainable social, cultural, and economic development, including escaping the trap of excessive dependency on the exportation of raw materials? What are the conditions under which Brazil's economy might take part in a national context based on the intelligent, sustainable management of its territory as a whole? Is it possible to take a historical leap to become a welfare society with a high level of scientific knowledge based on renewable energy sources, in biodiversity, and in the production of biomass? Is it possible to combine an economy geared toward the internal quality of society, in its regional diversity, which is also strong in the export of resources that are necessary for an international community amid a profound crisis of ecological scarcity?

In any case, the development of adequate political responses does not obviate the need for a deep and wide-ranging historical analysis. Despite its innumerable problems, Brazilian society is politically vibrant and is trying to find its own way forward. The path that Brazil will take will be relevant in defining the environmental future for humanity in the twenty-first century.

José Augusto Pádua is a professor of Brazilian environmental history at the Institute of History of the Federal University of Rio de Janeiro, where he also codirects the Laboratory of History and Nature. From 2010 to 2015, he was president of the Brazilian Association of Research and Graduate Studies on Environment and Society. As a specialist on environmental history and politics, he has lectured, taught, and done fieldwork in more than forty countries. He has published many books and articles in Brazil and abroad, including *Environmental History: As If Nature Existed* (edited with John R. McNeill and Mahesh Rangarajan).

Notes

This chapter was translated by Amy Chazkel and revised by John Soluri.

1. Fernand Braudel, *Ensaio sobre o Brasil do século XVI* (unpublished manuscript, 1943), cited in Luís Corrêa Lima, *Fernand Braudel e o Brasil* (São Paulo: Editora da Universidade de São Paulo, 2009), 172.

2. Milton Santos and Maria Laura Silveira, *O Brasil: Território e sociedade no início do século XXI* (Rio de Janeiro: Record, 2004), 12.

3. "Country Comparison: Population," (Demographics: Population density), Index Mundi, accessed 8 October 2017, http://www.indexmundi.com/g/r.aspx?t=0&v=21000&l=en

4. José Augusto Pádua, "As bases teóricas da história ambiental," *Estudos Avançados* 24, no. 68 (2010): 95

5. Thereza Santos and João Câmara, eds., *Geo Brasil 2002—Perspectivas do meio ambiente no Brasil* (Brasília: IBAMA, 2002), 32; Olivier Dabène and Fréderic Louault, *Atlas du Brésil* (Paris: Autrement, 2013), 38.

6. IBGE (Instituto Brasileiro de Geografia e Estática), *Mapa de biomas do Brasil* (Brasília: IBGE, 2004); Conservation International, *Biomas brasileiros: Retratos de um país plural* (Rio de Janeiro: Casa da Palavra, 2012).

7. Antonio Robert de Moraes, *Geografia histórica do Brasil* (São Paulo: Annablume, 2011), 77

8. "Povos indígenas isolados e de recente contato," Fundação Nacional do Índio, Ministério da Justiça, accessed 8 October 2017, http://www.funai.gov.br/index.php/nossas-acoes/povos-indigenas-isolados-e-de-recente-contato.

9. Antonio Robert de Moraes, *Geografia histórica do Brasil* (São Paulo: Annablume, 2011), 87.

10. Darcy Ribeiro, *The Brazilian People: The Formation and Meaning of Brazil* (Miami: University Press of Florida, 2000), ch. 4.

11. Nísia Lima, *Um sertão chamado Brasil* (São Paulo: HUCITEC, 2013.)

12. Candice Vidal e Souza, *A pátria geográfica: Sertão e litoral no pensamento social brasileiro* (Goiânia: Editora da UFG, 1997).

13. José Augusto Pádua, "European Colonialism and Tropical Forest Destruction in Brazil," in *Environmental History: As If Nature Existed*, ed. John R. McNeill, José Augusto Pádua, and Mahesh Rangarajan (New Delhi: Oxford University Press, 2010).

14. Arquivo diplomático da independência (Brasília: Ministério da Educação e Cultura, 1972), 1, 47, cited by Cid Valle, *Risonhos lindos campos: Natureza tropical, imagem nacional e identidade brasileira* (Rio de Janeiro: Senai, 2005), 156.

15. José Murilo de Carvalho, *Teatro de sombras: A política imperial* (Rio de Janeiro: Vértice, 1988), ch. 1.

16. Renato Peixoto, *A máscara de Medusa: A construção do espaço nacional brasileiro através das corografias e da cartografia no século XIX* (Ph.D. diss., História Social, UFRJ, 2005); Lilia Schwarcz, *The Emperor's Beard: Dom Pedro II and the Tropical Monarchy of Brazil* (New York: Hill and Wang, 2004).

17. Warren Dean, *With Broadax and Firebrand: The Destruction of the Brazilian Atlantic Forest* (Berkeley: University of California Press, 1995), ch. 2; Berta Ribeiro, *O índio na cultura brasileira* (Rio de Janeiro: Revan, 1991).

18. John Monteiro, "Rethinking Amerindian Resistance and Persistence in Colonial Portuguese America," in *New Approaches to Resistance in Brazil and Mexico*, ed. John Gledhill and Patience Schell (Durham: Duke University Press, 2012); Manuela Cunha, ed., *História dos índios no Brasil* (São Paulo: Companhia das Letras, 1992).

19. Diogo Cabral, *Na presença da floresta: Mata Atlântica e história colonial* (Rio de Janeiro: Garamond, 2014), 125.

20. Warren Dean, *With Broadax and Firebrand*, 263; Bert Barickman, *A Bahian Counterpoint: Sugar, Tobacco, Cassava and Slavery in the Recôncavo, 1780–1860* (Palo Alto: Stanford University Press, 1998).

21. Warren Dean, *With Broadax and Firebrand*, ch. 8; Steven Topic and Allen Wells, *Global Markets Transformed: 1870/1945* (Cambridge: Belknap Press, 2012), 224.

22. José Augusto Pádua, *Um sopro de destruição: Pensamento político e crítica ambiental no Brasil escravista* (Rio de Janeiro: Jorge Zahar, 2002).

23. Guilherme Capanema, *Agricultura: Fragmentos de um relatório dos comissários brasileiros à exposição universal de Paris* (Rio de Janeiro, 1858), 4.

24. Christian Brannstrom, "Coffee Labor Regimes and Deforestation on a Brazilian Frontier, 1915–1965," *Economic Geography* 76, no. 4 (2000).

25. Zuleika Alvim, "Imigrantes: A vida privada dos pobres do campo," in *História da vida privada no Brasil*, vol. 3, ed. Nicolau Sevcenko (São Paulo: Companhia das Letras, 2001), 220.

26. Silvio Correa and Juliana Bublitz, *Terra de promissão: Uma introdução à eco-história do Rio Grande do Sul* (Passo Fundo: Editora da Universidade de Passo Fundo, 2006).

27. Warren Dean, *With Broadax and Firebrand*, ch. 5; Robert Wilcox, "The Law of the Least Effort: Cattle Ranching and the Environment in the Savanna of Mato Grosso, Brazil, 1900–1980," *Environmental History* 4, no. 3 (1999), 338–368.

28. Warren Dean, *With Broadax and Firebrand*, 249; Regina Duarte, *Activist Biology: The National Museum, Politics and Nation Building in Brazil* (Tucson: The University of Arizona Press, 2016).

29. Marcos Gerhardt, "Extrativismo e transformação na Mata Atlântica meridional," in *Metamorfoses florestais: Culturas, ecologias e as transformações históricas da Mata Atlântica*, ed. Diogo Cabral and Ana Bustamante (Curitiba: Prismas, 2016).

30. Warren Dean, *Brazil and the Struggle for Rubber* (Cambridge: Cambridge University Press, 1987); Barbara Weinstein, *The Amazon Rubber Boom: 1850/1920* (Palo Alto: Stanford University Press, 1983).

31. José Augusto Pádua, "Biosphere, History and Conjuncture in the Analysis of the Amazon Problem," in *The International Handbook of Environmental Sociology,* ed. Michael Redclift and Graham Woodgate (Cheltenham: Edward Elgar, 1997).

32. IBGE, *Tendências demográficas* (Rio de Janeiro: IBGE, 2001); IBGE, *Sinopse do censo demográfico* (Rio de Janeiro: IBGE, 2011).

33. Jorge Barbosa, "Olhos de ver, ouvidos de ouvir: Os ambientes malsãos da capital da república," in *Natureza e sociedade no Rio de Janeiro,* ed. Maurício Abreu (Rio de Janeiro: Biblioteca Carioca, 1992); Janes Jorge, *Tietê: O rio que a cidade perdeu—São Paulo, 1890–1940* (São Paulo: Senac, 2006).

34. Antonio Leite, *A energia do Brasil* (Rio de Janeiro: Campus, 2015), ch. 2.

35. Martine Droulers, *Brésil: Une Géohistoire* (Paris: PUF, 2001), chs. 5 and 6.

36. Christian Brannstrom, *Coffee Labor Regimes,* 327.

37. John R. McNeill and Peter Engelke, *The Great Acceleration* (Cambridge: Belknap Press, 2014).

38. João Maia, *Estado, território e imaginação espacial: O caso da Fundação Brasil Central* (Rio de Janeiro: Fundação Getúlio Vargas, 2012).

39. Martine Droulers, *Brésil,* 254; Hervé Théry, "Retrato cartográfico e estatístico," in *Brasil: Um século de transformações,* ed. Ignacy Sachs, Jorge Wilheim e Paulo Pinheiro (São Paulo: Companhia das Letras, 2000); Departamento Nacional de Trânsito, Frota de veículos 2016, accessed 8 October 2017, http://www.denatran.gov.br/estatistica/261-frota-2016.

40. Christian Brannstrom, "Was Brazilian Industrialisation Fuelled by Wood? Evaluating the Wood Hypothesis, 1900–1960," *Environment and History* 11 (2005).

41. *Resenha energética brasileira* (Brasília: Ministério de Minas e Energia, 2015); Comitê Brasileiro de Barragens, Sistema informatizado do Cadastro Nacional de Barragens Brasil, accessed 8 October 2017, http://www.cbdb.org.br/5-69/Cadastro%20Nacional%20de%20Barragens; Carlos Vainer, "Águas para a vida, não para a morte: Notas para uma história do Movimento de Atingidos por Barragens no Brasil," in *Justiça Ambiental e cidadania,* ed. Henri Acserald, Selene Herculano and José Augusto Pádua (Rio de Janeiro: Relume-Dumará, 2004).

42. Site Reforma Agrária em Dados, Agricultura familiar, accessed 8 October 2017, http://www.reformaagrariaemdados.org.br/realidade/1-agricultura-familiar.

43. Sandro Dutra e Silva, José Pietrafesa, José Franco, José Drummond, and Giovana Tavares, eds., *Fronteira Cerrado: Sociedade e natureza no oeste do Brasil* (Goiânia: Editora da PUC-Goiás, 2013).

44. Ministério da Agricultura, Estatística e dados básicos de economia agrícola – Agosto 2017, accessed 8 October 2017, http://www.agricultura.gov.br/assuntos/politica-agricola/todas-publicacoes-de-politica-agricola/estatisticas-e-dados-basicos-de-economia-agricola/PASTADEAGOSTO.pdf; Companhia Brasileira de Abastecimento, *Acompanhamento da safra brasileira—grãos* (Brasília, Conab, 2015).

45. Raquel Rigotto, Dayse Vasconcelos, and Mayara Rocha, "Uso de agrotóxicos no Brasil e problemas para a saúde pública," *Cadernos de Saúde Pública* 30, no. 7 (2014).

46. Alfredo Sirkis, *Ecologia urbana e poder local* (Rio de Janeiro: TIX, 2010).

47. Susanna Hecht and Alexander Cockburn, *The Fate of The Forest: Developers, Destroyers and Defenders of the Amazon* (Chicago: Chicago University Press, 2010).

48. Fundação SOS Mata Atlântica and INPE (Instituto Nacional de Pesquisas Espaciais), *Atlas dos remanescentes florestais da Mata Atlântica* (São Paulo, 2014); Antonio Nobre, *O futuro climático da Amazônia* (São José dos Campos: ARA, INPE, INPA, 2014); Robert Buschbacher, *Expansão agrícola e perda da biodiversidade do Cerrado* (Brasília: WWF-Brasil, 2000).

49. Kathryn Hochstetler and Margaret Keck, *Greening Brazil: Environmental Activism in State and Society* (Durham: Duke University Press, 2007); Angela Alonso and Débora Maciel, "From Protest to Professionalization: Brazilian Environmental Activism after Rio-92," *The Journal of Environment and Development* 19, no. 3 (2010); José Augusto Pádua, "Environmentalism in Brazil: A Historical Perspective," in *A Companion to Global Environmental History,* ed. John R. McNeill and Erin Mauldin (Oxford: Wiley-Blackwell, 2012).

50. Joan Martinez Alier, *The Environmentalism of the Poor* (Cheltenham: Edward Elgar, 2003).

51. Clinton Jenkins and Lucas Joppa, "Expansion of the Global Terrestrial Protected Area System," *Conservation Biology* 142 (2009).

52. José Drummmond, José Franco, and Daniela de Oliveira, "Uma análise sobre a história e a situação das unidades de conservação no Brasil," in *Conservação da biodiversidade: Legislação e políticas públicas,* ed. Roseli Ganem (Brasília: Edições Câmara dos Deputados, 2011); José Augusto Pádua, "Tropical Forests in Brazilian Political Culture: From Economic Hindrance to Ecological Treasure," in *Endangerment, Biodiversity and Culture,* ed. Fernando Vidal and Nélia Dias (London: Routledge, 2015).

53. Instituto Trata Brasil, Situação do saneamento no Brasil, accessed 8 October 2017, http://www.tratabrasil.org.br/saneamento-no-brasil.

54. Instituto Brasileiro de Mineração, *Information and Analyses on the Brazilian Mineral Economy,* 7th edition (Brasília: IBRAM, 2012).

 CHAPTER 5

From Threatening to Threatened Jungles

Claudia Leal

In 1989, Sting, the charismatic lead singer of The Police, and Raoni Metuktire, one of the main leaders of the Kayapó indigenous group of the Amazon, met with a number of European leaders of stature: Pope John Paul II; François Mitterrand, the president of France; and King Juan Carlos of Spain. These meetings were part of a tour of seventeen countries that aimed at making people aware of deforestation in the Amazon Basin and the uncertain future of its indigenous people. Shortly before, Jean Pierre Dutilleux, a Belgian film director, invited Sting, who was visiting Brazil, to meet Raoni and see for himself the threats hanging over those millenary forests and their inhabitants. Alarmed by the destruction of tropical nature and its cultures, Sting and Dutilleux started the Rainforest Foundation, writing a book, *Jungle Stories: Fight for the Amazon,* to promote and finance their cause.

Many others joined the crusade to save the tropical rainforests, one half of which are found in Latin America. (The other half are located mainly in Africa's Congo river basin and in Southeast Asia.) Since 1987, for example, the Rainforest Alliance, a New York-based NGO with sixteen offices now around the world, has certified dozens of jungle products to be sustainably produced. According to its 2012 report, it "has protected hundreds of millions of acres of forests . . . and improved social conditions for millions of people" in an effort to counteract the effects of "slash-and-burn agriculture, overharvesting, reckless extraction and illegal logging."[1] Other organizations, such as the environmental website Mongabay, have tried to change the mentality of the younger generations, teaching them that these ecosystems help to stabilize the world's climate; are home to countless plants and animals; maintain the water cycle; protect lands from floods, droughts, and erosion; and guarantee the subsistence of tribal groups.[2]

The origin of this global idea of endangered ecosystems, which stresses the strategic importance of rain forests on a planetary level, lies in the dramatic increase of deforestation that began in the mid-twentieth century. But the history of jungles over the past two hundred years is more than a story about their gradual disappearance. In the second half of the nineteenth century, demand for tropical products, from rubber and precious hardwoods to bananas, stimulated the extraction of forest products and the expansion of monoculture crops, both of which spurred an incipient drive into Latin America's jungle regions, though without, paradoxically, causing significant deforestation overall. The widespread destruction, which occurred one hundred years later, was spurred by state efforts to develop these regions and the massive influx of landless peasants. The main result has been the opening of wide swaths of pastureland for cattle ranching, which in turn gave impulse to the creation of conservation areas as well as reserves for indigenous and peasant communities whose ways of life were considered to be compatible with the forests. In this long process, governments, local inhabitants, activists, and many others have struggled over these areas and their resources, which has weakened the idea—commonplace a century ago—of a green hell.

In a way that was characteristic of that time, a prominent traveler described the jungle as a space of slavery, degeneration, violence, and death. Arturo Cova, the protagonist and autobiographical narrator of *The Vortex*, a novel published in 1924, flees from Bogotá with his lover only to disappear into the dark world of Amazonian rubber extraction. Entrapped within an alien world, he is filled with despair: "Oh, jungle, wedded to silence, mother of solitude and mists! What malignant fate imprisoned me within your green walls? . . . You are a cathedral of sorrows. . . . Let me flee . . . from your sickly shadows formed by the breath of beings who have died in the abandonment of your majesty. You yourself seem but an enormous cemetery, where you decay and are reborn."[3]

The Vortex earned its author, José Eustacio Rivera, a leading place in the pantheon of Colombian literature. Like Rivera, other outstanding Latin American writers of that period wrote about the anxiety caused by the jungles. In the novel *Camaima*, published in 1935 by Rómulo Gallegos, the best-known Venezuelan writer of the twentieth century, protagonist Marcos Vargas travels to and is swallowed up by the jungle just like Cova.[4] The stories of the Brazilian writer Alfredo Rangel, titled *Inferno Verde* (1907), and some of the essays of his compatriot, the renowned Euclides da Cunha, published posthumously in *A margem da história* (1909), also convey the pessimistic and tragic view of the tropical forest. Their titles are telling: the jungle is the natural realm of eternal torture where progress never arrives. In these books, nature serves as a metaphor for a cruel and unjust society. Such social criticism is the axis of the six so-called jungle novels written by B. Traven in the 1930s. In these works, Traven, a German anarchist who had settled in Mexico, denounced the oppression of debt-enslaved indigenous workers in the Chiapas timber industry.

These books helped their readers, many of them Venezuelans or Brazilians, to become acquainted with the jungle and make it their own. A distinguished

Colombian man of letters expressed this feeling in praising Rivera: "No other poet has specialized, as this one has, in drawing a portrait of what is *our own.* [Rivera] is the unrivaled bard of *the tropics.*"[5] This interpretation and the book that inspired it would have been impossible half a century earlier. At the beginning of the nineteenth century, many of the nascent Latin American republics inherited extensive jungles from the colonial empires but no actual dominion over them. So nearly half the landmass of Latin America remained outside of state control: the challenge of incorporating these areas was enormous. Eventually, the extraction of a few forest products, and the expansion of plantations on the ashes of tropical vegetation, awakened Latin American states to the enormous potential wealth underlying these regions, encouraging them to define their national territories.

In fact, both Rivera and da Cunha became familiar with the Amazon jungle precisely as public officials responsible for defining international boundaries. In 1922, Rivera, in his capacity as a lawyer, became a member of the Colombo-Venezuelan Border Commission. During nearly a year of travel through the Amazon, he suffered from malaria, visited unexplored areas on his own account, listened to gruesome stories, and felt the anxiety he described in his novel.[6] In the same way, da Cunha spent various months in 1905 traveling up the river Purús, the subject of a dispute between Peru and Brazil. One of the boats in which the Brazilian commission traveled sunk, drowning half of their provisions, so that, in addition to fever and discomfort, its weakened team suffered hunger as they sought to fulfill their patriotic duty. Nevertheless, da Cunha thought that their trials were minor compared to the fate of those who had made the jungle their home: "In the forest, days of mourning are human beings' entire existence, monotonous, painful, obscure and anonymous—oppressive rounds of bitter and unalterable paths, without beginning and without end, inscribed in the close circuit of the rubber trails."[7]

Da Cunha was speaking of a part of the biggest expanse of jungle in the world—Amazonia—which includes the Amazon and Orinoco river basins and is shared by nine countries, among them Brazil, which has sovereignty over 60 percent of this forest. There are two other expanses of rainforest in Latin America that, since their area is smaller and they border the Atlantic Ocean and the Caribbean Sea, have been decimated more rapidly. One of them extends from the Pacific coast of northern South America to southern Mexico, crossing Central America along the Caribbean coast: it is now highly fragmented. The other jungle was part of the larger Atlantic Forest, which bordered the coast of Brazil from the northeast to Uruguay. The rest of that forest, of which barely 8.5 percent survives, was not what is typically called a rainforest: the climate in its southern tract is too cold and dry, while inland higher altitudes or meager rainfall also contribute to the formation of other kinds of forests (See map 5.1).

Map 5.1. Tropical rainforest cover.

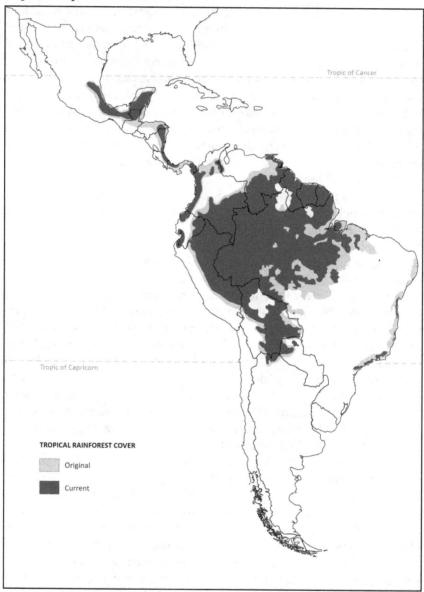

Map made by Camilo Uscátegui, based on United Nations Environment Program.

Both heat (between 25 and 27 degrees Celsius on average), the product of elevations under a thousand meters, and constant rainfall (between two thousand and three thousand millimeters annually on average) characterize tropical rainforests and help explain their amazing diversity. Trees that grow there never lose their leaves and some easily reach a height of twenty-five or thirty meters. But not all jungles are the same: those in Latin America have more species of trees, birds, bats, and butterflies than those of others regions. The abundance of epiphytic plants, which germinate in trunks and branches, gives them a unique appearance, as do the narrow trails formed by leaf-cutter ants as they tirelessly carry the leaves they use to produce their food. No elephants, orangutans, or gorillas live in these jungles, as they do in those of Africa and Asia, but they are the home of jaguars, anacondas (snakes that grow up to ten meters long), and a great variety of small primates, like the spider monkey and the black-faced capuchin. Let us now examine their past.

Extractive Economies and Plantations

In 1925, a few months before retiring from his position as secretary of public education in postrevolutionary Mexico, José Vasconcelos foresaw a rebirth of humankind centered in the tropics: "The world of the future will belong to whoever conquers the Amazon region. Universopolis will rise by the great river, and from there the preaching, the squadrons, and the airplanes propagandizing the good news will set forth."[8] With these words Vasconcelos challenged a long Western tradition that conceived of forests as the "shadow of civilization."[9] His illusions had been roused by the spectacular growth of Manaus, a city with two hundred thousand inhabitants situated 1,450 kilometers from the mouth of the Amazon River, and that, twenty-eight years before, had inaugurated a luxurious opera house: perhaps it was possible to triumph over the barbarism of the jungle.

Manaus and its famous opera house were the best-known symbols of a way of relating to the jungle that, beginning in the colonial period, rested on the selective extraction of parts of plants and animals that could be sold on the world market. The colonial appropriation of the jungles was meager, however, with the notable exception of parts of the Atlantic Forest. Up to the mid-nineteenth century, jungles were still almost exclusively inhabited by indigenous groups, even though the numbers of some of them had been reduced by diseases and slavery. Those groups gathered a great variety of natural products for self-consumption and barter, and practiced an itinerant agriculture based on crops such as manioc and maize. Their management of the environment over thousands of years left deep marks on the composition of soils and vegetation—the creation of scattered *terras pretas* (highly fertile black soils) in Amazonia, and an abundance of useful species, such as palms with edible fruits or materials for construction or crafts, whose reproduction they favored—leading specialists to refer to domesticated forests or wild gardens.[10]

On the margins of some jungle areas Europeans began procuring products for commercial use with the forced or negotiated aid of aboriginal inhabitants. In 1750, for example, the Caribbean coast of what is now Nicaragua exported mahogany and sarsaparilla, a root with medicinal properties.[11] In the region of the lower Amazon River, a much greater variety of wild products were exploited: in addition to sarsaparilla, they included vanilla, species similar to clove and cinnamon, cacao, annatto dye (a food coloring), manatee meat and oil, and even decorative products such as feathers.[12] The Portuguese term for these goods, *drogas do sertão* (drugs of the backlands), highlights the difference between extraction and cultivation. The *sertão* or uncultivated lands were the source of *cacao bravo*, "wild" cacao, as opposed to the *cacao manso* or "tame" cacao grown on haciendas.[13] Jungle regions were also the source of Nueva Granada's (present-day Colombia's) alluvial gold. In the eighteenth century, mines in the extremely humid jungles of the Pacific coast produced the main export of the viceroyalty of that name, anticipating current mineral prospection and exploitation in many jungle areas.[14]

The incipient extractive economy of the colonial period became much stronger in the second half of the nineteenth and first half of the twentieth centuries. The industrial revolution created a strong demand for raw materials, some of which were made from plants found in rainforests. The rubber boom is the best example. The Amazon was its main scenario though it also included the forests of Central America and the horrors of Belgian colonization in the jungles of the Congo. There were other booms that left their mark on specific regions, such as the sale of vegetable ivory, the seeds of a number of palms that grow in the forests of the Pacific coast between Panama and Ecuador. These seeds are the size of a small chicken egg and have a color and texture similar to ivory, which is why they were used to make buttons until they were replaced by plastic in the 1940s.[15] In the same period, synthetic materials replaced *chicle,* the sap of the *Manilkara zapota* tree, in the manufacture of chewing gum. Since the 1870s, U.S. companies had exploited this tree from the jungles of Guatemala and Mexico, where it was abundant.[16] The extraction of sarsaparilla continued in the jungles of the Caribbean coast of Central America and timber extraction intensified. Belize, which had been a center for the export of logwood, became a leading exporter of mahogany, along with southern Mexico, for more than a century.[17] In a similar way, China, India, and Europe increased their imports of timber from the forests of Southeast Asia.[18]

The ways in which rubber was extracted allows us to understand how those forests survived the boom. *Hevea* trees, the source of white rubber, the best on the market, are, like *chicle* trees, scattered throughout the forest. To obtain their latex, rubber tappers opened trails through the jungle that connected one hundred or two hundred trees and made daily rounds to collect the sap that oozed from incisions made in their trunks. In addition, rubber tappers cared for the plantlets of *Hevea* and occasionally planted some themselves.[19] The case of black rubber was very different, but it too explains the survival of the forest. This latex, derived from trees of the *Castilla* genus, which are found in the Amazon as well as the jungles of the Pacific and Central America, dries and turns black on contact with the air. Since the exudations were meager and the rubber was less valuable than its white counterpart, the tappers cut down the tree to "bleed" its latex all at once. Thus, in a short time, the stock of black rubber drastically fell. Due to rising prices, peasants and businessmen planted black rubber trees in Mexico, Central America, and Colombia. But when the prices fell in 1913, due to the opening of Asian plantations, those initiatives ended. However, thanks to the trees they planted and the ones that had replaced their progenitors in the open clearings of the forest, there are still black rubber trees in the jungles where they were once felled.

Not all economic activities responsible for the push into the jungles were extractive ones, nor did they maintain the forests. The Spanish colonizers also developed monocultures, for which they felled and burned the vegetation that served as fertilizer. Nevertheless, the initial environmental impact of this form of agriculture on the jungle was limited. Even though the economy of northeastern Brazil in the sixteenth and seventeenth centuries revolved around sugar production in areas of rainforest (and, to a lesser extent, dry tropical forest), by 1700 the sugarcane plantations had barely deforested twelve thousand hectares, less than 1 percent of the territory of Sergipe, the smallest state in the northeast.[20] The sugar mills only planted sugarcane in the valleys, close to rivers, though they also felled trees for fuel and to make the boxes in which sugar was shipped.[21] In the region of the lower Amazon and in the Guianas, especially Surinam, a number of plantations producing sugar, cotton, and even a little cacao also left their mark near the coast and along the banks of a few rivers.

In the nineteenth century, and especially in the early decades of the twentieth century, the number of plantations producing tropical products destined for U.S. and European markets increased, as did deforestation. Something similar occurred in the jungles of Southeast Asia with the establishment of rubber and oil palm plantations. In Latin America, these new plantations were concentrated in the coastal zones of Central America (especially in the Caribbean) and the states of Rio de Janeiro and São Paulo in Brazil. In the latter case, coffee, beginning in the 1820s, spread for more than a century through the Atlantic Forest as the woodlands of the valleys and low mountains of the littoral were cleared and burned. As the soils became exhausted, the plantations opened up new areas, giving rise to a kind of itinerant agriculture. Population growth and urbanization, which accompanied the economic rise of the southeast region, exerted even more pressure on the forest and sped up its destruction.[22]

In the coastal forests of Panama, Costa Rica, Honduras, and the Mexican regions of Tabasco and Veracruz, thousands of hectares were devoted to bananas so that consumers in the United States could enjoy a high-caloric fruit packed in a natural wrapper, thereby enriching the shareholders in foreign firms such as the United Fruit Company.[23] These monocultures required the installation of railways to make export commodities viable in previously distant lands. In Costa Rica, the railway built to ship coffee from the central-valley highlands led to the establishment of plantations in the Caribbean region of Limón in the 1880s. A similar story occurred in Honduras, where the government awarded land grants to the companies that built new railways but whose main goal was to grow bananas for export.[24] The infrastructure built for petroleum drilling—towers, pipelines, and camps—caused the elimination of the jungles of la Huasteca on the coast of the state of Veracruz.[25] By contrast, extractive economies mainly used rivers—the natural roads of the jungles—and trails to find latex or valuable seeds and thus minimized their impact on the geography of the forests.[26]

The susceptibility of monocultures to pests and diseases exacerbated deforestation, as exemplified by the strategy used in Honduras to deal with Panama disease. Given the difficulties of developing a disease-resistant variety of banana that would also satisfy consumer tastes, the companies chose to abandon infested areas and plant in new terrain, which in turn was abandoned when it too became infected.[27] A similar story occurred in the cacao plantations on the coast of Ecuador: over forty-two thousand hectares cleared between 1885 and 1910 had to be abandoned two decades later when they were infected by the Monilia fungus and witches' broom disease.[28] Monocultures continued to be a factor in the annihilation of tropical rainforests, as shown by the development of banana plantations in the Ecuadorian coast in the 1950s and the Urabá region of Colombia in the 1960s.

The incorporation of these jungle areas into the world economy through the extraction of natural resources and by growing tropical products made it easier for states to nationalize these territories. For instance, while Colombia finally extended its sovereignty over the Darien Gap (the link between South and Central America), long inhabited by the fiercely independent Cuna or Tule indigenous group, other countries strove to exert control over parts of the Amazon. Similar processes occurred in the Congo river basin and in Southeast Asia, though in the name of European colonialism. In Latin America, the national appropriation of such jungles found its most effective agents in settlers who considered themselves to be Venezuelan, Mexican, and the like, in contrast with the indigenous groups, who did not have national identities and who suffered a new wave of disease-induced deaths. Costa Rica's banana economy, for example, attracted migrants from the more populated highland zones to the Caribbean, a region with barely twenty thousand inhabitants in 1850. Similarly, impoverished peasants from Brazil's northeast wagered their future on rubber extraction in the Amazon. Along with those settlers often came state and national institutions, such as new forms of property rights to land and resources, judges and mayors, and even the Spanish and Portuguese languages. The migrants founded towns and cities in the midst of the jungle. Manaus is the best known example, but there are many others, like Iquitos in the Peruvian Amazon, Tumaco in the Colombian Pacific, and San Pedro Sula in Honduras. All of these towns became the administrative centers from which the national states consolidated their hold over jungle regions.

The conquest of these areas caused disputes between countries over the location of international borders. The rubber economy led to the war over Acre between Bolivia and Brazil (1899–1903) and the Colombo-Peruvian War (1932–1933). After drawn-out disputes, the Amazon Basin was pierced by imaginary lines marking clearer claims of sovereignty; another line was traced to separate the Guatemalan Petén from the Lacandon Jungle in Mexico. With their territories defined, countries were able to create and disseminate definitive national maps, the shapes of which have formed a kind of logos that now serve as strong national referents.

Thanks to those maps, the jungles were finally thought of as integral parts of the national territory of each country, but only the first-hand accounts from those areas, with their descriptions and intrigues, allowed people's imaginations to be filled with their sights and smells. We have seen how the best writers depicted the jungles as formidable geographies, where visitors risked their hides at every step. When those writers, as well as many other outsiders, spoke of the dark side of the jungle, they blended fiction with reality. For example, in the construction of the 350-kilometer-long railway between Madeira and Marmoré (1907–1912), on the frontier between Brazil and Bolivia, more than six thousand laborers are said to have died; and the building of the Panama Canal (1881–1914) cost more than twenty-five thousand lives. The message of the chronicles was clear: both menacing nature and unchecked injustices meant the civilizing mission still beckoned.

In this way, imaginaries about pestilent and cruel jungles helped to reinforce the hierarchical character of national geographies that had been coming into being since the colonial period. Just as the contradictory images of the Asian and African tropics—as beautiful but hellish—helped legitimize European colonial enterprises, the ideas that gained ground in the American tropics strengthened the mandate to domesticate areas regarded as savage.[28] The new regions entered into each nation's territory and self-definition in a subordinate position. They stopped being *terra incognita* and were now defined by their relationship with the central zones of the countries into which they were incorporated. The formal inclusion of these areas into the administrative framework of the state made their secondary status evident. The administrative divisions of the Colombian Amazon, for example, were called *intendancies* or *commissariats,* which denoted a position inferior to the *departments* into which most of the country's territory was divided.

The southeastern part of Brazil and the Sula Valley in Honduras are exceptions to the rule. Coffee growing, along with the mining and cattle ranching of Minas Gerais and the growth of manufacturing in São Paulo, helped propel the former region into the political and economic center of the country: a thriving territory, symbolized by two great cities, Rio de Janeiro and São Paulo. But these coastal regions stopped being jungle areas along the way.

Other jungles continued to be places that the center had not subjugated completely. For that reason, some of them—the Pacific region of Colombia and Ecuador, parts of the lower Amazon, and the interior of Surinam—turned into free territories where Maroons, former slaves, and descendants of slaves remade their lives.[29] The marginal character of these regions was (and still is) related to the fact that they were zones of contact: between settlers and indigenous inhabitants, domestic and wild animals, private property and commons. Even though the onslaught that began in 1960 has led to the previously unimaginable annihilation of millions of hectares of jungle, the territories thus transformed are still regarded as geographically and culturally distant: "the antithesis of the nation," to borrow Margarita Serje's expression.[30]

Settlers and Pastures

In the second half of the twentieth century, deforestation took on surprising momentum: thousands of peasants migrated to the jungles and cleared the vegetation in search of a better life (See illustration 5.1). A man known as "Mico" (monkey) Hernández, one of so many, recalled his arrival in Vistahermosa, in the Colombian Amazon, in 1972: "We went with my wife and began to work . . . . We cleared the forest together, I with the axe and she with the machete; I felled the trees and she cleared the scrub. . . . In those days everything that is now open was full of trees like *guáimaro, guacamayo, balso, tambor.* No one lived there. When we felled the trees and burnt the undergrowth people began to arrive . . . in those early days there were lots of animals . . . a lot of howler monkeys, curassows, wild turkeys, wild pigs. You came across tapirs that could weigh over 180 kilos. There were spider monkeys and coatis. What didn't you find there! If you didn't hunt, it was because you were too lazy to

Illustration 5.1. The history of jungles in Latin America is not dominated by state-managed forestry, as in Southeast Asia or Europe, nor is it a relentless tale of destruction, as this picture of a village in the Pacific coast of Colombia suggests. This village—located on the Napi River, a tributary of the Micay River, and photographed by geographer Robert C. West in the mid-1950s—was inhabited by descendants of slaves who mined gold in the eighteenth century.

Photograph R9 N11 from the photographic archive "Robert West: The Pacific Lowlands of Colombia," housed at the Sala de Libros Raros y Manuscritos, Biblioteca Luis Angel Arango, and the Main Library at Universidad de los Andes, both in Bogotá. See https://robertwest.uniandes.edu.co/.

pull the trigger."[31] With the arrival of peasants the landscape changed and the animals "fled" to the interior of the jungle and were no longer seen.

Hernández migrated in the context of accelerated population growth. The number of Latin Americans rose from barely 70 million in 1900 to 175 million in 1950 and to 515 million inhabitants in the year 2000. In the second half of the twentieth century many peasants tried their luck in cities, turning the region into an urban one. A few preferred to look for a piece of land they could call their own in tropical forests. Thus, the Lacandon Jungle, which could claim only a thousand inhabitants in 1950, had 150 thousand in 1990, while in the neighboring jungle of Petén, the figure rose from 25 thousand inhabitants in 1962 to around 300 thousand in 1986. This growth rate greatly surpassed the national average for each country. Between 1950 and 2000, the population of the Ecuadorian Amazon multiplied by a factor of ten, while that of Ecuador as a whole tripled. In the Brazilian Amazon, the population grew by a factor of six (from less than two million to thirteen million), which was more than double the national rate.[32] Even this latter rainforest region became mostly urban: by 1991 it had 133 cities (eight of them with more than a hundred thousand inhabitants) containing 58 percent of its population.[33]

States played a fundamental role in this story of settling and destroying the jungle. They had become stronger and more solvent than they had been at the beginning of the century, and built roads to integrate the frontier, nourishing themselves on the dreams of development and economic growth that were staples to this postwar period. In no country were these policies as adamant as in Brazil, perhaps because the Amazon accounts for more than half of its territory. The strongest thrust came from the military government (1964–1985), which created the Superintendency for the Development of the Amazon and announced the economic development policy known as "Operation Amazon." The plans partly rested on the long-held desire to link the interior to the coast. As early as 1953, President Getúlio Vargas created an institution responsible for the development of the Amazon, whose crowning achievement was the construction, between 1956 and 1960, of the road linking Brasilia with Belém, the port at the mouth of the Amazon River. The military government intensified the development of infrastructure by paving that road and building others, including the 4,000-kilometer-long Trans-Amazon Highway, one of Brazil's longest.[34]

Roads have been crucial in promoting colonization. In the 1960s and 1970s, several governments thought of the vast jungle territories as safety valves that could defuse the acute tensions engendered by the highly concentrated ownership of land, which set off alarms after the Cuban Revolution in 1959. The United States government even encouraged the implementation of agrarian reforms, which were sometimes or partially circumvented by offering frontier lands instead of expropriating existing properties. Between 1963 and 1980, the Colombian state, for example, encouraged migration to the jungle by means of the Caquetá 1 and 2 projects, which ended up impacting three million hectares.[35] The military government of Brazil, for its part, tried to give a social slant to its Amazonian policy with the National Integration Program, which focused on colonization. The Ecuadorian government followed a similar path by creating, in 1978, the National Institute for the Colonization of the Amazonian Region of Ecuador. But instead of creating agricultural colonies, this institute devoted itself to formalizing the property rights of peasants who arrived there on their own account.[36] Something similar happened in other countries. However, as in the case of Mico Hernández, the great majority of peasants who migrated to the jungles settled there with little or no state support.[37]

The settlers and the state imposed a new model for the transformation of the jungle: the establishment of pastures for extensive cattle ranching. This development transformed the old monoculture model: no longer limited to coastal plantations that produced tropical fruits for export, the new estates grew grass on a far larger stage to raise and fatten cattle mostly destined for national markets. The success of cattle ranching is explained both by the policies that supported it and certain characteristics intrinsic to the activity. The Brazilian government granted lavish subsidies that stimulated the formation of enormous estates in the 1970s and 1980s. In Costa Rica subsidies designed to diversify the economy managed to boost beef production (near its ports) to third place in export value, a development known through much of Central America as the "hamburger connection."[38] Additionally, turning forest into pasture in both Central America and the Amazon helped to bring more land into the property market and encourage land speculation, especially in the context of road building. Frontier peasants also raised cattle because of the multiple advantages of stock raising over agriculture, as Wilcox and Van Ausdal show in this book. Cattle can walk to the market and can be sold at any time. Pastures extend the useful life of a jungle clearing beyond the three years in which crops are relatively productive and constituted a commodity that could easily be sold to ranchers who followed. Thus, many settlers kept transforming the jungle into pasture, often selling it to ranchers, thereby replicating the general pattern of unequal land ownership.[39]

Natural resource extraction also continued. While the booms were less significant—in hardwoods or animal skins—they facilitated processes of colonization and the establishment of pastures. Through 1960, the economy of the Brazilian Amazon revolved around extractive activities that depended on the maintenance of the forest. Those who remained when the rubber boom ended continued to extract rubber, although on a smaller scale; they also extracted Brazil nuts, hearts of palm, and *babasú* seeds, and kept small food plots for their own consumption and sale to local markets. As roads penetrated the region, it became possible to sell valuable hardwoods, especially mahogany and cedar, in national markets. The building of the Brasilia-Belém road sped up the extraction of mahogany, a phenomenon later replicated in other areas of the Brazilian Amazon. From coastal forests—like those of the Pacific coast of Colombia and Central America—a wider variety of timber species were exported.[40] Technological changes—such as the widespread use of chainsaws— helped to stimulate logging. The revenues earned in the 1960s and 1970s from the export of jaguar skins and, to a lesser extent, those of ocelots and giant otters, also helped to support frontier settlement. Highlighting the magnitude of this trade, rough estimates suggest that the total number of jaguar skins exported from Latin America during these years was equivalent to the entire jaguar population of the Amazon today.[41]

The extractive economy diversified with the exploitation of resources from the subsoil. As the price of gold rose in the 1980s, the mining frontier spread into various areas of the Amazon. Its main protagonists have been small-scale informal miners, known as *garimpeiros* in Brazil, who numbered half a million in 1990.[42] The recent gold rush has also affected the Colombian Pacific, where old placer mines have been exploited anew with backhoes, which, more than deforestation, has stripped away the soil and contaminated the ground and water with mercury. The panorama of mining also includes large-scale projects and other minerals, such as Brazil's Carajás complex, the world's largest open-pit iron mine, which began operations in the late 1960s.

The overall result of the above combination—of landless peasants, governments eager to exploit every corner of their national geography, markets for products as varied as skins and gold, and technological developments—was an unprecedented rise in the rates of deforestation that, in turn, generated alarm about the future of Latin America's rainforests. Central America had the highest rates of deforestation, but the largest deforested areas were and are still located in the Amazon. Unfortunately, we do not have precise data regarding the extent of forest loss. The best available information comes from the Brazilian Amazon, where the intense clearing of forests began in the 1970s and was concentrated in what is known as the "arc of deforestation" along the southern and southeastern edges of the basin. The intensity of this trend began to taper off in 2005, but the cleared area continues to be significant.[43] It was estimated that by 2003, 16.2 percent of the Brazilian Amazon had been deforested. (Just 15 percent of this area corresponds to deforestation that took place before 1970 and was concentrated in Pará and Maranhão.)[44] Nonetheless, this means that some 80 percent of Brazil's Amazonian forests are still standing. The panorama in Central America is quite different. The Petén lost half of its forests between 1970 and 1985, while those of Costa Rica were reduced by 65 percent between 1950 and 1990, making it, at one point, one of the most deforested countries in the world.

Similar processes occurred in Southeast Asia, turning tropical deforestation into a dramatic global problem, even though the loss of forests in Africa has, so far, been less intense. Between 1960 and 2000, forest cover (60 percent of which is made up of rainforests) fell from 55 to 20 percent in Thailand, 45 to 20 percent in the Philippines, and 70 to 50 percent in Indonesia.[45] The main cause was timber extraction: the Philippines, Malaysia, Indonesia, and Papua New Guinea have exported enormous amounts of timber, mainly to Japan. Another important cause has been the increase in land devoted to agriculture, related, in part, to growing demand in a region much more densely populated than Latin America.

Widespread forest destruction has helped erode the idea of menacing nature, part and parcel of colonization efforts, and partially replace it with that of an endangered ecosystem. The book, *Mi alma se la dejo al diablo* (I leave my soul to the devil), published in 1982 and still reminiscent of Rivera and da Cunha, tells the fatal, true story of a humble peasant whose body was found in an abandoned jungle encampment after dying of hunger: a few scrawled words he left became the book's title.[46] This tragic account fed into the perception that open pastures were the ideal landscape, not only because they allow you to see into the distance, but also because they clearly embody the fruit of human labor. The wish to civilize the land by destroying the forest still reverberates through tropical Latin America and coexists with its opposite: a conservation impulse that has produced a new rainforest cartography.

Protected Areas and Communal Territories

In 1989, the first Encounter of the Indigenous Nations of Xingú, which took place in Altamira, in the Brazilian Amazon, became a landmark in the protest movement—which is still active—against the construction of hydroelectric dams. The same area had been visited two decades before by the Belgian film director, Dutilleux, who, along with his Brazilian counterpart, made a documentary about Raoni (the Kayapó leader), which was nominated for an Oscar in 1979. Shortly after the above encounter, Dutilleux took Sting to the region, an initiative that, through the Rainforest Foundation, consolidated a partnership between people from the North and organizations of local inhabitants that has lasted for decades. Xingú thus exemplifies the intense struggles to define the future of the rainforest, as well as the ties that bind the wide network of environmental and indigenous movements and their shared recognition of ethnic rights, especially the right to territory.

A central node in the environmental movement is comprised of scientists, especially conservation biologists. Their field emerged in the 1980s in response to tropical deforestation. Its founders coined the term *biodiversity*—so popular now that we forget its recent origin—to refer to genetic, species, and ecosystem diversity, and singled out tropical rainforests as the most biodiverse areas of the world. These ideas were welcomed with alacrity partly because they were not that new. Inspired by romanticism, the nineteenth-century European naturalists who traveled to the jungle encountered sublime places and emphasized the innumerable plant and animal species they contained. For example, British naturalist Henry Walter Bates published *A Naturalist on the River Amazons* in 1863, which "oscillates between expressions of awe and the sort of meticulous description suggested by his rambling subtitle (*A Record of the Adventures, Habits of Animals, Sketches of Brazilian and Indian Life, and Aspects of Nature under the Equator, during Eleven Years of Travel*)."[47] Books of this kind established a lasting tradition that was foreshadowed by Alexander von Humboldt, one of the most influential exponents of this rapturous view.[48]

Conservation biology issued an urgent summons to study and protect nature in the tropics. By 1970, countries with Amazonian territories had established thirty-three national parks, but only five of them were located in Amazonia. Ironically, efforts to exploit the jungle had resulted in the development of cartographic surveys that were useful not only for conceiving mining projects but also for designing protected areas. Between 1970 and 1985, cutting-edge technology was employed to map part of Brazil, especially the Amazon. In 1974, with this project still in its initial stages, the first study to establish conservation priorities for the Brazilian Amazon was undertaken. On the Colombian side of the border, between 1972 and 1978, a similar initiative arose to study and integrate the Amazon region through maps. The widespread use of airplanes greatly aided these mapping and early conservation efforts. These projects were part of a wider reappraisal of Amazonia that turned it into the "lungs of the planet" due to the great amounts of carbon dioxide it absorbs, not to mention its role in the global water cycle, since it discharges 20 percent of the fresh water that the oceans receive. There were more than enough reasons and tools to designate areas to preserve tropical forests.

Thus, between the early 1970s and early 1980s, some of the most emblematic protected areas in the rainforests of Latin America were established. Among them is Manú National Park in Peru, founded in 1973, which was elevated to the status of a UNESCO Biosphere Reserve in 1977 and a World Heritage Site in 1987. The Montes Azules Biosphere Reserve, which covers a large tract of the Lacandon Jungle, dates to 1979. The biggest protected areas in Honduras and Panama were established in 1980 and 1981 respectively: the Río Plátano Biosphere Reserve in the rainforests of Mosquitia, and Darien National Park, also a World Heritage Site. The vast areas still covered by rainforests, along with the premise that the effectiveness of a protected area is proportional to its size, meant that these new conservation units were enormous.

The efforts to demarcate areas where mining, cattle ranching, and colonization would be prohibited did not end there. In the 1980s, Colombia created six parks in the Amazon; prior to this there had only been two. This trend, which gained pace at the start of the new millennium, occurred throughout the countries that shared the Amazon Basin: by 2010 there were fifty-one parks covering nearly five hundred thousand square kilometers, and protecting some 10 percent of the forests still standing in the region.[49] When other types of conservation units are also taken into account, protected areas now cover about 22 percent of the Amazon.[50] They have ensured protection against businesses that bother asking for permits, but not against illegal mining or other activities that may even be seen locally as legitimate legal activities. In some cases, especially in Costa Rica, protected areas have been a boon to the economy through ecotourism.

In addition to natural reserves, the creation of communal territories, which acknowledge the rights of peoples who lived in the jungle before the onslaught against them, is another significant development. Some of these territories overlap with protected areas, largely because the ways of life of indigenous peoples, peasants who extract forest resources, and black communities are often considered to be sustainable. These two trends have thus reinforced each other. The famous struggle of the *seringueiros* of the state of Acre, in the Brazilian Amazon, is a good illustration of this collaboration, as well as of the conflicts—often violent—that these movements engender. In the second half of the 1970s, rubber tappers in Acre organized themselves to oppose the advance of cattle ranchers and land speculators who threatened their way of life. While they conceived of their struggle within the agrarian reform movement, anchored in land rights, the environmental dimension of their claims soon became apparent because without the jungle there would be no *seringueiros*. This conservationist slant, along with support of outside allies, helped the movement gain global recognition. Its leader, Chico Mendes, was invited to Washington, where his proposal to create extractive reserves won the support of the World Bank and the Inter-American Development Bank. While there are now more than eighty-seven such reserves, they have come at a high price, including the 1988 assassination of Mendes by cattle ranchers.[51]

As the *seringueiros* were organizing, something similar was happening within indigenous communities who, relatively few in number, have experienced a veritable avalanche of intruders inside their territories. Some groups, like the Miskitos and the Cunas of the Caribbean, had established their own ethnic organizations much earlier. In the 1970s, however, a new scheme of regional organizations emerged, such as that representing the varied communities of the Peruvian rainforest, founded in 1978. Extending the organizational scope to the international level, the First Congress of Indigenous Organizations of the Amazon Basin was held in 1984. The resulting coordinating group in turn stimulated the creation of regional organizations in the countries where they did not yet exist. This was the case in Colombia, where the first inter-ethnic, indigenous organization, embracing sixty-two Amazonian nations, dates from 1995. These new structures benefited from a favorable international context. The creation of the International Indian Treaty Council in 1974 and the passage of the International Labor Organization's Convention 169, both ratified by nearly all Latin American states with rainforests, laid the legal foundations for the recognition of the territorial rights of indigenous nations.

These indigenous organizations have teamed up with the many NGOs working to protect rainforests and their inhabitants since the 1980s. The Brazilian case indicates the dimension of this movement: the Amazon Working Group, created in 1992, represents no less than 602 organizations. While local and regional groups formed in each country, taking advantage of the growing funding possibilities, international NGOs opened offices in Latin America. Tropenbos International, begun in Holland in 1986, began implementing a program in the Colombian Amazon that same year. Greenpeace opened an office in Brazil in 1991, where it sponsored a campaign focused on the Amazon, adopting an atypical approach (for this NGO) that combined environmental and social issues.[52] In Colombia, the von Hildebrand brothers illustrate the close link between indigenous rights and nature conservation. Martín, an anthropologist, has been a key defender of indigenous rights in the Amazon, first as a public official and later with the Gaia Foundation, which he founded in 1991. Patricio, a biologist, established the Puerto Rastrojo Foundation in 1982 to research Amazonian ecology and through which he contributed to the creation of national parks.

The joint mobilization of the indigenous peoples and their partners has led to the formal recognition of many of these groups' territories, highlighting Latin America's position at the forefront of this trend in the tropical world. In the Amazon, officially recognized indigenous territories amount to 28 percent of the region.[53] As in Brazil, the process of recognizing such rights in Colombia took off at the end of the 1980s, resulting in the establishment of 156 indigenous reserves, which cover nearly twenty-six million hectares and 82 percent of officially recognized indigenous territories (despite the fact that these reserves amount to just a quarter of the total number in the country). In Central America, the creation of indigenous territories has been uneven. Panama recognized the Cuna's territory as early as 1930, and the indigenous peoples native to the Lacandon Jungle were awarded a reserve totaling more than six hundred thousand hectares in 1972 (which caused conflicts with indigenous people who had migrated into the forest).[54] But on the Mosquito Coast of Honduras and Nicaragua, the recognition of property rights held by indigenous communities only advanced in recent decades.[55] Whether officially recognized or not, many other parties have a stake in these territories, creating bitter disputes. One famous case involved the Awas Tigni community of the Mosquitia, which denounced the government of Nicaragua before the Inter-American Court of Human Rights for granting a timber concession in their territory. In 2001, the court ruled in their favor.

The experience of indigenous groups served as an example for black communities in the Pacific lowlands of Colombia. Since 1996, the government has granted 163 collective titles covering more than five million hectares. Added to the two and a half million hectares covered by indigenous reserves, the vast majority of this jungle region is now comprised of ethnic territories. In a similar manner, since 2003 the Brazilian government has constituted "Maroon territories," the largest of which lie in the Amazonian states of Pará and Maranhão. These novel territorial rights fostered the formation of black social movements. Even though communities of black people have gained rights to the areas in which they live, they still face many problems of governance, related to the persistence of external threats (like coca growing and the presence of armed groups, in the Colombian case) as well as internal disputes.

Taken as a whole, these efforts have reduced the pace of deforestation in recent years. There are even some areas in which forests have recovered. While this process started more than a century ago in the Panama Canal Zone and Rio de Janeiro's Tijuca, elsewhere, such as in El Salvador and Costa Rica, the recovery has been more recent. Additionally, there are enormous areas in the Amazon still inhabited by uncontacted tribes. But deforestation persists, as do conflicts over who may use the jungle and how. For that reason, many of these areas are known for being violent: the longest-lasting and strongest guerrilla group in the region—the Revolutionary Armed Forces of Colombia (FARC)—maintained its traditional strongholds in the piedmont of the Colombian Amazon. Even though this is an exceptional case, it exemplifies the many other tensions that have arisen from the accelerated changes of recent decades and the conflicting interests at stake.

This somber picture leads many to espouse the pessimistic view expressed in the final pages of Gabriel García Márquez's *Love in the Time of Cholera*:

> The river became muddy and narrow, and instead of the tangle of colossal trees that had astonished Florentino Ariza on his first voyage, there were calcinated flatlands stripped of entire forests that had been devoured by the boilers of the riverboats, and the debris of godforsaken villages. . . . At night they were awakened not by the siren songs of manatees on the sandy banks but by the nauseating stench of corpses floating down to the sea. . . . Instead of the screeching of the parrots and the riotous noise of invisible monkeys, which at one time had intensified the stifling midday heat, all that was left was the vast silence of the ravaged land.

However, the captain of the vessel "looked at Florentino Ariza, his invincible power, his intrepid love, and he was overwhelmed by the belated suspicion that it is life, more than death, that has no limits."[56] In the end, instead of a long and irreversible path toward certain demise, the twists and turns in the story of Latin America's jungles highlight a degree of resilience: their tangled vegetation still covers two fifths of the region and accounts for 85 percent of its forests.[57]

Claudia Leal holds a Ph.D. in geography from the University of California at Berkeley and is associate professor in the department of history at the Universidad de los Andes in Bogotá. She has been a fellow of the Rachel Carson Center for Environment and Society in Munich and visiting professor at Stanford University and Universidad Católica de Chile. She is the author of *Landscapes of Freedom: The Building of a Postemancipation Society in Western Colombia*, and coeditor (with Carl Langebaek) of *Historias de Raza y Nación en América Latina* and *The Nature State: Rethinking the History of Conservation* (with Wilko Graf von Hardenberg, Mathew Kelly, and Emily Wakild).

Notes

This chapter was translated by Jimmy Weiskopf and Shawn Van Ausdal.

1. Rainforest Alliance, *Protecting Our Planet: Redesigning Land Use and Business Practices—25 years of Impacts,* 17 September 2012, accessed 27 May 2015, http://issuu .com/rainforest-alliance/docs/anniversary_120917_b/3?e=3062032/2615611.

2. Rhett Butler, "Why Are Rainforests Important?," Mongabay, 24 June 2004, accessed 27 May 2015, http://kids.mongabay.com/elementary/401.html#FJXsBMH92ZxaKpBZ.99.

3. José Eustacio Rivera, *The Vortex,* trans. E. K. James (Bogotá: Panamericana Editorial, 2001), 155–156.

4. Lúcia Sá, *Rain Forest Literatures: Amazonian Texts and Latin American Culture* (Minneapolis: University of Minnesota Press, 2004), 72–73.

5. Agustín Nieto Caballero, quoted in "Nota sobre el autor" (Note about the author), in *La vorágine,* by José Eustacio Rivera (Bogotá: Biblioteca Popular de Cultura Colombiana, 1924), my italics.

6. Eduardo Neale-Silva, *Horizonte humano: Vida de José Eustacio Rivera* (Mexico City: Fondo de Cultura Económica, 1960).

7. Quoted in Susanna B. Hecht, *The Scramble for the Amazon and the "Lost Paradise" of Euclides da Cunha* (Chicago: University of Chicago Press, 2013), 374.

8. José Vasconcelos, *The Cosmic Race: A Bilingual Edition,* trans. Didier T. Jaen (Baltimore: Johns Hopkins University Press, 1997), 25.

9. Robert Pogue Harrison, *Forests: The Shadow of Civilization* (Chicago: University of Chicago Press, 1993).

10. Johannes Lehmann, Dirse C. Kern, Bruno Glaser, and William I. Woods eds., *Amazonian Dark Earths: Origin, Properties, Management* (Dordrecht: Kluwer Academic Publishers, 2003).

11. Karl H. Offen, "The Geographical Imagination: Resource Economies and Nicaraguan Incorporation of the Mosquitia, 1838–1909," in *Territories, Commodities, and Knowledges: Latin American Environmental History in the Nineteenth and Twentieth Century,* ed. Christian Brannstrom (London: Institute of Latin American Studies, 2004).

12. David Cleary, "An Environmental History of the Amazon: From Prehistory to the Nineteenth Century," *Latin American Research Review* 36, no. 2 (2001): 64–96.

13. Dauril Alden, "The Significance of Cacao Production in the Amazon Region during the Late Colonial Period: An Essay in Comparative Economic History," *Proceedings of the American Philosophical Society* 120, no. 2 (1976): 103–135.

14. Claudia Leal, *Landscapes of Freedom: Building a Postemancipation Society in the Rainforests of Western Colombia* (Tucson: University of Arizona Press, 2018).

15. Leal, *Landscapes of Freedom.*

16. Norman B. Schwartz, *Forest Society: A Social History of Petén, Guatemala* (Philadelphia: University of Pennsylvania Press, 1990).

17. Lara Putnam, *The Company They Kept: Migrants and the Politics of Gender in Caribbean Costa Rica, 1870–1960* (Chapel Hill: University of North Carolina Press, 2002); O. Nigel Bolland, *Colonialism and Resistance in Belize: Essays in Historical Sociology* (Benque Viejo del Carmen: Cubola Productions, 2003 [1988]); Jan de Vos, *Oro verde: La conquista de la Selva Lacandona por los madereros tabasqueños, 1822–1949* (Mexico City: Fondo de Cultura Económica, 1988).

18. Peter Boomgaard, *Southeast Asia: An Environmental History* (Santa Barbara: ABC-CLIO, 2007).

19. Barbara Weinstein, *The Amazon Rubber Boom, 1850–1920* (Stanford: Stanford University Press, 1983).

20. Warren Dean, *With Broadax and Firebrand: The Destruction of the Brazilian Atlantic Forest* (Berkeley: University of California Press, 1995), 79.

21. Thomas D. Rogers, *The Deepest Wounds: A Labor and Environmental History of Sugar in Northeast Brazil* (Chapel Hill: University of North Carolina Press, 2010).

22. Dean, *With Broadax and Firebrand*.

23. Carolyn Hall, Héctor Pérez Brignoli, and John V. Cotter, *Historical Atlas of Central America* (Norman: University of Oklahoma Press, 2003); Stan Ridgeway, "Monoculture, Monopoly, and the Mexican Revolution: Tomás Garrido Canabal and the Standard Fruit Company in Tabasco (1920–1935)," *Mexican Studies* 17, no. 1 (2001): 143–169.

24. Putnam, *The Company They Kept*; Ronny Viales Hurtado, "La colonización agrícola de la región atlántica (Caribe) costarricense entre 1870 y 1930. El peso de la política agraria liberal y de las diversas formas de apropiación territorial," *Anuario de Estudios Centroamericanos* 27, no. 2 (2001): 57–100.

25. Myrna Santiago, "The Huasteca Rainforest: An Environmental History," *Latin American Research Review* Special Issue 46 (2011): 32–54.

26. André Vasques Vital, "A força dos varadouros na Amazônia: O caso da comissão de obras federais do território do Acre e as estradas de rodagem (1907–1910)," *Fronteiras: Journal of Social, Technological and Environmental Science* 6, no. 1 (2017): 23–44.

27. John Soluri, *Banana Cultures: Agriculture, Consumption, and Environmental Change in Honduras and the United States* (Austin: University of Texas Press, 2006).

28. Stuart McCook, "Las epidemias liberales: Agricultura, ambiente, y globalización en Ecuador, 1790–1930," in *Estudios sobre historia y ambiente en América Latina*, vol. 2, *Norteamérica, Sudamérica, y el Pacífico,* ed. Bernardo García Martínez and María del Rosario Prieto (Mexico City: El Colegio de México, Instituto Panamericano de Geografía e Historia, 2002), 223–246.

29. Flávio dos Santos Gomes, "A 'Safe Haven': Runaways Slaves, Mocambos, and Borders in Colonial Amazônia, Brazil," *The Hispanic American Historical Review* 82, no. 3 (2002): 469–498; Leal, "Landscapes of Freedom"; Richard Price, *First-Time: The Historical Vision of an Afro-American People* (Baltimore: Johns Hopkins University Press, 1983).

30. Margarita Serje, *El revés de la nación: Territorios salvajes, fronteras y tierras de nadie* (Bogotá: Ediciones Uniandes, 2005); Germán Alfonso Palacio, "An Eco-Political Vision for an Environmental History: Toward a Latin American and North American Research Partnership" *Environmental History* 17, no. 4 (2012): 725–743.

31. Alfredo Molano, Darío Fajardo, and Julio Carrizosa, *Yo le digo una de las cosas . . . La colonización de la Reserva La Macarena* (Bogotá: Fondo FEN Colombia y Corporación Araracuara, 1989): 17–19.

32. These figures correspond to the Northern Region (states of Acre, Amapá, Amazonas, Pará, Rondônia, Roraima, and Tocantins).

33. John O. Browder and Brian J. Godfrey, *Rainforest Cities: Urbanization, Development and Globalization of the Brazilian Amazon* (New York: Columbia University Press, 1997).

34. Shelton Davis, *Victims of the Miracle* (Cambridge: Cambridge University Press, 1977).

35. Teófilo Vásquez, *Territorios, conflicto armado y política en el Caquetá: 1900–2010* (Bogotá: Ediciones Uniandes, 2015).

36. Thomas K. Rudel, *Tropical Deforestation: Small Farmers and Land Clearing in the Ecuadorian Amazon* (New York: Columbia University Press, 1993).

37. Claudia Leal, *A la buena de Dios: Colonización en La Macarena, ríos Duda y Guayabero* (Bogotá: Fescol-Cerec, 1995).

38. Mary Pamela Lehmann, "Deforestation and Changing Land Use Patterns in Costa Rica," in *Changing Tropical Forests: Historical Perspectives on Today's Challenges in Central and South America,* ed. Harold K. Steen and Richard P. Tucker (Durham: Forest History Society, 1992).

39. Susanna Hecht and Alexander Cockburn, *The Fate of the Forest: Developers, Destroyers and Defenders of the Amazon* (New York: Harper Perennial, 1990); Marianne Schmink and Charles H. Wood, *Contested Frontiers in Amazonia* (New York: Columbia University Press, 1992).

40. Claudia Leal and Eduardo Restrepo, *Unos bosques sembrados de aserríos: Historia de la extracción maderera en el Pacífico colombiano* (Medellín: Universidad de Antioquia, Universidad Nacional sede Medellín, Instituto Colombiano de Antropología e Historia, 2003).

41. Esteban Payán and Luis A. Trujillo, "The Tigrilladas in Colombia," *CAT News* 44 (2006): 25–28.

42. David Cleary, *Anatomy of the Amazon Gold Rush* (Iowa City: University of Iowa Press, 1990).

43. Totaling 121,990 square kilometers between 2004 and 2014, according to the Prodes Project of the Brazilian Ministry of Science, Technology and Innovation.

44. Philip M. Fearnside, "Deforestation in Brazilian Amazonia: History, Rates, and Consequences," *Conservation Biology* 19, no. 3 (2005): 680–688.

45. Boomgaard, *Southeast Asia.*

46. Germán Castro Caycedo, *Mi alma se la dejo al diablo* (Bogotá: Plaza y Janés, 1982).

47. Candace Slater, *Entangled Edens: Visions of the Amazon* (Berkeley: University of California Press, 2002), 41.

48. Nancy Leys Stepan, *Picturing Tropical Nature* (London: Reaktion Books, 2002).

49. Figures estimated on the basis of the websites of the national parks service of each country.

50. "Amazonia 2016 Protected Areas & Indigenous Territories," RAISG: Amazon Geo-Referenced Socio-Environmental Information Network, Amazonia SocioAmbiental, accessed 2 October 2017, http://raisg.socioambiental.org/en/mapas/#areas_protegidas.

51. Kathryn Hochstetler and Margaret E. Keck, *Greening Brazil: Environmental Activism in State and Society* (Durham: Duke University Press, 2007); José Augusto Pádua, "The Politics of Forest Conservation in Brazil: A Historical View," *Nova Acta Leopoldina* 114, no. 390 (2013): 65–80.

52. Hochstetler and Keck, *Greening Brazil.*

53. RAISG, accessed 2 October 2017, http://raisg.socioambiental.org/en/mapas/#areas_protegidas.

54. Jan de Vos, *Una tierra para sembrar sueños: Historia reciente de la Selva Lacandona, 1950–2000* (Mexico City: Fondo de Cultura Económica, 2002).

55. "Nicaragua" and "Honduras," Portal Territorio Indígena y Gobernanza, sponsored by the Right and Resources Institute and HELVETAS Swiss Intercooperation, accessed 15 February 2016, http://www.territorioindigenaygobernanza.com/.

56. Gabriel García Márquez, *Love in the Time of Cholera,* trans. Edith Grossman (New York: Alfred A. Knopf, 1999), 336, 348.

57. Food and Agriculture Organization of the United Nations, *State of the World's Forests* (Rome: FAO of the UN, 2011); Michael Williams, *Deforesting the Earth: From Prehistory to Global Crisis* (Chicago: University of Chicago Press, 2002).

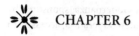 CHAPTER 6

The Ivy and the Wall
Environmental Narratives from an Urban Continent

Lise Sedrez and Regina Horta Duarte

> *Se va enredando, enredando,*
> *como en el muro la hiedra*
> Tangling, intertwining,
> like the ivy on the wall
> —Violeta Parra

The Environmental City

Latin America today is a predominantly urban region. Around 80 percent of its population lives in cities, towns, or urbanized areas. Within these places, four out of every five Latin Americans negotiate their access to food, water, air, land, and green spaces; they dispose of the waste they produce; they confront storms and earthquakes; they live side by side with populations of insects, rats, pigeons, dogs, and cats. These urban residents' day-to-day encounter with nature occurs on the beaches of Havana; with the rains that stream down the buildings toward the manholes in the streets of Bogotá; in the *arroyos*—underground streams—that run beneath Buenos Aires and overflow during the seasonal heavy rains of the *Sudestada*; with the trees on Lima's public squares, where neighborhood children play; in the smoky air in the center of Mexico City; with the erosion of the hills that dominate the city of Rio de Janeiro; as well as in the smells of animals, people, and machines ingrained in the colonial-era sidewalks that have survived into the twenty-first century. The city also constitutes a built environment, where different living beings have coexisted in a modified landscape and live alongside objects and technologies that are the products of the processing of natural resources; this processing, in turn, implies new alterations of the natural environment. The construction of this *urban nature*, which unites trees and buildings, rivers and streets, animals and automobiles, food and trash, is just as much a part of Latin American environmental history as its mountains, forests, deserts, and mines.

Latin American cities took shape as extremely complex networks characterized by resonances between urban centers and the biophysical environments in which they were built and that surrounds them. The networks include large and small cities within Latin America, and links with other cities and regions of the world. In forming these networks, cities and towns have both tested the limits and possibilities of the biosphere and have transformed it as well. This urban transformation has taken neither a uniform trajectory nor, in many cases, a linear one. Demands and problems emerged over time that were unforeseeable. Residents in sixteenth-century Mexico City were unable to imagine the dilemma that pollution would present in the twentieth century, even though this pollution was the result of an urban trajectory shaped by four centuries of its history. Nor could they conceive the size their city would reach. Among the thirty-six metropolises in the world with over ten million inhabitants, according to the census taken in 2015, four are Latin American: Mexico City, São Paulo, Buenos Aires, and Rio de Janeiro (see illustration 6.1). Latin America's complex process of urban globalization continues to expand, and cities with fewer than a half a million inhabitants show the highest rates of growth, such as Tijuana (Mexico), Temuco (Chile), or Belém (Brazil).[1]

If cities dominate modern Latin American landscapes, they were also important in the past, even when the majority of the population lived outside them. Urban life flourished in pre-Columbian societies, as the vestiges of Mayan cities (e.g., El Caracol and Tikal), as well as Toltec (e.g., Tula), Aztec (e.g., Tenochtitlán), and Incan (e.g., Cusco) ones, demonstrate. The city was an important feature of the colonial experience, although its logic and structure differed significantly from the urban aggregations that had existed prior to

Illustration 6.1. Smog around the Pico de Orizaba, in Mexico City, the second largest city of the Americas, 2015.

Photograph by Gogadicta, iStock.

Iberian colonization. The arrival of the Europeans in the fifteenth century and the process of colonization demanded that the connections between the New and the Old Worlds had strong urban foundations. The crossing of the seams of Pangea, as Alfred Crosby called the great colonial seafaring enterprise, depended upon coastal cities to stitch these two worlds tightly together.[2] Ships full of plants, animals, and germs entered these ports, and departed them carrying precious metals, sugar, tobacco, lumber, medicinal commodities harvested in the hinterlands, leather, and animals. For centuries to come, waves of humans from other continents also came ashore at the region's port cities, particularly Europeans and enslaved Africans. Demographic growth heightened demand for land and resources.

Latin American colonial cities were the main points of articulation in the region for the dynamics of economic mercantilism. The urban network was designed from the perspective of the colonizers. Either by appropriating traditional indigenous centers and indigenous roads, or by building brand new cities and roads in forests and river banks, colonizers sought, above all, constant communication with the metropole. Good ports were a priority, protected from enemy attacks, and near waterways that facilitated access to the wealth of the hinterland. Many of the earlier urban constructions were concentrated near bays, capes, and estuaries. São Salvador, founded on the Bay of All Saints, in Brazil, emerged in 1549 as a city, a fortress, and the capital of the new Portuguese colony, quickly becoming an important port for the export of sugar and for the import of enslaved Africans until the nineteenth century.

The colonizer also valued proximity to precious natural resources, particularly silver and gold. Potosí (in modern-day Bolivia) became one of the largest and wealthiest cities in the world in the seventeenth century, with approximately two hundred thousand inhabitants. The city was home to a production complex for the extraction of silver, with a system of aqueducts, artificial lakes, and watermills. The mercury used in the process came from the mines of Huancavelica. The circulation of labor, capital, and ore created channels of communication and transportation between urban centers including Potosí, Chuquisaca, La Paz, Cuzco, Lima, and the port in El Callao. Products that inhabitants of Potosí did not produce but wanted to have, such as grains, domestic animals, firewood, coca leaves, fruits, textiles, and beverages arrived by way of these same channels.[3]

Urban growth also benefitted from the concentration of the labor force and colonial control over it. Many cities established hierarchies of domination in spaces that had already been domesticated and transformed by indigenous societies, which facilitated colonial control over the human element that was necessary for the transformation of nature into wealth. For example, Mexico City emerged from the ruins of former Aztec capital Tenochtitlán; and Jesuits established their missions alongside indigenous settlements. Puebla, in New

Spain, took advantage of both indigenous labor and privileged natural conditions, including fertile soils and the San Francisco River, which supplied the city with high-quality water. The flow of the river made possible the use of the hydraulic energy in the perimeter of the city, while it eliminated the city's wastes. The surrounding areas also provided abundant wood for construction and fuel to supply a population that reached over fifty thousand inhabitants in 1777.[4]

Many urban areas emerged less from the experience of conquest or as a result of a direct connection with the metropole, but rather from the colonies' internal economy. Towns grew along the edges of subsistence routes, at outposts for the exploration of the hinterlands, and especially along rest stops for mule trains. These population centers were all connected, however tenuously, to the capitals and to the colonial project. For example, Mendoza, founded in 1561 in an area that had been inhabited by indigenous Huarpes people, was a place where traders coming from the Río de la Plata area en route to Santiago de Chile would stop to rest and resupply themselves before crossing the Andes. In that semiarid climate, indigenous systems of irrigation using waters from an oasis formed by water coming from the Cordillera guaranteed Mendoza's water supply. Towns, many of which were ephemeral, also sprouted in the shadows of Mendoza's large rural properties, born out of the many types of sociability that rural workers sought as they came together to pray, dance, and seek companionship and love.[5]

Colonial cities also coexisted with hurricanes, earthquakes, floods, and droughts, or with the constant necessity to renew the water supply or to drain flooded areas (as in Mexico City or Rio de Janeiro). If, for example, hurricanes were almost unknown to the European colonizers who settled in Havana, Cuba, they are regular phenomena in the region and, according to Stuart Schwartz, provide the basis for a historical metanarrative of the Caribbean with respect to both slavery and the region's plantations. During hurricane season, cities flooded, mosquitos multiplied, and yellow fever and malaria exploded, leading to shortages of food and the intensification of existing social conflicts. Far from the metropoles, the colonies were obligated to disregard colonial enmities and exclusive trade arrangements, in exchange for mutual protection. The concentration of human populations and buildings in cities carried a particular risk of disaster during hurricanes.[6] Earthquakes, which are even less predictable than hurricanes, similarly helped form Latin American urban history during the colonial era. Lima and Callao suffered a massive earthquake in 1746, the sixth since the arrival of the Europeans, but these cities' strategic position guaranteed their reconstruction. However, the memory of the 1746 catastrophe deeply marked the conflicts and negotiations of this period.[7]

The colonial experience generated a way of reacting to environmental challenges that would endure after the creation of independent, nation states. Thus, cities affected by hurricanes and earthquakes were reconstructed in places and under conditions that failed to secure them against natural disasters, and swampy areas were drained to make way for urban expansion, despite constant flooding.

Even colonial urban regulations concerning access to water and to land in the region lasted after independence, defining the contours of urban expansion. In Spanish America, the establishment of communally titled territories (*ejidos*) that provided indigenous communities with access to water and pasturage in the sixteenth century persisted for centuries. These arrangements created both legal and traditional limitations that configured how cities might grow and use these resources, until the *ejidos* finally succumbed to the urban planning of the liberal generation of *científicos* at the end of the nineteenth century.[8]

Pasturage and water, earthquakes and tropical storms, metals and mountains, swamps and bays: the biophysical sphere has always been a protagonist, and not just the stage on which the history of Latin American cities has played out. The creation, organization, and expansion of cities are processes in which human inhabitants interact with the natural environment; and the natural setting of a city, in turn, can be radically transformed by human action. In *The Lettered City*, Ángel Rama argues that Latin American cities were ordered, literate, bureaucratic centers of power and poles of transformation.[9] We would add that the Latin American city has also been, and continues to be, an environmental city.

The "Second Conquest of America"[10] and its Urban Interface

At the beginning of the nineteenth century, the young, independent nations of Latin America already contained a network of interconnected cities and towns that was a legacy of the colonial experience. From small villages to large cities, from the capitals to the frontier outposts, from the interior to the coast, all were interconnected as a result of political decisions, economic pressures, and environmental demands. Municipal administrations made use of a significant collection of rules and ordinances that regulated the use of common water and land and waste disposal, even though these rules were unevenly applied.[11]

During the first half of the nineteenth century, Latin America experienced a period of urban decline. Only in rare cases did cities surpass their colonial contours, with the notable exception of Havana, Cuba, which benefitted from a booming sugar economy. Argentina, Brazil, and Mexico each had one or two primary cities, channels of communication between the region and global markets, and innumerable towns and villages that were pulled into the orbit of these major cities, whether from near or far. The conflicts of the period had the effect of diminishing the attraction of these larger urban centers.[12]

This tendency reversed itself in the second half of the nineteenth century. With the greater integration of Latin American nations in global commodity markets, cities took on a new dynamic. Extractive and agroexport activities fed an industrial sector hungry for raw materials, fertilizers, and tropical products. This process transformed not only landscapes characterized by their forests, fields, and mountains, but also urban landscapes almost to the same degree as the arrival of the Europeans had in the fifteenth century. Coffee, guano, cotton, tobacco, sugar, cacao, fruits, and rubber circulated through the region's port cities. Here, contracts and prices were negotiated, and roads and railways were designed that would open up new areas to exploitation. Overseas commerce, particularly in the Atlantic world, fed such new commercial centers as Barranquilla (Colombia), which eclipsed other colonial coastal cities, such as Cartagena, in importance.[13]

By the end of the nineteenth century, modernization and the expansion of monoculture and productive rural areas oriented toward export markets had brought about waves of internal migration from the countryside to the city. In Mexico, Central America, Venezuela, and Colombia, this modernization exacerbated the process through which traditional forms of land use had been breaking down, and expelled peasants and indigenous people from their rural homes to the outskirts of cities, where they lived in miserable conditions.

Despite the precariousness of the services and housing available there, cities promised new opportunities for social mobility to recently arrived migrants, both from the region and overseas. Countries such as Brazil and Argentina received thousands of European and Asian immigrants, which brought about massive demographic changes. The population of Buenos Aires increased from 230 thousand to 677 thousand inhabitants between 1875 and 1895, surpassing 1.5 million residents in 1914. Rio de Janeiro's population rose from five hundred thousand to 1.2 million between 1890 and 1920. Mexico City had a half a million inhabitants in 1900, and it became the most populous city in the Americas over the course of the twentieth century. More modest yet still significant growth took place in Montevideo, Uruguay, which reached a hundred thousand residents in 1890, and Bogotá, Colombia, which reached a hundred thousand in 1905.[14]

The region's national governments invested in audacious urban renewal projects. These included the widening, paving, and landscaping of avenues and main streets, the installation of water and sewage systems, electrical illumination, trolley lines, and the construction of public buildings, squares, and urban parks. Urban reforms expressed the anxieties of native elites, whose wealth came from the export sector, and who were fascinated by the European lifestyle. From Mexico City to Buenos Aires, including Rio de Janeiro, São Paulo, Lima, and Havana, political leaders ordered the opening of avenues, demolished hills, and had swamps turned into landfill. The unequal distribution of these benefits of urbanization unfolded along the lines of race and class, and wealthy areas of the city made use of the parks, promenades, and improved sanitary conditions. The obsession with straight lines ignored the underlying biosphere on which these cities were built. Defined by the engineer's compass, urban grids were imposed without accounting for hills, fields, rivers, or lagoons, "in an exercise of arrogance" with environmental consequences that often have endured until the present day.[15]

Meanwhile, over the nineteenth century, Latin American cities became places that nurtured the efforts of both European and North American naturalists, hired by local governments, as well as the pioneering works of Latin American researchers. Knowledge about the natural world was systematized in scientific institutions in urban settings. Museums of natural history were established in Buenos Aires (1812), Rio de Janeiro (1818), Santiago (1822), Bogotá (1823), Mexico City (1825), Lima (1826), and Montevideo (1837). In the course of the nineteenth century, museums also emerged in Caracas (1875) and Costa Rica (1887). In these institutions, fauna, flora, mineral resources, and anthropological and archeological characteristics of each nation were studied and inventoried, in the context of the postcolonial opening of these areas to the world and the parallel desires to impose civilization and to affirm national identities. Financed by their respective governments, the museums opened their doors to the public, and thousands of visitors were exposed to a pedagogy of nature as they wandered through the exhibitions. Natural history museums frequently served as the main sites on which Latin American countries organized their participation in the era's world's fairs and expositions. Brazil, Venezuela, Chile, Argentina, and Mexico, for example, anxiously sought to integrate themselves into the "civilized" world, particularly by exhibiting samples of its exuberant, rich natural bounty.[16] Urban areas became centers for learning about the natural world well beyond their own city limits.

The adoption of the European model of urbanism, demographic increases, the increasing complexity of the service sector, and the growth of the middle classes and of consumerism greatly increased consumption of animal energy and biomass in Latin American cities. The environmental impact, both within the city and outside it, was profound: Rio de Janeiro and São Paulo owe their growth, in part, to the systematic conversion of the Atlantic Forest into firewood and charcoal.[17]

Local geographies and politics alike conditioned the options that presented themselves to Latin American cities. Bogotá, strangled by a ring of *haciendas* that were only really broken up in the second decade of the twentieth century, built vertically without expanding horizontally. Even in a rather humid area crossed by the San Francisco, San Augustin and Arzobispo Rivers, this spatial limitation created critical problems for supplying Bogotá with water. In contrast, Buenos Aires and Mexico City spread outward, in the direction of neighboring *pueblos,* incorporating new communities.

The belts of municipalities that formed around capital cities often grew faster than the urban centers. In Rio de Janeiro, between 1911 and 1921, the population of so-called rural districts grew by 71 percent, while that of the city center grew just 16 percent. In Buenos Aires, the yellow fever epidemic of 1871 drove a large part of the city's elite to the rural areas surrounding the city, giving rise to the neighborhoods Recoleta and Palermo. The growth of peripheral regions, both rich and poor, pressured governments to provide more public services including drainage, sanitation, river channelizations, water, energy, sewage, and transportation service. Trains and streetcars were necessary to connect urban centers with peripheries that supplied labor and perishable foods. In 1879, Buenos Aires had 146 kilometers of railways for a population of two hundred thousand inhabitants, joining distant districts and neighboring municipalities to the center.[18] The suburbs were, therefore, part of a continuum between the countryside and the city. Alongside workers' housing, small family farms and rural properties supplied city centers with vegetables, fruits, cereals, fodder for draft animals, pigs, and chickens.[19]

The old colonial urban network had turned more complex by the early twentieth century, becoming an integrated system in which each city or town had a direct impact on the dynamics of other, related, settlements. Such characteristics as size, population, political importance, and location were the result of both the specific historical trajectory of each city and the product of its position in the system as a whole. From capital cities to peripheral villages, the urban experience can be explained less by the opposition between the city and the countryside than as a continuum: an image that accounts for the unequal and disorderly nature of this relationship, but which stresses the integration of cities in rural economies, in communities whose livelihoods are based on extractive industries, and in the Latin American landscape in general.

A fundamental element in the creation of these urban connections in Latin America from the mid-nineteenth century on, railways expanded the urban reach into what were still remote areas, allowing the emergence of hundreds of frontier towns. Thanks to railways, an initial wave of industrialization strengthened the creation of new regional centers such as Atlixco (in Puebla, Mexico), which almost tripled its population between 1890 and 1910. The installation of textile factories in the region would be a result of the arrival of the railway line Ferrocarril Interoceánico in 1893.[20] In Colombia, the economic conjuncture rearticulated the Andean axis with the Atlantic economy, a relationship formed by the importance of the Magdalena River. Bogotá's dominion over the hot and also the temperate climates of central Colombia's high plateau was fundamental for the country's insertion in the new global economy and for the consolidation of the city's position as the capital of the nation: investments in railroads and highways connected the Magdalena and Cauca Rivers to the country's new productive areas, and thus to Caribbean ports. These new transportation networks linked the towns that sustained themselves by harvesting coffee from the Colombian Andes with the port cities of Maracaibo, Barranquilla, and Buenaventura, from which a large part of the coffee was exported.

On the coastline, the era of steam power allowed ships to forge interoceanic connections more regularly during the export boom at the end of the nineteenth century, bringing dynamism to the region's port cities. Tumaco, Colombia, "the pearl of the Pacific," demonstrates the contradictions of these processes. Tumaco centralized the activities related to the harvesting and commercialization of *tagua*, the seed of a palm tree (*Phytelephas tumacana*), which, because of its similarity to ivory, was broadly used in the United States and Europe for the production of buttons. The palm was common in the area surrounding the port, and its exploitation transformed Tumaco into a city with elegant buildings and shops filled with merchandise. Because of the tensions between white elites involved in the export businesses in the center of the city and the black populations who ventured into the forest to gather the seeds, different perceptions and representations of nature were layered over the existing racial conflicts that divided that society: white elites strongly associated cities with civilization in contrast to the jungle, a place of savagery and blackness.[21]

In the Amazon forest, North American businessman Percival Farquhar directed the construction of the Madeira-Mamoré railway following an agreement between the Bolivian and Brazilian governments. The finished railway facilitated the connection of small towns in the heart of the Amazon all the way to Porto Velho (founded in 1907), the city from which it was possible to navigate all the way to the Atlantic, making possible the export of rubber. The venture attracted laborers from throughout the world, involving them in a history of environmental destruction, the expansion of international capital, and the displacement of human populations. With the decline in the export of rubber and the opening of the Panama Canal in 1914, the railway lost importance, and the small urban centers that had grown up around the rubber trade became useless and emptied out, forming a string of "ghost towns" lost among the chimeras of modernity.[22]

Cities, however, wove together not only stories of countries, diverse populations, and ecologically distinct regions; they were also the stage on which social conflicts played out, in which race and class conditioned people's varied experiences of the natural world. This urban network, more diversified and more extensive than during the colonial era, captured natural resources and transferred them to the city, and thereby to the voracious system of global commerce, which required constant investments of energy to sustain its dynamic: steam, firewood, hydraulic energy, coal, and draft animals. The expansion of services, the concentration of the labor force, and the growth of cities' administrative and political importance intensified urban areas' impact on the biosphere. More than just points of transit, cities were spaces where resources were fought over and transformed. Even before the expansion of industrialization from the 1930s on, cities both old and new became important consumers of natural resources. Once this standard of consumption was established, it would increase spectacularly starting in the second half of the twentieth century, requiring constant investments of energy to sustain its dynamic.

The Modernizing City and Environmental Challenges

If until now we have examined the environmental history of the Latin American city by considering its place in an urban network—with our attention also focused on the areas around the city, and its connections with extractive and productive areas—the city's internal dynamics also need to be explored. By the late nineteenth century, the internal processes that regulated the urban metabolism, the creation of green spaces, and the rise of new environmental sensibilities were marked by intense conflicts, inserted in broader political and social contexts. Cities were thus the stage on which social conflicts played out, in which race and class conditioned people's varied experiences of the natural world.

The new urban model, as an integral part of modern agroexport commerce, required that cities make significant morphological adjustments. Cities that functioned in the sixteenth century with hundreds or thousands of inhabitants had to attend to unprecedented demands for transportation, housing, territorial expansion, and access to food in the twentieth century. Urban shantytowns and forms of collective housing for the poor that went by a variety of names—*favelas, conventillos, cortiços,* and *barrios,* to name a few—proliferated on the peripheries of Latin American cities. Municipal governments took pains to limit the circulation of this "suspicious" population as part of a demand for social order that was further legitimated by a need to control epidemics and contagious diseases.

The same logic that inserted Latin American cities in global dynamics also facilitated the international diffusion of fatal microorganisms, particularly under such favorable socioenvironmental conditions as poor nutrition among the lower classes, increasing population density, problems with the supply of potable water, and unhealthy working and living conditions. The Spanish influenza epidemic of 1918–1919 was a tragic, exemplary case. In Latin America, Brazil was the country that was most deeply affected. The flu arrived in the Brazilian cities of Recife, Salvador, Rio de Janeiro, and Belo Horizonte, among others, leaving a death toll of around three hundred thousand. The cities of Caracas and Santiago de Chile were also deeply affected. In Argentina, the influenza virus probably entered by way of the port of Buenos Aires and spread to other cities connected to the capital by railroads, such as Salta and Tucumán, with disastrous consequences.[23]

Epidemics set systematic sanitation campaigns and urban renewal projects in motion. Doctors and engineers contracted by the state undertook spatial transformations with significant environmental implications—and with a certain opportunism, as a means to establish greater social control. In Barracas, a neighborhood in Buenos Aires, the traditional *saladeros* that produced salted meats along the banks of the Riachuelo River were removed following a long polemic in 1870, having been accused of poisoning the river with their activities. The refrigerated meat producers that took their place, were, within the modernizing logic famously expressed by the Argentine president Domingo Sarmiento, better suited to this new way of occupying space, but probably were neither more hygienic nor less polluting than the small producers.[24] In Santos, a coastal city in the state of São Paulo, bubonic plague arrived in 1899 and generated sanitation campaigns. The disease reached Caracas in 1908, giving rise to public policies that included demolitions and the disinfection of homes, as well as the renovation of the aqueduct that sent clean and abundant water to the city's residents. In 1904, the Brazilian doctor and public health pioneer Oswaldo Cruz combatted yellow fever in Rio de Janeiro by renovating residential areas and imposing vaccinations on the population of the city's collective housing. Strong opposition to forced vaccinations led to a popular uprising, the *Revolta da Vacina,* fuelled by anger at the authoritarian form by which these public health measures were carried out.[25] Rio de Janeiro's inflexible prefect, Francisco Pereira Passos, flattened hills, filled in the city's marshy areas, and demolished *cortiços,* the crowded collective housing where the city's poorest classes resided.

Urban reforms not only transformed capitals, but also reverberated throughout the urban network. In Honduras, Puerto Cortés, a town that was virtually ignored until the twentieth century, became one of the principal ports for the export of bananas by the United Fruit Company. The town expanded over the mangrove swamps and estuaries of the region.[26] This adaptation of the landscape, carried out in rapid and summary fashion, also reveals the critical environmental vulnerability of the way urbanization often unfolded in Latin American history. Cities quickly expanded along flood areas. Intense rains, which had previously fallen on prairies and forests, now caused disasters for growing urban populations.[27] In 1911, strong rains in the Río de la Plata basin brought on dramatic floods in the riverside cities in Buenos Aires province, such as in Avellaneda.

The occupation of new urban areas, like the modernization of old capital cities, was marked by a pattern of inequality. The poor population, whether native-born, slave, or immigrant, settled in the least expensive places to live and generally built their own homes with precarious materials and structures. Shantytowns, such as Rio de Janeiro's favelas, rose up the hillsides or spread across marshes and atop stilts, concentrated in and forming areas of the highest environmental risk, in a confluence of social and environmental vulnerability.

Larger and more extensive than their predecessors, twentieth-century cities also placed more pressure on surrounding ecosystems, and even those at a distance. Cities of the early twentieth century demanded more construction materials, more water, and more land than ever before, proportional to their demographic growth and the growing sophistication of their more privileged inhabitants. In Bogotá, the mountains to the east, which for centuries had supplied the city with wood and potable water, became a distressing landscape for the city's residents. The overexploitation of wood for fuel and construction caused soil erosion and desiccation. Engineers, doctors, and local authorities quickly drew up plans for reforestation, privileging exotic species, such as eucalyptus and pine, to the detriment of the native species. These elite professionals, using a discourse of utilitarianism, attributed the degradation of mountains to the presence of precarious settlements inhabited by poor populations who used mountain resources to live. Local authorities removed poor inhabitants, a decision that evidences the political and social aspects involved in disputes over nature.[28]

Urban growth in nineteenth-century Rio de Janeiro also caused the deforestation of the surrounding areas, compromising the volume of water in the springs that supplied the city. This prompted Rio's municipal government to re-create Tijuca Forest, planting over a hundred thousand seedlings of native trees, as well as exotic ones, with the help of the Brazilian monarchy. In addition these restoration efforts, the suspension of logging enabled the Tijuca forest to reestablish itself, according to José Pádua, in a case of "synergy between human activity, and the actions of nature." The forest would later go through a process of landscaping, with the construction of roads, trails, fountains, and lakes geared toward the forest's use for the leisure of the urban population, which bequeathed Rio with a splendid green space of 3,953 hectares in the heart of one of the largest metropolitan areas in Latin America.[29]

Urban forests such as Tijuca are part of the history of Latin America, but the coexistence of city and forest has not always been easy. Disputes over the woodlands to the south of Mexico City exemplify the complexity and the non-linear nature of the relationships between the city and nature. At the beginning of the twentieth century, urban reformers and conservationists brought attention to the dangers of the destruction of the forests, arguing that the forests were important for the wellbeing of the city's inhabitants. As spaces of conflict between *pueblos* and economic elites, the use of woodlands would be the object of intense negotiations following the Mexican Revolution. In 1926, a law established a precarious balance between the conservation of woodlands and their use by populations in the areas surrounding the city. However, under the administration of President Lázaro Cárdenas, these peasant communities would witness the disappearance of not only their main channels of political communication but also the government's interest in guaranteeing the hard-won rights to the woodlands, whose trees were increasingly coveted for the production of energy and paper.[30]

The modern city conferred new value on green spaces, whether they were landscaped spaces inspired by European models, or the remains of ancient woodlands. Green public areas established a new urban aesthetic. In Bogotá, Centenary Park, inaugurated in 1883 to commemorate the one hundredth anniversary of the birth of Simón Bolívar, occupied an area that had already served as a park twenty years earlier.[31] Chapultepec Park in Mexico City is one of the oldest areas of preservation in the Americas. Sprawling over 850 hectares in the middle of Latin America's largest metropolitan region, the park attests to the historical resilience of urban green areas as they are transformed, adapted, and renovated.[32]

The survival of these green areas stands in contrast to the growing appetite for woodlands both in and outside the urban perimeter. Cities widely consumed firewood and charcoal. Growing numbers of homes and manufacturing and industrial establishments significantly intensified the process of deforestation, and the pace at which these resources were harvested soon became insufficient to satisfy the demand for them. In Brazilian cites in the first half of the twentieth century, textile mills, steel furnaces, and railroads all needed a constant supply of wood. In addition to these industries, the combined consumption of household wood stoves, small pottery kilns, lime kilns, dairy farms, and sugar mills devoured a massive volume of wood. The consumption of firewood was not the only source of urban energy; in the state São Paulo, industrial growth also relied on the use of hydroelectric energy and fossil fuels.[33]

The production of organic and chemical wastes was a perilous dimension of the urban metabolism. Garbage, sewage, odors, and industrial waste were produced more quickly than waterways could dissolve and/or carry them away. Santiago de Chile, for instance, discovered that the Mapocho River was no longer sufficient to clean the city due to the new volume of waste. Moreover, in the first decades of the twentieth century, the elites of Santiago developed a growing intolerance for the putrid smell of garbage that permeated the city streets and, in effect, symbolically identified the range of urban odors with social hierarchy. These new sensibilities and the yearning for the symbols of modernity led Santiago elites to invest in sanitation projects, although the results were far from satisfactory.[34]

More successful, Rio de Janeiro had taken the lead by installing a comprehensive sewer system as early as 1871, second only to the cities of Hamburg and Brooklyn (New York). Collected sewage was compacted in a processing station near the Guanabara Bay and then carried in a sludge transport vessel, or *navios lameiros,* which dumped its contents in the middle of the bay. Until the middle of 1920, solid wastes were taken to a sanitary dump that was also located at the edge of the bay to be later carried away by the water, as industrial wastes also were.[35] Like the Mapocho River and the Guanabara Bay, other rivers, bays, and estuaries had a limited capacity to absorb or dissolve cities' increasingly voluminous quantity of organic and inorganic wastes.

Other cities, such as Bogotá and Buenos Aires, burned their garbage. In Buenos Aires, special train cars carried garbage—to the great discomfort of residents who lived near rail stations—to large open spaces, where it was incinerated. In 1912, the system was decentralized, shortening the time it took to travel between the trash collection site and the facility where it was burned. Communities of informal street workers called *papeleros* (paper men), *basureros* (garbage men) or *catadores de papel* (paper gatherers) also participated in the processing of urban refuse, whose ultimate destination was either incineration or simply dumping in open spaces. Men, women, and children sought metals, paper, and food—anything of which one could make further use—from the city's refuse. A municipal report from 1904 in Buenos Aires describes the difficult experiences of these families, who lived in the garbage with "1,500 pigs, lots of dogs, and thousands of rats," with humans and nonhumans sharing the "same occupation and the same routines."[36] Companies hired by municipalities to process refuse sometimes hired *papeleros,* who were generally from cities' most marginal populations and considered, like the materials that they scavenged, a sort of "social refuse" of the urban system. In 1878, the *basureros* in the garbage dump south of Buenos Aires's Los Mataderos neighborhood were mostly indigenous people who had emigrated from the *pampa.*

These new times were also times of new sensibilities, and the urban scene was fertile ground in which these could flourish. For example, people began to see animals with a new affection. Societies for the protection of animals protested against vivisection, the use of animals in circuses and bullfights, cockfighting, and the mistreatment of pack animals, which all came to be seen as cruel practices. In Havana, the Cuban Society for the Protection of Animals and Plants (Sociedad Cubana Protectora de Animales y Plantas) was created in 1882. Similar organizations in Caracas, Buenos Aires, São Paulo, San José de Costa Rica, and Rio de Janeiro carried out campaigns and published illustrated pamphlets and magazines.[37] Debates arose around what a model civilization should look like and how relations between humans and animals served as a key theme with which to reflect on relations between human beings themselves.

Despite the peculiarities of each place, from the 1860s to the mid-twentieth century, most Latin American cities confronted shared challenges of adjusting to new economic demands, confronting the costs of the acceleration of the urban metabolism, and managing common spaces in cities with profound social inequalities. The ways in which cities approached these challenges established conditions that would make these same questions even more pressing later during Latin America's industrial era.

The Age of Acceleration

Considering that the "Second Conquest of Latin America" fed the development of agroexports and extractive development, the post–World War II context implied important transformations with the advance of industrialization in Latin America. Latin America's industrial production reached considerable levels as compared with the standards of earlier periods, bringing expanded and increasingly complex environmental challenges. A second wave of urbanization, beginning in 1950, effectively transformed a majority of Latin Americans into urbanites. For the most part, this process was related to industrialization, which had begun in the late nineteenth century and consolidated with the import substitution industrialization policies that most Latin American governments adopted to a greater or lesser extent in the mid-twentieth century.[38] The region's industrialization, therefore, concerned the acceleration of much older processes involving the occupation of territory and the transformation of natural resources, with a significant impact on the entire region, including areas and landscapes that until then had been little exploited. Urban Latin American thus participated in what John McNeill characterizes as an "age of acceleration."[39]

Industrialization ramped up the demand for energy, natural resources, and labor—and consequently, for water, land, and urban services. Local sources of firewood and charcoal, already in short supply for a modernizing city, reached an unsustainable limit for the growth of São Paulo. The industrial growth of this city increased its energy demand exponentially, and from 1950 on the hydroelectric plants in the region multiplied, transforming large forested areas into dams and reservoirs. In the 1970s, the Brazilian military government initiated negotiations with Paraguay for the construction of the Itaipu hydroelectric plant, which upon completion was the largest in the world, providing cities in southeastern Brazil with energy produced over eight hundred kilometers to the west of São Paulo. The Itaipu Dam submerged the great waterfall Sete Quedas, which was made up of nineteen waterfalls grouped in seven levels in the Paraná River. In just a few days following an official ceremony closing the floodgates, the immense cataracts surrounded by vast forested areas disappeared, under dramatic protest from environmentalists. The same energy demands justified the construction of nuclear plants in Angra dos Reis, between

Rio de Janeiro and São Paulo, in operation since 1985, imposing challenges that remain unresolved concerning the final destination of radioactive waste. The environmental footprint of the industrial city has left a wide mark, in both time and space.

Urban networks are increasingly dense, and the spaces between cities are transformed into corridors, forming a continuous landscape everywhere marked by urban occupation. In the highways that knit together the metropolitan conurbation, the automobile became one of the characteristic features of the urban landscape and the daily lives of its inhabitants, with all the environmental consequences that accompany the vertiginous increase in their number, such as noise and atmospheric pollution and colossal traffic congestion.

The importance of the city as a space of territorial transformation on a broader scale cannot be underestimated, and environmental questions have merged with cultural ones. This process of change to the human geography of the region concerns not only the amount of foodstuffs that cities consume, but also the appropriation of traditional products by reproducing them in cities with new technologies for consumption by the urban masses, far from their original cultural contexts. For example, urbanites reinvented *fogo de chão*, the traditional barbeque from the prairies and plains of southern Brazil, now served in popular, and often elegant, restaurants called *churrascarias*, while the consumption of meat transforms landscapes of native vegetation into cattle pastures. *Tortillas* and *arepas*, duly industrialized and packaged in plastic, fill supermarket shopping carts in Mexico City and in Bogotá—and even in the great centers of the Latin American diaspora, such as New York and London. Mexican fast-food chains have spread throughout the world. Urban consumption thus shapes the urban landscape as well as the nonurban one. Latin American cultural products are exported en masse, just as its agricultural products and its immigrant population are, for a global urban society, hungry for novelties.

Access to such basic services as sanitation and water is provided in an unequal and precarious manner. At the end of the twentieth century, almost 60 percent of the growth of Mexico City derived from housing constructed without any planning, by men and women who were part of the informal labor market. Today, Mexico City holds the dubious honor of being home to what is perhaps the largest shantytown in the world, Neza-Chalco-Itza, with over four million inhabitants. In São Paulo, favelas represented almost 20 percent of the city's homes in 1993, and their numbers grew by more than 16 percent per year in the 1990s.[40] In the Caribbean, the enthusiastic use of DDT from the 1950s on made it possible to settle areas that until then had been unhealthy and unviable for human habitation, resulting in unregulated occupation of these lands and increasing pressure on natural resources. In the northern Brazilian state of Amazônia, at the outskirts and inside the largest remaining tropical forest in the world, accelerated deforestation accompanied an increase in areas covered with precarious, informal housing; almost 80 percent of the urban growth in the region occurs in favelas and *barrios* without even the minimum of urban services. Residents of cities in the interior of the forest, like Letícia (Colombia) and the bordering city of Tabatinga (Brazil), suffer from the absence of sewage treatment, the contamination of the water supply, and urban flooding, which are likewise common dilemmas for inhabitants of peripheries of megacities.

How has this diverse population coexisted with urban nature? There is a continuous struggle for greater access to natural resources—water, soil, air— for different sectors of the city. Large urban areas are not homogenous, and different groups seek to insure the availability of these resources, in quantity and quality. The development of *barrios,* favelas and *villas* represents a decisive part of the environmental history, and not just the political history, of the city. The city that consumes and transforms resources also generates large quantities of wastes, including domestic sewage, trash, and atmospheric contaminants, which place these necessary resources at risk. Noise, chemical, and atmospheric pollution are problems that affect the daily lives of most Latin Americans.

Atmospheric pollution has afflicted Latin American societies since the postwar period. In 1950 in Poza Rica, Mexico, twenty-two people died and 320 were hospitalized because of an accidental leak of hydrogen sulfide (H_2S). In Mexico City, topographical and meteorological conditions led to the concentration of pollutants, and smog came to blot out the urban landscape since the end of the 1950s. In 1961 in Lima, sanitation engineers complained that emissions from the fishing industry led to critical contamination episodes that threatened public health. In Santiago de Chile, since the end of the 1950s, atmospheric pollution grew considerably worse; in addition to emissions generated by industry, automobiles, the heating of buildings, and trash incineration, the city's topographic and meteorological conditions favored the concentration of pollutants and thermic inversions.[41] In the 1970s, alarming atmospheric conditions sparked a debate about "urbanicide," mortality caused by the city itself.[42]

Governments have sought to approach the problem from a regional perspective. In Buenos Aires, the Primera Conferencia Latinoamericana de Contaminación del Aire took place in 1962, followed by the creation of the Asociación Argentina contra el Aire Contaminado. Studies emerged on the topic in Latin American universities, which in 1961 had fifty-seven sanitary engineering programs, often with international consultants.[43] In 1967, the Red Panamericana de Muestreo Normalizado de la Contaminación del Aire (REDPANAIRE), headquartered in Lima, Peru, initiated its activities in Latin America with the objective of identifying and proposing measures to control air pollution. With this institution, the council director of Oficina Panamericana de Saude (OPAS)/Oficina Sanitária Panamericana (OSP), an agency connected to World Health Organization (WHO) undertook projects that consolidated some forms of action that had already been underway in Latin America since the beginning of that decade.[44]

REDPANAIRE involved Argentina, Bolivia, Brazil, Chile, Colombia, Costa Rica, Cuba, El Salvador, Guatemala, Jamaica, Mexico, Peru, Uruguay, and Venezuela. Until 1974, twenty-nine cities from these countries participated in the program whose activities included the installation of sampling stations, training programs led by specialists, and the distribution of equipment needed to carry out measurements. Each station sent collected data to the OPAS, which organized the comparative analysis of these data and disseminated them in detailed reports. The project confronted obstacles in several cities due to a shortage of specialized technical personnel, structurally precarious facilities, and difficulties in the use of regular and methodical measurements needed to make local diagnostics and comparisons viable.[45] The declared objective of the program was to create a set of practices for the measurement, oversight, control, and prevention of atmospheric pollution, such that the industrial and urban growth expected in the following decades in Latin America and the Caribbean would occur with minimum levels of emissions. The ultimate aim was to make development and environmental management compatible with each other, thus avoiding a repetition of the grave problems that industrialized countries already had been experiencing. The program failed miserably in meeting this goal. Latin American governments feared that any attempt to control pollution would present an obstacle to progress, as the Brazilian representatives made clear at the United Nations Conference on the Human Environment in Stockholm in 1972.[46]

In the decades to follow, the developmentalist imperative hindered these timid preventative actions, and pollution remains a central problem in Latin American cities. In 1992, the air quality in Mexico City reached levels considered acceptable for just eleven days out of the year. The region's urban environmental dilemmas are not limited, therefore, to demands for access to water and housing, but also include the quality of these resources, including potable water, proper sanitation, and healthy air. In recent decades, some cities have made notable advances in this respect. For years, the Brazilian city of Cubatão, which housed a plethora of chemical plants, oil refineries, steel mills, and chemical manufacturing plants, stood as a tragic symbol of pollution. Suffering from acid rain, the loss of soil and vegetation, and an extremely high infant mortality rate, Cubatão was a perverse reminder of the costs of the so-called Brazilian miracle and its deregulation of industry, promoted during the height of the military regime. Pressure applied by diverse social groups and energetic public policies changed part of this dismal picture in the late 1990s, when measures put in place improved air quality significantly.[47] Mexico City achieved more modest, but significant, success: it cut in half the amount of ozone in the last decade of the twentieth century, and managed to cut another 25 percent of ozone emissions in the first decade of the twenty-first century. However, its fleet of automobiles has increased by 35 percent during this same period. Environmental conditions in Santiago de Chile, despite considerable improvements, are still frightening; on 22 June 2015, in the middle of the Americas' Cup soccer tournament, city authorities declared a state of emergency, removing 40 percent of vehicles from circulation and interrupting the operations of over nine hundred factories for over twenty-four hours.[48]

Following the proliferation of environmental agencies that monitor pollution, erosion, and urban water and air quality, political negotiations and conflicts related to socioeconomic inequality in urban areas began to incorporate questions of health, pollution, and waste disposal. After the United Nations Conference on the Environment and Development in 1992 in Rio de Janeiro, these debates have multiplied throughout the region. In general, old disputes for urban justice—basically sanitation and housing—have returned, reinvigorated by environmental perspectives. There is a new emphasis on the role of the state and its public policies and their preeminent importance in the urban environment, in particular in capital cities, since the practices developed there reverberate throughout the rest of Latin America's interurban networks.

The city is also the privileged locus for the creation of innovative solutions. At times, acute crisis situations favor policy experimentation. In living through the drama of extreme food scarcity in 1989 brought about by the combination of the international blockade imposed by the allies of the United States on the one hand and the end of the Soviet bloc on the other, the residents of Havana, Cuba, turned their cityscape green with a myriad of small organic gardens. There is no doubt that the Cuban government played an essential role in stimulating the organization of urban cultivation, but nothing would have been possible without the creativity and knowledge of urban residents. As Adriana Premat affirms, it would be wrong to ignore that the "greening" of Havana was also a social action "from below."[49] The creativity of urban communities was also essential for the emergence of the association of trash pickers called Asmare in the Brazilian city of Belo Horizonte. In the 1980s, the mayor's office prohibited the trash pickers from working in the center of the city. When they ignored the prohibition and gathered in an area near the city center, the police set fire to their belongings and dwellings. Leaders of the trash collectors sought out the mediation of a religious organization, the Pastoral de Rua da Igreja Católica, and founded an association in 1990. In the following years, the trash pickers made great gains, obtaining warehouses in which to work, new carts to transport the materials they collected, and an agreement with the municipal government that officially recognized their importance to the city. Currently, Asmare is a model for the collection and recycling of tons of reusable material, combining public service with urban sanitation, and working toward social inclusion.[50]

Especially in this age of acceleration, Latin American urban nature is a frequent object of negotiation, whether between diverse actors as in the examples described above, or between private and public use. There has been, for example, an increase in the number of "green condominiums"—enclosed residential areas for urban elites that claim to provide bucolic contact with woodlands and pure air—basically, a romanticized rural experience inside the city. Whether in Mexico City or in Belo Horizonte, Brazil, the privatization of urbanized nature was only possible with the exclusion of communities that historically occupied these areas, who now must witness the enclosure and privatization of these areas.[51]

Privatization also includes the commodification of urban landscapes by way of the tourist industry, which reached an unprecedented scale in the second half of the twentieth century. The pressure that the floating population of tourists exerts on urban resources is significant, especially in fragile ecosystems such as urban coral reefs or dunes. If Caribbean cities were the principal point of contact with Atlantic markets for agricultural commodities through the nineteenth century, these cities became a bridge for an international tourist market in the twentieth century, as points of arrival for millions of tourists annually (see chapter 2 in this volume).

Growing social conflicts also emerged around the commodification, privatization, and contamination of water in Latin America, one of the richest regions of the world in fresh water resources. Alongside commercial agriculture, urbanization, industries, and tourism, these conflicts have entailed the growing use of water resources and increases in the discharges of contaminating wastes. International investor groups apply pressure for the privatization of water and sewage systems, generating social and political impasses that are decisive for the future of these societies.[52]

In the dispute between privatized, commodified, or public spaces, urban populations are increasingly demanding areas of social interaction in the context of urban nature. Tree-lined squares and parks are spaces of leisure for games and family gatherings, sociability, and political rallies and protest. For example, New Year's Eve on the famous Copacabana Beach, with fireworks and offerings for the Afro-Brazilian deity Yemanjá, the lady of the waters, has become one of the high points of Rio's tourist calendar; in 2015, over two million people circulated between the ocean, the sand, and the asphalt (see illustration 6.2). Mexico City's Chapultepec Park is visited by some thirteen million people per year.[53] These communitarian, natural spaces are among the most egalitarian expressions of Latin American urban nature. These spaces are at times worn out by use, and they alternate as spaces of violent conflict and conviviality and cultural resilience. And they are also, simply, proud celebrations of urban nature.

Illustration 6.2. Celebration of New Year's Eve, Copacabana Beach, 2012.

Photograph by Lise Sedrez.

The urban environmental history of Latin America should not be limited, therefore, to the city limits. This history demands a global narrative, both continental and local, a narrative that unites the uses of local, human, and natural resources with the transformation of global standards of consumption. Nor can cities be seen in isolation. Diverse as a group and internally unequal, Latin American cities resist being forced into a single model. Indeed, the various possible models for the study of the region's cities are interdependent: large and small cities, tropical and temperate, coastal and frontier, island and Andean cities, are connected in that they share the common experience of occupying the vast Latin American territory, and in undergoing a common process of transforming nature. Even when the urban population was a minority on the continent, the centrality of cities in directing the occupation of space makes environmental history of Latin America a narrative that is also urban. And nonetheless, as is evident in the periodization proposed in this chapter, these models have demands and trajectories in common, sharing the environmental, social, and political processes upon which they are placed. As in the music of Violeta Parra, the ivy and the wall are entangled and intertwined.

Lise Sedrez is a professor of history of the Americas at the Universidade Federal do Rio de Janeiro. She earned her doctorate in history from Stanford University in 2005. She is currently the coeditor, with Chris Boyer, of Arizona University Press's Latin American Landscapes series. Her book, *A History of Environmentalism: Local Struggles, Global Histories,* coauthored with Marco Armiero, was published by Continuum in 2014. She also edits the Online Bibliography on Latin American Environmental History (boha.historia.ufrj.br). Her research is focused on urban environmental history and the agency of urban nature in Latin American societies.

Regina Horta Duarte is full professor of history at the Universidade Federal de Minas Gerais, Brazil. She is the author of the book *Activist Biology: The National Museum, Politics, and Nation Building in Brazil* (2016). She also has published articles in international journals such as *Latin American Research Review, Environmental and History, Journal of Latin American Studies, Luso-Brazilian Review, ISIS,* and several Brazilian journals. She is a founding member of the Sociedad Latinoamericana y Caribeña de Historia Ambiental (SOL-CHA). Her current research interests focus on the environmental history of zoos in Latin America, in the first half of the twentieth century.

Notes

The authors thank Amy Chazkel for the translation of this chapter into English, the editors and the anonymous reviewers, and Brazil's Ministry of Science, Technology, Innovation and Communications (CNPq) for its support. Regina Horta Duarte also thanks the Minas Gerais State Agency for Research and Development (FAPEMIG).

1. Mike Davis, *Planet of Slums* (London: Verso, 2006), 8.
2. Alfred Crosby, *Ecological Imperialism* (Cambridge: Cambridge University Press, 2004), 44.
3. Pierre Villar, *A History of Gold and Money, 1450 to 1920* (Atlantic Highlands, NJ: Humanities Press, 1976), 112–133.
4. Rosalva Loreto, *Una vista de ojos a una ciudad Novohispana* (Puebla: BUAP/CONACYT, 2008), 19–33.
5. Fania Fridman, "Breve história do debate sobre a cidade colonial brasileira," in *Cidades latino-americanas,* ed. Maurício de Abreu and Fania Fridman (Rio de Janeiro: FAPERJ/Casa da Palavra, 2010), 43–73. On Mendoza and the water supply, see Jorge Ponte, "Historia del regadío: las acequias de Mendoza, Argentina," *Scripta Nova* 10, no. 218 (2006), retrieved 20 June 2017 from http://www.ub.edu/geocrit/sn/sn-218-07 .htm .
6. The author defines *metanarrative* as "a general organizing theme that would allow me to examine the past of the region over the long course of its history." Stuart Schwartz, *Sea of Storms* (Princeton: Princeton University Press, 2015): xi, 33–69.
7. Charles Walker, *Shaky Colonialism* (Durham: Duke University Press, 2008).

8. Vera Candiani, *Dreaming of Dry Land: Environmental Transformation in Colonial Mexico City* (Palo Alto: Stanford University Press, 2014), "Note" and ch. 8, Kindle edition.

9. Ángel Rama, *The Lettered City* (Durham: Duke University Press, 1996).

10. Stuart McCook, "The Neo-Columbian Exchange: The Second Conquest of the Greater Caribbean, 1720–1930," *Latin American Research Review* 46, no. 4 (2011): 11–31.

11. Candiani, *Dreaming of Dry Land.*

12. Richard Morse, "The Development of Urban Systems in the Americas in the Nineteenth Century," *Journal of Interamerican Studies and World Affairs* 17, no. 1 (1975): 4–26.

13. Germán Palacio, "Urbanismo, naturaleza y territorio en la Bogotá Republicana, 1810–1910," in *Ciudad y naturaleza: tensiones ambientales en Latinoamérica, siglos XVIII–XXI,* ed. Rosalva Loreto (Puebla, Mexico: ICSyH/BUAP, 2012), 165–187.

14. James Scobie, "The Growth of Latin American Cities, 1870–1930," in *The Cambridge History of Latin America,* vol. 4, ed. Leslie Bethell (Cambridge: Cambridge University Press, 1986), 233–266.

15. Ramón Gutiérrez, "Reflexões sobre o urbanismo do século XIX," in *Cidades do Novo Mundo: Ensaios de urbanização e história,* ed. Fania Fridman (Rio de Janeiro: Garamond, 2013), 139–162. See also Henry Lawrence, *City Trees: A Historical Geography from the Renaissance through the Nineteenth Century* (Charlottesville: University of Virginia Press, 2006), 250–253.

16. Maria Lopes, "Cooperação científica na América Latina no final do século XIX," *Interciencia,* 25, no 5 (2000): 228–233; Leon López-Ocón, "La exhibición del poder de la ciencia : La América Latina en el escenario de las Exposiciones Universales del siglo XIX," in *O mundo ibero-americano nas Grandes Exposições,* J. Mourão ed. (Evora: Vega, 1998), 67–89. See also Beatriz González and Jen Anderman eds., *Galerias del progreso* (Rosário: Beatriz Viterbo Editora, 2006).

17. Christian Brannstrom, "Was Brazilian Industrialization Fuelled by Wood? Evaluating the Wood Hypothesis, 1900–1960," *Environment and History* 11, no. 4 (2005): 395–430.

18. Antonio Brailovsky, *Historia ecológica de la ciudad de Buenos Aires* (Buenos Aires: Kaicron, 2012), 195.

19. Leonardo dos Santos, "Zona, sertão ou celeiro? A constituição do cinturão verde da cidade do Rio de Janeiro e seus impasses, 1890–1956," In *História urbana: memória, cultura e sociedade,* ed. Gisele Sanglard, Carlos Araújo, and José Siqueira (Rio de Janeiro, RJ: FGV Editora, 2013), 251– 278.

20. Victor Genaro Luna Fernández, Mario Aliphat Fernández, and Laura Caso Berrera, "La transformación rural-urbana de la región de Atlixco, Puebla del periodo prehispánico al siglo XX," in *Ciudad y naturaleza: tensiones ambientales en Latinoamérica, siglos XVII–XXI,* ed. Rosalva Loreto (Puebla, Mexico: ICSyH/BUAP, 2012), 121.

21. Claudia Leal, "Un Puerto en la selva: Naturaleza y raza en la creación de la ciudad de Tumaco, 1860–1940," *Historia Critica* 30 (2005): 39–66.

22. Fernández, Genaro, Aliphat, and Caso, "La transformación rural-urbana"; Palacio, "Urbanismo, naturaleza y territorio en la Bogotá republicana," 167; Francisco Hardman, *Trem fantasma* (São Paulo: Cia das Letras, 1988), 117–153.

23. Dora Dávila, *Caracas y la gripe española de 1918* (Caracas: Universidad Catolica Andrés Bello, 2000), 34–40; Adrián Arboneti, "Historia de una epidemia olvidada: La

pandemia de gripe española en la argentina, 1918–1919," *Desacatos* 32 (2010): 159–174; Liane Bertucci, *Influenza, a medicina enferma* (Campinas: Unicamp 2004).

24. Graciela Silvestri, *El color del río: historia cultural del paisaje del Riachuelo* (Buenos Aires: Universidad Nacional de Quilmes, 2003), 173.

25. Nicolau Sevcenko, *A Revolta da Vacina* (São Paulo: Brasiliense, 1984); Dilene Nascimento, "La llegada de la peste al estado de São Paulo en 1899," *Dynamis* 31, no. 1 (2011): 65–83.

26. John Soluri, *Banana Cultures* (Austin: University of Texas Press, 2005).

27. Lise Sedrez and Andrea Maia, "Narrativas de um dilúvio carioca: Memória e natureza na grande enchente de 1966," *Revista História Oral* 14, no. 2 (2011): 221–254.

28. The choice of eucalyptus to reforest the area caused physical damage to Bogotá. Luis Jiménez, "Unas montañas al servicio de Bogotá: imaginarios de naturaleza en la reforestación de los cerros orientales 1899–1924" (Master's thesis in history, Universidad de Los Andes, Bogotá, 2011), 36–57; Julián Osorio, "Los cerros y la ciudad," *Anuario de Ecología, Cultura y Sociedad* 5, no. 5 (2005): 129–142.

29. Claudia Heynemman, *Floresta da Tijuca: natureza e civilização no século XIX* (Rio de Janeiro: Prefeitura do Rio de Janeiro, 1995). On the synergy between human action and environmental recuperation, see José Augusto Pádua, "Tempo de oportunidades," *Revista de Historia da Biblioteca Nacional do Rio de Janeiro* no. 82 (2012): 55–61.

30. Matthew Vitz, "La ciudad y sus bosques: La conservación forestal y los campesinos en el Valle del México, 1900–1950," *Estudios de historia moderna y contemporánea de México* 43 (2012): 135–172.

31. Maria Lucía Guerrero, "Pintando de verde Bogotá: Visiones de la naturaleza a través de los parques del Centenário y de la Independencia, 1880–1920," *HALAC* 1, no. 2 (2012): 112–139.

32. Emily Wakild, "Parables of Chapultepec: Urban Parks, National Landscapes and Contradictory Conservation in Modern Mexico," in *A Land Between Waters: Environmental Histories of Modern Mexico,* ed. Christopher Boyer (Tucson: University of Arizona Press, 2014), 192–217.

33. Marcos Martins, "A política florestal, os negócios de lenha e o desmatamento, Minas Gerais, 1890–1950," *HALAC* 1, no. 1 (2008): 33–48; Brannstrom, "Was Brazilian Industrialization Fuelled by Wood?," 397.

34. Fernando Ramírez, "Pestilencia, olores y hedores en el Santiago del Centenario," *HALAC* 1, supp. (2012), 58–59.

35. Lise Sedrez, "The Bay of All Beauties: State and Environment in Guanabara Bay, Rio de Janeiro, Brazil, 1875–1975" (Ph.D. dissertation, Stanford University, 2004).

36. Brailovsky, *Historia ecológica,* 225.

37. Reinaldo Funes, "Los orígenes del asociacionismo ambientalista en Cuba," in *Naturaleza en declive,* ed. Reinaldo Funes (Valencia: Centro Francísco Tomás y Vaiente, 2008), 267–310. See, for example, the magazine "El Zoofilo Venezolano," founded in 1896, in Caracas, or "O Zoófilo Paulista," founded in 1919.

38. Orlandina Oliveira and Bryan Roberts, "Urban Growth and Urban Social Structure in Latin America, 1930–1990," in *The Cambridge History of Latin America,* vol. 4, ed. Leslie Bethell (Cambridge: Cambridge University Press: 1994), 253– 324.

39. John McNeill and Peter Engelke, *The Great Acceleration: An Environmental History of the Anthropocene since 1945* (Cambridge: Harvard University Press, 2016).

40. Davis, *Planet of Slums,* 28.
41. Carolina Riveros, "El problema de la contaminación atmosférica en Santiago de Chile, 1960–1972" (Master's thesis, Pontifícia Universidad Católica, Santiago de Chile, 1997), 75–84, 137–142.
42. Biblioteca Nacional de Chile, "La contaminación atmosférica de Santiago," *Memoria Chilena,* 2014, accessed 20 June 2017 from http://www.memoriachilena.cl/602/w3-article-3507.html.
43. Humberto Bravo, "Variation of Different Pollutants in the Atmosphere of Mexico City," *Journal of the Air Pollution Control Association* 10, no. 6 (1960): 447–449; N. Henry, "Polluted Air: A Growing Community Problem," *Public Health Reports* 68, no. 9 (1953): 859; J. Kretzschmar, "Particulate Matter Levels and Trends in Mexico City, São Paulo, Buenos Aires and Rio de Janeiro," *Atmospheric Environment* 28, no. 19 (1994): 3188; Ricardo Haddad and John Bloomfield, "La contaminación atmosférica en América Latina," *Boletín de la OSP* 57, no. 3 (1964): 241–249; John Bloomfield, "La Ingenieria santiaria frente al proceso de industrialización," *Boletín de la OSP* 65, no. 6 (1968): 549–561; Honório Botelho, *O ensino de engenharia sanitária da UFMG* (Belo Horizonte: Ed. Engenharia, 1972), 34.
44. *Seminario Latinoamericano de contaminación del aire* (Washington, DC: OPAS, 1970), 2–3.
45. REDPANAIRE, *Resultados obtenidos Junio 1967–Diciembre 1970* (Lima: Centro Pan-americano de Ingenieria Sanitaria y Ciencias del Ambiente, Série Técnica, 1971), 10; REDPANAIRE, *Report 1967–1974* (Lima: Pan American Center for Sanitary Engineering and Environmental Sciences, 1976).
46. Regina Duarte, "'Turn to pollute': poluição atmosférica e modelo de desenvolvimento no 'milagre' brasileiro (1967–1973)," *Tempo* 21, no. 37 (2015): 64–87.
47. John McNeill, *Something New under the Sun: An Environmental History of the Twentieth-Century World* (New York, London: WW Norton & Company, 2001), 82; Lúcia Ferreira, *Os Fantasmas do Vale: questão ambiental e cidadania* (Campinas: Ed. Unicamp, 1993).
48. "Chile Declares First Environmental Emergency since 1999 over Air Pollution," *Time,* 22 June 2015, accessed 20 June 2017 from http://time.com/3930737/santiago-air-pollution-emergency/.
49. Adriana Premat, "State Power, Private Plots and the Greening of Havana's Urban Agriculture Movement," *City & Society* 21 (2009): 28–57. See also Catherine Murphy, *Cultivating Havana: Urban Agriculture and Food Security in the Years of Crisis* (Oakland: Food First Institute, 1999); Fernando Funes et al., *Sustainable Agriculture & Resistance* (Oakland: Food First Books, 2002).
50. Aparecido Gonçalves et al., "Dezoito anos catando papel em Belo Horizonte," *Estudos Avançados* 22 (2008): 231–238.
51. Regina Duarte, "'It Does Not Even Seem Like We Are in Brazil': Country Clubs and Gated Communities in Belo Horizonte, Brazil, 1951–1964," *Journal of Latin American Studies* 44 (2012): 435–466.
52. Carlos Porto-Gonçalves, "Água," in *Enciclopedia Contemporânea da América Latina e do Caribe,* ed. Emir Sader et. al. (São Paulo: Boitempo, 2006), 55–59.
53. Wakild, "Parables of Chapultepec," 216.

CHAPTER 7

Home Cooking
Campesinos, Cuisine, and Agrodiversity

John Soluri

In 1934, a campesino organization in the Colombian county of Quipile signed a labor pact with coffee estate owners. The contract's first clause listed the food that landowners should provide their workers: "Breakfast: two cups of maize soup with peas and arracacha [a tuber]; Lunch: two cups of hominy or vegetable soup; cabbage; and a 1/5 pound of beef; dinner: arracacha; yucca, plantain, beans, balú [a legume] in sufficient quantities; and a 1/5 pound of beef. All of the food should be properly prepared and seasoned. Plantains should not be cooked in iron pots that are not enameled nor in copper pots."[1] For this particular group of campesinos, the quantity and quality of the food that they ate apparently mattered a great deal. Although the degree of culinary detail was probably unusual for labor contracts signed in early-twentieth-century Latin America, the list suggests that cuisine—not mere caloric intake, but the suite of foods and their preparation that constitute a meal in any given place and time—can be an important dimension of popular culture and political negotiation.

The Quipile campesinos' "menu" also provides a glimpse of the agrodiversity that provided for a varied diet. Maize, manioc (yucca), common beans, *arracacha,* and *balú* were native to the Americas. In contrast, beef and plantains reached the Americas via the transatlantic crossings of Iberians and Africans, as did the crop that generated employment for Quipile's campesinos: arabica coffee, the most important agroexport in modern Latin American history. This agrodiversity has a history: millions of people—growing, eating, and exchanging crops over millennia in a variety of geographical settings—played a leading role in creating modern Latin America's agrodiversity.[2] Agrodiversity is not limited to the variety of crops cultivated in a given location, but also incorporates intraspecific (genetic) diversity and the life-forms that thrive on or near a site of cultivation.[3] Diverse agricultural systems are not only central to human identities and cuisines, they are also vital for maintaining resilient and productive ecosystems.[4]

Finally, the form in which the Quipile campesinos articulated their culinary preferences—a labor contract—reminds us that campesino agriculture and foodways evolved in relationship to both markets and governments. Far from static, the emergence and reemergence of campesinos in modern Latin America/Caribbean have been linked to the abolition of slavery, the rise of agroexport trades, European immigration, land reform, and urbanization. I define "campesinos" in a literal sense: as people who produce food in the *campo* or countryside. Throughout modern Latin America, campesinos (*camponês* in Portuguese) engaged in both self-provisioning and market exchange, relying heavily, but not exclusively, on household members for labor.[5] Campesinos sometimes held titles to the land that they farmed, but they also farmed land as tenants or squatters. The term therefore encompasses a range of social relationships, ethnicities, and legal identities.

Campesinos have been central to growing foods consumed in Latin America. As late as the 1920s, approximately 80 percent of agricultural production in Latin America was oriented toward domestic markets; this production generated much more employment and overall economic activity (GDP) than did agroexport sectors.[6] As the twentieth century advanced, the growth of factory and service work, along with state investments in public health, transportation, and education, stimulated flows of rural-to-urban migration. Between 1929 and 1980, the percentage of Latin Americans living in urban areas more than doubled (from 32 percent to 65 percent) while the percentage of the working population engaged in agriculture fell from 49 percent to 32 percent between 1960 and 1980.[7]

Paradoxically, the number of smallholders increased between 1950 and 1970 in most countries (including Argentina, Brazil, Chile, Colombia, Costa Rica, Ecuador, Guatemala, and Peru). In Mexico, the implementation of agrarian reform led to a sharp increase in smallholders between 1930 and 1950. Several other Latin American/Caribbean governments initiated land reforms of varying scale and scope during the mid-twentieth century. All told, campesino farms in Latin America rose from approximately nine million in the mid-1960s to around fifteen million in the mid-1980s.[8] Most of these smallholders grew food for both subsistence and urban markets, accounting for an estimated 40 percent of the food consumed domestically in the mid-1970s.[9]

Amid this changing demography, urban labor organizations gained political power; there were not only more mouths to feed, but mouths that had the ears of politicians committed to industrialization. By the end of the 1950s, most Latin American governments had established some controls over urban food provisioning.[10] Governments also became increasingly concerned about nutrition and diets of the urban poor; for example, in Mexico City, the government of Ávila Camacho experimented with public dining halls intended to serve (and educate) working-class Mexicans.[11] To what extent did the growth of both cities and government intervention in nutrition and food prices in the mid-twentieth century change campesino livelihoods, and what effects did expanding urban markets have on agrodiversity?

This chapter does not offer a definitive answer to that question, but begins to trace the entangled pasts of campesinos, agrodiversity, cuisine, and urbanization by making excursions into the histories of four crops widely cultivated by campesinos: maize (*Zea mays*), potatoes (*Solanum tuberosum*), common beans (*Phaseolus vulgaris*), and coffee (*Coffea* spp.). I examine campesino production of maize in Mesoamerica and potatoes in the Andean highlands, two crops domesticated in the Americas thousands of years ago that continue to be tied to indigenous cultures (see chapters 1 and 3 in this volume). Common beans are also native to the Andes and Mesoamerica, but here I explore their production and consumption by largely nonindigenous smallholders in Brazil in order to challenge the tendency to conflate modern indigenous cultures and agrodiversity. Finally, I look at coffee to show that export-oriented agriculture is not always incompatible with agrodiversity; in fact, coffee farms throughout the tropics relied on agrodiversity and campesino family labor to sustain themselves. Collectively, these four examples seek to explain the persistence of biocultural diversity in the context of rapidly changing societies by considering the significance of cuisine—"home cooking."

Maize and Campesinos in Mesoamerica

Important throughout the Americas, maize has special meanings in Mexico, where people have cultivated it for several thousands of years. Maize is renowned for its proclivity for outcrossing (cross-pollinating); in Mexico, researchers working since the 1940s have documented some fifty-nine "landraces"—populations of crop plants that have "morphological integrity and geographic identity."[12] How has this agrodiversity persisted through the numerous political, social, and technoscientific revolutions described in chapter 1 of this volume by Boyer and Cariño Olvera? A partial answer lies in the very agrarian structures that both spawned and resulted from Mexico's revolutions. Historians have demonstrated that the family farm or *rancho* did not proliferate in Mexico until the early nineteenth century. Haciendas could not compete with campesino maize production on *temporal* (rain-fed) soils until the late-nineteenth century when rising grain prices and surplus labor enabled haciendas to expand onto temporal lands through the use of sharecroppers, a system that became widespread during the Porfiriato, when urban demand for maize expanded, along with the railroads needed to move grain in bulk. Some sharecropping contracts stated that hacienda owners would supply seed for planting, a provision that if put into widespread practice might have limited varietal diversity. However, sharecroppers probably grew additional varieties for self-provisioning.[13]

The Mexican Revolution resulted in the formation of a powerful central government committed both to economic nationalism and to ensuring the livelihoods of its citizens. Under President Lázaro Cárdenas (1934–1940), more than 10 percent of the country's total surface area was redistributed to some eight hundred thousand Mexicans.[14] The Cárdenas administration also created government agencies to purchase, store, and sell wheat, corn, rice, and beans. The Partido Revolucionario Institucional (PRI) governments that followed shifted priorities from land reform to boosting yields and subsidizing urban consumers. These policies primarily benefited affluent farmers who incorporated hybrid seeds, fertilizers, and irrigation to boost production; industrialists who were able to suppress wages; and urban workers who were the primary targets of government food subsidies.[15]

Under the often contradictory and unevenly implemented policies of the PRI, smallholder cultivation more than doubled between 1930 and 1991 from 5.4 million hectares to 12.1 million hectares; as late as 1970, 60 percent of agricultural units were smaller than five hectares.[16] This expansion should not be equated with improved economic or social standing for most campesinos; in fact, many people abandoned farming and moved to cities or migrated to the United States. Those who continued to farm often worked rain-dependent *temporal* soils, leaving them highly susceptible to drought. In all likelihood, this spatial dimension of campesinos' economic marginality contributed to the maintenance of agrodiversity by reducing the utility of "Green Revolution" crop varieties, most of which benefited irrigated farms.

Campesinos often cultivated and ate several varieties of maize; a study carried out in northern Mexico in the mid-twentieth century found that cooks used distinct varieties for tortillas, soups, beverages, and desserts.[17] The Mexican Revolution's repudiation of the Eurocentric cultural politics of Porfirio Díaz included public fiestas that featured tortilla soup and mole poblano.[18] In the 1940s, after decades of advocating wheat-based diets over maize, cookbooks and nutritionists targeting middle-class women vindicated the cultural and nutritional value of maize, beans, and chili peppers, which were staples of rural and urban workers (see illustration 7.1). In addition, women generally accepted the arrival of corn-grinding mills that created time to engage in other activities, including income-generating ones.

Fieldwork carried out in Jalisco and Chiapas in the 1980s demonstrates the tight relationship between cuisine and agrodiversity. In the mountainous region of Cuzalapa, Jalisco, indigenous and mestizo campesinos grew white maize for tortillas; purple maize for roasting cobs; and yellow varieties for animal feed.[19] They intercropped maize with squash (during the rainy season) or beans (on irrigated lands). All told, they planted twenty-six varieties of

Illustration 7.1. Construction workers gathered around a large comal with tortillas, Mexico City, 1953.

Photo by Nacho López, copyright 380016 Secretaria de Cultura, Inah, Sinafo, Fn, Mexico.

maize, six of which they considered to be local varieties related to a single landrace of white maize (*Tabloncillo*). Two white varieties accounted for more than 60 percent of maize plantings; the twenty "exotic" varieties made up less than 14 percent of cultivated maize. Campesinos obtained new varieties from both neighbors and farmers outside of Cuzalapa via nonmonetary exchanges. Overall, most Cuzalapan cultivators were keen to try out new varieties, but they were not quick to replace local varieties. This kind of pragmatic experimentation also occurred in Chiapas, where the availability of purchased seed and agrochemicals seemed compatible with maize diversity in the late 1980s.[20] Campesinos in Chiapas chose varieties based on a number of characteristics related to yields, labor demand, and agroecological conditions. Although they strongly associated superior taste with landraces, campesinos sold and consumed a mixture of improved varieties and landraces.[21]

By the end of the twentieth century, several varieties of white corn grew on more than seven million acres of temporal lands, an indication of the centrality of white-corn tortillas in Mexican diets.[22] In the state of Mexico, white kernel varieties accounted for 96 percent of total maize production in the early twenty-first century. Although much of this white maize was used for self-provisioning or sold on local markets, a significant amount was sold to industrial flour mills, masa and tortilla makers, or wholesale outlets in Mexico City. Industrial consumers secured much of their white maize from regions where hybrid varieties predominated. This is suggestive of a market dynamic that has been more thoroughly documented for export commodities than for domestic ones, revealing the power of processors to influence the crop varieties cultivated by farmers. On the other hand, in Mexico City's principal wholesale market, the Central de Abastos, demand for "specialty varieties," including large-grained maize preferred for making *pozole* (a soup), and blue corn for making *antojitos* (savory snacks), suggests that urban markets may have some capacity to sustain agrodiversity.[23] However, much more research is needed to understand the history of campesinos' articulation with urban markets, including wholesalers and processors, in order to understand how these middle spaces between farm and table have shaped, or been shaped by, taste preferences.[24]

This history of maize in modern Mexico indicates that both the successes and failures of government policies directed toward agriculture and urban food provisioning help to explain the persistence of smallholder maize production. The collapse of the PRI and subsequent shift to neoliberal policies have further weakened the ability of smallholders to sustain agrarian livelihoods, raising the possibility that off-farm earnings, including remittances from the United States, may be playing an indirect role in sustaining maize diversity. The many uses of maize in Mexican cookery makes possible—but by no means ensures—that campesino farms will remain repositories of agrodiversity. The hand-made tortillas that have become an object of nostalgic desire for middle-class urban eaters may help to preserve agrodiversity in the short term, but they also reflect the bitter irony that "peasant diets" have become a marker of affluence.

Potatoes and Campesinos in Peru

In the central Andes, millions of campesinos cultivate a remarkable variety of crops, including maize, potatoes, and other tubers (e.g., *ulluco, mashua,* and oca) rarely consumed outside of the Andes, in addition to quinoa, a pseudo-grain that has recently become part of a transnational culinary chic. Historically, Andean campesinos have valued eating a mix of potatoes of distinct colors, shapes, and flavors: both Quechua and Aymara speakers have conveyed this idea with a single word—*chalo*—since at least the sixteenth century.[25] One early twentieth-century observer recorded some fifteen preparations for maize that utilized ten different landraces.[26] However, Andean campesinos, like their counterparts in Mexico, experienced significant changes from the late nineteenth century to the present that often undermined their ability to maintain agrodiversity.

Following independence, Peru emerged as a divided republic: a criollo-dominated coast with a regional cuisine based on wheat, and the largely indigenous highlands, whose foodways were based on tubers, *choclo* (maize), and *ají* (hot peppers).[27] In Peru, the postindependence period (1826–1880) did not bring any dramatic changes in the crops cultivated by highland campesino communities.[28] However, government, business, and campesino initiatives brought about significant shifts during the twentieth century. The construction of transportation infrastructure (highways and railroads) helped to revive commercial agriculture during the first half of the twentieth century. Hacienda owners, like their counterparts in prerevolutionary Mexico, directly planted their best lands and granted campesinos usufruct rights to marginal lands in return for their labor.

The growth of business and population in the highland city of Cusco and other urban areas prompted campesinos to devote more land to a high-yielding landrace potato (*qompis*) that became an important commodity in regional markets (see illustration 7.2). In addition, barley cultivation prospered following the establishment of a beer brewery in Cusco that paid high prices for a particular kind of malting barley. At times, market demand for certain crops undermined agrodiversity by creating new demands on land and labor. For example, *qompis* potatoes, barley, and quinoa all prosper at similar elevations; the mid-twentieth-century expansion of fields planted in *qompis* and barley led to a reduction in the area planted in quinoa. Around the same time, hacienda workers cut back on their cultivation of an early potato known locally as *chawcha* on account of conflicting labor demands and limitations on irrigation water, which was increasingly diverted to fields planted in *qompis*.[29]

The pressures on agrodiversity in the Peruvian highlands heightened in the second half of the twentieth century. In 1969, the government of General Juan Velasco instituted sweeping land reform measures that did away with haciendas and servile labor relations in the highlands. Some twenty thousand campesinos in Pacaurtambo gained usufruct rights to land. Velasco simultaneously implemented policies to promote industrialization, urbanization, and large-scale agriculture in Peru's coastal regions. The results were contra-

Illustration 7.2. Ezequiel Arce and his potato harvest, Cusco, Peru, 1934.

Photo by Martín Chambi. Courtesy of Archivo Fotográfico Martín Chambi, Cusco, Peru, www.martinchambi.org.

dictory: the promise of land reform brought immigrants into some farming regions, but government policies designed to control urban inflation drove down prices for wheat and potatoes. At the same time, rising demand for beer and rice spurred campesinos to plant more fields in barley and upland rice. New educational and employment opportunities, along with novel consumer desires, contributed to increases in migration.[30] Finally, government agencies and credit programs promoted new crop varieties that provided high yields when grown with fertilizers.

The combined effects of these myriad changes on agrodiversity are not easy to generalize. Campesinos in the Paucartambo province maintained foodways based on various landraces and diverse ecosystems even as they incorporated commercial crops such as barley, fava beans, and peas into their fields.[31] Highland farmers, often women, recognized the culinary value of landraces to be superior to that of "improved" varieties.[32] For example, campesinos described landraces as "floury" (*harinosas*) potatoes in contrast to many commercial varieties considered to be "watery" (*aguanosas*). Some cultivators associated watery varieties with the use of chemical fertilizers.[33] Nevertheless, by the late 1980s, 35 percent of Paucartambo's campesinos did not cultivate one or more of the following native crops: potato, maize, *ulluco,* and quinoa.[34] Moreover, the cultivation of floury potato varieties declined significantly. Comparative fieldwork in two communities (Paucartambo and Tulumayo) found that modern potato varieties constituted the bulk of the potatoes sold in both regions, but that cultivation and consumption of *chalo* potatoes (mixed local varieties) persisted as well.

The inroads made by nonnative, commercial crops in these regions of highland Peru did not necessarily signify a loss of cultural identity or an economic devaluation of native crops. Prosperous campesino families acquired local prestige by cultivating and cooking meals based on crops such as floury potatoes, quinoa, and maize-based chicha beer.[35] A similar connection between prestige and indigenous foods exists in Ecuador.[36] In the central Andes, food and drink are intimately related to practices of reciprocal labor: the quality of work often depended upon the quality of food provided by the owners of the fields. Finally, some landraces of potatoes adapted to very high altitudes (around four thousand meters) fetched higher prices in urban markets due to culinary qualities considered superior to those of the introduced potatoes cultivated at lower altitudes.[37]

The modern history of agrodiversity in highland Peru appears to resonate with that of southern Mexico: the persistence of indigenous farmers in centers of crop plant domestication goes a long way toward explaining the extraordinary diversity of tubers and maize, among other crops. At the same time, many indigenous and nonindigenous campesinos in highland Peru became increasingly tied to regional markets for their labor and crops. Twentieth-century land reforms did away with servile labor, but did not eliminate campesinos' social marginalization and economic precariousness. Urban elites also sought out "artisanal" varieties of potatoes and other crops, creating niche markets similar to those for blue maize in Mexico City. This suggests how changing tastes and meanings inscribed in foods can help to explain the persistence of agrodiversity. Of course, countervailing forces, including demand for low-cost, high-yielding varieties can simultaneously undermine diversity.

Beans and Brazilian Camponês

The history of common beans (*Phaseolus vulgaris L.*) in Brazil presents a contrast to maize in Mexico or potatoes in Peru because there is no evidence that people domesticated beans in what is today Brazil.[38] However, this has not prevented bean-based dishes such as *feijoada* (usually a black bean stew flavored with pork) from becoming a culinary expression of Brazilian national identity.[39] In Brazilian cities, beans and rice accompanied by another dish (*mistura*) constitute a "full meal."[40] However, rural households consume nearly double the amount of beans (23.5 kilograms) eaten in urban households (12.9 kilograms), a reflection of the crop's vital role in cropping systems in a variety of environments throughout most of Brazil. Common beans also differ from maize and tubers on account of their symbiotic relationship with soil-dwelling cyanobacteria that enables bean plants to capture (or "fix") atmospheric nitrogen. Beans and other legumes therefore provide eaters with vegetable proteins while enriching soils.[41]

Ancient cultivators domesticated beans in both Mesoamerica and the Andean region, giving rise to two distinct gene pools. Beans reached Brazil prior to the arrival of Europeans, but their precolonial history remains murky.[42] The Portuguese crown recognized the value of beans as early as 1707 when it decreed that its subjects should plant beans along the routes they traversed in the colony.[43] In culinary lore, beans became "married" to rice following the relocation of the Portuguese court to Rio de Janeiro in 1808. One early-nineteenth-century French traveler described black beans as "an dispensable dish on the tables of the rich and . . . almost the only gourmet food of the poor."[44] An 1836 census of rural districts in São Paulo indicated that beans, along with maize and manioc, were an important source of household income: more than two thousand farmers grew beans, "from the poorest squatter to the most capitalized planter."[45]

The ubiquity of beans did not diminish even when the state of São Paulo became a center of coffee production in the early twentieth century. In 1905, when more than twenty thousand farms produced coffee, beans grew on more than thirty-one thousand farms.[46] Coffee accounted for a far greater percentage of the total value of the state's agricultural production (64 percent) than did beans (7 percent), but bean production was far less concentrated than coffee production; many growers produced relatively small harvests of beans. As I discuss below, the form that coffee production took created spaces for food crops like common beans. Coffee profits would also stimulate the growth of the city of São Paulo, which in turn created new urban markets for beans, corn, and other foodstuffs.

In the second half of the twentieth century, both northeast and southeast Brazil underwent rapid deruralization due in part to the legal disintegration of the resident labor system on large estates in 1963.[47] Many rural people responded to new political rights but persistent landlessness by migrating to cities or Amazonia. Beans followed them. In the 1960s, urban wholesale and retail markets in Brazil stocked at least seven varieties of common beans, including Preto (black), Manteigão, Mulatinho, Roxinho, Rosinha, Pardo, and Amarelo.[48] Between 1955 and 1973, bean production in the Amazonian state of Rôndonia increased at an annual rate of more than fifteen percent. By the 1980s, upland rice, beans, and maize occupied nearly 60 percent of cultivated land in Rôndonia in spite of the government's promotion of perennial crops.[49] Bean harvests also increased sharply in the states of Sergipe, Bahia, and Maranhão during the 1970s.[50] All told, the area planted in beans increased from 2.36 million hectares in 1950 to 4.36 million hectares in 1980, when only corn, soy, and rice occupied more land. However, between 1980 and 2006, land planted in beans fell sharply (2.19 million hectares) while corn, soy, and sugar all increased.[51]

Camponês cultivated several hundred varieties of common beans in the twentieth century. Institutional bean-breeding programs began in Brazil in the 1930s, but apparently few new varieties were distributed prior to 1970. As late as 1979, Brazil's Ministry of Agriculture estimated that 87 percent of the country's bean production came from farms that relied on their own seed.[52] However, this changed over the final two decades of the twentieth century, when government programs distributed more than one hundred varieties of selected or hybridized beans. Many of these improved varieties contained germplasm collected in other countries and stored in the International Center for Tropical Agriculture (CIAT) in Colombia.[53]

Recent efforts to assess the genetic, morphological, and agronomic diversity of common beans in Brazil indicate that Mesoamerican landraces predominate.[54] This is noteworthy since the Andes are much closer in a geographical sense. On the other hand, agroecological conditions in many parts of Brazil more closely resemble those found in Mesoamerica than Andean regions. An examination of landrace varieties collected from 279 municipalities stretching from the northeast to the southeast revealed significant diversity in terms of seed colors, sizes, and growth habits. Although the genetic diversity was not as great as that found in Mesoamerica or the Andes, many hybrid bean varieties exist in Brazil that are not found elsewhere. Brazil can therefore be considered a secondary center of bean domestication.[55]

Little historical research has been carried out that can account for the diversity of Brazil's beans, but institutional collections provide some clues. For example, bean varieties that share common names are rarely duplicates; instead, they often present clear differences in shape, agronomic features, and genetic makeup. This variation points to the historical dimensions of agrodiversity: the small farmers from whom collectors procured seeds did not classify beans in ways that reflected a genotype. In fact, there is a considerable amount of genetic diversity within accessions (or samples) held by Brazilian seed banks, which further reflects the diversity of the smallholder farms that often cultivated multiple varieties and seldom put a premium on seed purity.[56] As already noted in the case of maize and potatoes, the historical role of smallholders in contributing to the agrodiversity of beans in Brazil cannot be easily exaggerated, nor can the centrality of beans in Brazil's foodways.

Campesinos and Coffee "Forests"

Latin American agroexports often conjure images of large-scale plantations, estates, haciendas, or ranches. There is more than a little truth to this association: the production of bananas, cattle sugar, wheat, oil palm, soy, and even agave destined for tequila has often been based on large-scale monocultures. However, historical research has demonstrated that this model can be over-drawn: mixed scales of production prevailed in different times and places in many agroexport sectors, including bananas, cacao, and sugar.[57] Coffee is unique among export crops for the prevalence and persistence of small- and medium-scale farms characterized by comparatively high levels of planned and associated biological diversity.

Coffee plants are perennial, shade-tolerant woody plants. In contrast to herbaceous annual crops, coffee plants require four or five years before they begin to yield substantial harvests. Historically, small-scale coffee growers in Latin America have overcome this lag by intercropping coffee with food-producing plants and shade trees whose yields of food and/or fuelwood could be used for subsistence or sale. Wealthy planters throughout much of tropical Latin America granted workers access to land for planting food crops as a key means of recruiting and retaining their labor forces.

Many literate travelers remarked on the agrodiversity found on nineteenth-century coffee estates. In 1844, John Wurdemann marveled over a prosperous coffee farm in Cuba: "Imagine more than three hundred acres of land planted in regular squares with evenly pruned [coffee] shrubs each containing about eight acres, intersected by broad alleys of palms, oranges, mangoes and other beautiful trees; the interstices between which are planted with lemons, pomegranates, Cape jessamines, tuberose, lilies, and various other gaudy and fragrant flowers."[58] Wurdemann added, in a more matter-of-fact tone, that the planter also grew maize and plantains for sale to sugar estates and yams, yucca, sweet potatoes, and rice for autoconsumption. In the Brazilian coffee district of Vassouras, fazendas established in the 1840s and 1850s also cultivated a variety of crops. Intercropping maize, beans, and manioc with young coffee plants was a common practice that supplied both shade for coffee plants and food for the slave labor force whose diets included maize porridge, beans cooked with bacon, manioc flour (sprinkled liberally on beans), and jerked beef. Fruits and vegetables, both native and introduced, were also part of fazenda diets, along with sugar and coffee.[59] In addition, slaves planted maize, beans, and coffee on *roças*, plots of land that planters customarily ceded for subsistence production.[60]

In the aftermath of slave emancipation in Brazil, the organization of labor in coffee production changed. In the important coffee-growing state of São Paulo, a system known as the *colonato* emerged that combined wages, piecework, and access to food plots. The *colonato* relied on the subsidized immigration of nearly one million, mostly Italian, immigrants. Hired in family units, the immigrants played a central role in increasing both coffee production and food crops in São Paulo during the first half of the twentieth century.[61] The *colonos'* expectations that they had the right to intercrop maize, beans, or rice while they tended to young coffee farms became a source of social tension in coffee districts. Planters attempted to use contracts to restrict *colono* intercropping once coffee farms matured, but when faced with the prospect of disgruntled workers quitting, wealthy planters often opened up new coffee farms in order to control labor costs and quality, a socioecological dynamic that likely accelerated rates of deforestation and erosion.[62]

Campesinos in twentieth-century Colombia, Central America, Puerto Rico, and Venezuela also combined coffee production with self-provisioning, blurring the lines between subsistence and commercial agriculture.[63] For example, in Tarrazú, Costa Rica, campesinos used banana or plantain to provide both shade for coffee plants and sustenance for people and animals. They also planted shade trees that bore fruits or firewood. Tarrazú farmers often cultivated yucca, *tipiquizque*, or *ñampí*, so-called minor crops that were central to regional cuisine.[64] This biologically diverse landscape was not static; in some places, soil fertility declined and erosion increased due to annual plantings of maize and beans on sloping lands.

Beginning in the 1970s, Latin America coffee production started to intensify. The introduction of new "sun" varieties (including *caturra*), selected for their ability to grow in farms with limited shade and higher planting densities, led to an increase in the use of manufactured fertilizers, fungicides, and herbicides. Prosperous campesino households adopted these practices more quickly and thoroughly than their less prosperous neighbors, who held small plots of land on which old and new coffee varieties mingled.[65] This "technification" of coffee served to undermine agrodiversity in some growing regions.

However, at almost the same time, the rising popularity of high-priced "specialty coffees" in Europe and the United States created niche markets for shade-grown arabica coffees. Shifting consumer desires and alternative commodity chains (including organic and fair trade) created modest economic opportunities for smallholders to cultivate shade coffee farms.[66] Campesino coffee farms in contemporary El Salvador and Nicaragua have a rich diversity of trees in addition to several varieties of coffee, maize, beans, and medicinal plants.[67] These farms account for significant percentages of household incomes (50 to 100 percent) and subsistence crops (at least 40 percent) in addition to providing fuel. Finally, smallholder coffee farms provide habitats for diverse dense populations of birds, bats and other small mammals, and insects (arthropods).[68]

The history of agrodiversity in Latin American coffee systems is not generalizable: extensive deforestation to create large estates worked by slaves (in Brazil) or indigenous debt peons (in Guatemala and Nicaragua) existed alongside smallholder farms that often consisted of polycultures.[69] In most coffee-growing regions, including those in southern Brazil, local crops and export coffee reinforced one another. Landowners' practice of granting laborers the right to cultivate food crops among young coffee plants facilitated an overall expansion of coffee by helping to mobilize scarce labor resources. In reality, many so-called coffee farms were polycultures worked by campesino family labor. However, the persistence of such diverse agroecosystems in the early twenty-first century must be viewed in conjunction with the entrenched poverty and out-migrations of young people that shape the lives of most coffee smallholders.[70]

Latin Fusion: Campesinos, Cuisine, and Enduring Agrodiversity

After 1980, the number of small and mid-sized farms declined sharply throughout Latin America. Nevertheless, campesinos remained vital to food production at the start of the twenty-first century: in Brazil, four million farms produced nearly half of the nation's maize, 70 percent of its beans, and almost 90 percent of its manioc; in Mexico, 2.1 million farmers grew maize on 6.5 million hectares of land, and some five hundred thousand farms cultivate 1.7 million hectares of beans.[71] Finally, in Colombia, over four hundred thousand smallholders grew coffee at the start of this century. Not bad for actors sometimes dismissed by economic historians as "unproductive" or "traditional."[72]

Campesinos have also played a crucial, if seldom acknowledged, role in the "modern" conservation of agrodiversity via institutional plant breeding programs. Seed banks established in Mexico (1966), Colombia (1967), and Peru (1970) formed the foundation for plant breeding programs that developed "improved" varieties in the late twentieth century.[73] What is often forgotten is that campesino fields provided most of the specimens conserved in these collections. More research is needed in order to understand the creation and maintenance of in situ (or "living") diversity. I have identified some potentially important, broad contexts such as cuisine and urbanization, but very little has been documented about how and when campesinos articulated notions of "agrodiversity"—a term that did not become popular among scholars until the 1990s.[74]

The histories of Mesoamerican maize and Andean tubers confirm a tight connection between enduring indigenous cultures and the persistence of agrodiversity, which Nicolas Cuvi discusses in chapter 3 of this volume. Local and regional cuisines often valued landrace varieties of maize and potatoes as conveyed by the enduring Andean notion of *chalo*. These food preferences sensibly carried implications for the kinds of crops cultivated. These taste preferences were expressed in both market and communal relationships; indigenous campesinos grew landraces for regional markets and to build social prestige in their communities. Neither communal nor market pressures impeded indigenous campesinos from experimenting with new crops or "improved" seeds, but they appear to have rarely abandoned landraces by choice.

The rich diversity of common beans found in Brazil—a secondary center of diversity—suggests that dense populations of indigenous communities are not essential for agrodiversity. The myriad culinary variations based on rice and beans found in Brazil and throughout Latin America contributed, along with persistent rural poverty, to promote varietal diversity. Similarly, the fact that coffee, an introduced crop grown primarily for export, gave rise to some of Latin America's most diverse agroecosystems, points to the historical dynamism and multiple pathways taken by campesinos and agrodiversity alike.

The social and geographical marginality endured by many campesinos also promoted agrodiversity insofar as Green Revolution hybrids and land reform programs often failed to reach poor farmers. Campesinos forced to cultivate on marginal soils or unfavorable climates were far less likely to use "improved" varieties of crops or synthetic fertilizers. In addition, limited cash income probably compelled many campesinos to engage in self-provisioning and nonmonetary exchanges of seeds. Finally, the same broken, mountainous landscapes that discouraged investment in large-scale farming also created ecological and microclimatic conditions that favored plant diversity.

In the second half of the twentieth century, urbanization and rapid population increases affected campesino lives throughout much of Latin America. The growth of small, medium, and large cities (see chapter 6 by Horta Duarte and Sedrez) created new demands for food crops. I have presented evidence that suggests urban markets can at times promote, and at other times discourage, crop plant diversity. The effects of specific government policies and rural-to-urban food chains on campesinos and agrodiversity represent fertile terrain for future historical research.

The late-twentieth-century rise of neoliberal economic policies, the sharp decline in the number of smallholders, unabated urbanization, and significant shifts in urban diets make this research all the more pressing. As the Quipile campesinos conveyed, their political marginality did not prevent them from trying to fulfill a desire to eat on their own terms. In so doing, campesinos and urban dwellers alike have played a vital role in Latin America's food systems, a linkage that persists today with far-reaching implications for the future of global agrodiversity.

John Soluri is associate professor and director of global studies in the Department of History at Carnegie Mellon University. He has published *Banana Cultures: Agriculture, Consumption and Environmental Change in Honduras and the United States* (Texas, 2005), winner of the Elinor Melville Prize. His research and teaching focus on transnational environmental histories of agriculture, food, energy, and the commodification of nonhumans in Latin America. In a humbling effort to practice what he preaches, he is a long-time board member of Building New Hope, an NGO that works in solidarity with urban and rural people in Central America.

Notes

1. Marco Palacios, *El café en Colombia, 1850–1870. Una historia económica, social y política* (México: El Colegio de México/El Ancora Editores, 1983 [1979]).
2. C. Levis, F. R. C. Costa, F. Bongers, M. Peña-Carlos, et al., "Persistent Effects of Pre-Columbian Plant Domestication on Amazonian Forest Composition," *Science* 355 (3 March 2017): 925–931; Karl S. Zimmerer, "Conserving Agrodiversity Amid Global Change, Migration, and Nontraditional Livelihood Networks: The Dynamic Uses of Cultural Landscape Knowledge," *Ecology and Society* 19 (2014): 1; Stephen B. Brush, "The Issues of In Situ Conservation of Crop Genetic Resources," in *Genes in the Field: On-Farm Conservation of Crop Diversity,* ed. Stephen B. Brush (Boca Raton, FL: Lewis Publishers, 2000), 3–28.
3. John Vandermeer and Ivette Perfecto, *Breakfast of Biodiversity: The Truth about Rain Forest Destruction* (Oakland: Food First Books, 1995), 130–136. The term "agrodiversity" gained popularity among scholars in the 1990s; see Harold Brookfield and Michael Stocking, "Agrodiversity: Definition, Description and Design," *Global Environmental Change* 9 (1999): 77–80; and David Wood and Jillian M. Lenné, "The Conservation of Agrobiodiversity On-Farm: Questioning the Emerging Paradigm," *Biodiversity and Conservation* 6 (1997): 109–129.
4. Bradley J. Cardinale, J. Emmett Duffy, Andrew Gonzalez, David U. Hooper, Charles Perrings, Patrick Venail, et al., "Biodiversity loss and its impact on humanity," *Nature,* 486 (7 June 2012): 59–67; Marta Astier, Erika N. Speelman, Santiago López-Ridaura, Omar R. Masera, and Carlos E. Gonzalez-Esquievel, "Sustainability Indicators, Alternative Strategies and Trade-offs in Peasant Agroecosystems: Analyzing 15 Case Studies from Latin America," *International Journal of Agricultural Sustainability* 9 (2011): 409–422; and V. Ernesto Méndez, Christopher M. Bacon, Meryl Olson, Katlyn S. Morris, and Annie Shattuck, "Agrobiodiversity and Shade Coffee Smallholder Livelihoods: A Review and Synthesis of Ten Years' Research in Central America," *The Professional Geographer* 62, no. 3 (2010): 357–376.
5. The Portuguese word *camponês* has multiple meanings in Brazil; Cliff Welch, *The Seed Was Planted: The São Paulo Roots of Brazil's Rural Labor Movement 1924–1964* (University Park: Pennsylvania State University Press, 1999), 8.
6. Luís Bertola and José Antonio Ocampo, *The Economic Development of Latin America since Independence* (Oxford: Oxford University Press, 2012), 99–100.
7. Bertola and Ocampo, *The Economic Development of Latin America,* 139.
8. *Peasant Agriculture in Latin America and the Caribbean* (Santiago, Chile: ECLAC/FAO, 1986), 13–14.
9. Census data from the 1970s indicate that campesinos were central to maize production in Brazil (52 percent), Colombia (47 percent), Costa Rica (60 percent), Ecuador (45 percent), Chile (44 percent), and Panama (80 percent); *Peasant Agriculture in Latin America and the Caribbean,* 18–19.

10. Thomas C. Wright, "The Politics of Provisioning in Latin American History," in *Food, Politics and Society in Latin American History*, ed. John C. Super and Thomas C. Wright (Lincoln, NE: University of Nebraska Press, 1985), 24–45; Enrique C. Ochoa, *Feeding Mexico: the Political Uses of Food since 1910* (Wilmington: Scholarly Resources, 2010), 9; and Paulo Drinot, "Food, Race, and Working-Class Identity: *Restaurantes Populares* and Populism in 1930s Peru," *The Americas* 62 (2005): 245–270.

11. Sandra Aguilar-Rodríguez, "Cooking Modernity: Nutrition Policies, Class, and Gender in 1940s and 1950s Mexico City," *The Americas* 64 (2007): 177–205.

12. Stephen B. Brush, *Farmers Bounty: Locating Crop Diversity in the Contemporary World* (New Haven: Yale University Press, 2004), 87; and "Maize: From Mexico to the World," International Maize and Wheat Improvement Center (CIMMYT), accessed 4 April 2017, http://www.cimmyt.org/maize-from-mexico-to-the-world/.

13. Simon Miller, "The Mexican Hacienda between the Insurgency and the Revolution: Maize Production and Commercial Triumph on the Temporal," *Journal of Latin American Studies* 16 (1984): 309–336.

14. Ochoa, *Feeding Mexico*, 41.

15. Ochoa, *Feeding Mexico*, 99–126 and 157–176; and John Richard Heath, "Constraints on Peasant Maize Production: A Case Study from Michoacan," *Mexican Studies/Estudios Mexicanos* 3 (1987): 263–286.

16. Roger Bartra, *Agrarian Structure and Political Power in Mexico*, trans. Stephen K. Ault (Baltimore: Johns Hopkins University Press, 1993), 93.

17. Brush, *Farmers Bounty*, 168.

18. Jeffrey Pilcher, *¡Qué vivan los tamales!: Food and the Making of Mexican Identity* (Albuquerque: University of New Mexico Press, 1998), 163–164.

19. Dominique Louette, "Traditional Management of Seed and Genetic Diversity: What is a Landrace?" in *Genes in the Field*, ed. Stephen B. Brush (Ottawa, Canada: International Development Research Center Books, 2000), 109–142.

20. Mauricio R. Bellon, "The Ethnoecology of Maize Variety Management: A Case Study from Mexico," *Human Ecology* 19 (1991): 389–418.

21. Bellon, "The Ethnoecology of Maize"; and Mauricio R. Bellon and Stephen Brush, "Keepers of Maize in Chiapas, Mexico," *Economic Botany* 48 (1994): 196–209.

22. Arturo Warman, *El campo mexicano en el siglo XX* (Mexico: Fondo de Cultura, 2001), 128–129.

23. Alder Keleman and Jonathan Hellin, "Specialty Maize Varieties in Mexico: A Case Study in Market-Driven Agro-Biodiversity Conservation," *Journal of Latin American Geography* 8 (2009): 147–174.

24. Rita Schwentesius and Manuel Ángel Gómez "Supermarkets in Mexico: Impacts on Horticulture Systems," *Development Policy Review* 20 (2002): 487–502; Nelly Velázquez, *Modernización agrícola en Venezuela: Los valles altos andinos 1930–1999* (Caracas: Fundación Polar, 2004); Aníbal Arcondo, *Historia de la alimentación en Argentina* (Córdoba: Ferreya Editor, 2002); Pierre Ostiguy and Warwick Armstrong, *La evolución del consumo alimenticio en la Argentina (1974–1984)* (Buenos Aires: Centro Editor de América Latina, 1987).

25. Brush, *Farmers Bounty*, 105.

26. Karl S. Zimmerer, *Changing Fortunes: Biodiversity and Peasant Livelihood in the Peruvian Andes* (Berkeley: University of California Press, 1996), 60.

27. Vincent C. Peloso, "Succulence and Sustenance: Region, Class, and Diet in Nineteenth-Century Peru." In *Food, Politics and Society in Latin America*, ed. John C. Super and Thomas C. Wright (Lincoln: University of Nebraska Press, 1985), 46–64.

28. Zimmerer, *Changing Fortunes*, 59.

29. Zimmerer, *Changing Fortunes*, 61.

30. Zimmerer, *Changing Fortunes*, 77–84; and Sarah A. Radcliffe, "Gender Relations, Peasant Livelihood Strategies and Migration: A Case Study from Cuzco, Peru," *Bulletin of Latin American Research* 5 (1986): 29–47.

31. John Murra developed the idea of ecological verticality in the early 1970s; for a critical discussion, see Enrique Mayer, *The Articulated Peasant: Household Economies in the Andes* (Boulder, CO: Westview Press, 2002), 239–277.

32. Brush, *Farmers Bounty*, 102–103.

33. Pierre Morlon, A. Hibon, D. Horton, M. Tapia, and F. Tardieu, "Qué tipo de mediciones y qué criterios para la evaluación?" in *Comprender la agricultura campesina en los Andes Centrales: Peru-Bolivia*, ed. Pierre Morlon (Lima: IFEA and CBC, 1996), 276–319.

34. Zimmerer, *Changing Fortunes*, 84.

35. Brush, *Farmers Bounty*, 108–109.

36. Mary Weismantel, *Food, Gender and Poverty in the Ecuadorean Andes* (Philadelphia: University of Pennsylvania Press, 1988).

37. Morlon et al., "Qué tipo de mediciones y qué criterios para la evaluación?" 297.

38. Marília Lobo Burle, Jaime Robert Fonseca, Maria José del Peloso, Leonardo Cunha Melo, Steve R. Temple, and Paul Gepts, "Integrating Phenotypic Evaluations with a Molecular Diversity Assessment of a Brazilian Collection of Common Bean Landraces," *Crop Science* 51 (2011): 2668–2680; and M. I. Chacón, S. B. Pickersgill, and D. G. Debouck, "Domestication Patterns in Common Bean (*Phaseolus vulgaris* L.) and the Origin of the Mesoamerican and Andean Cultivated Races," *Theoretical and Applied Genetics* 110 (2005): 432–444.

39. Carlos Alberto Dória, "Beyond Rice Neutrality: Beans as Patria, Locus, and Domus in the Brazilian Culinary System," in *Rice and Beans: A Unique Dish in a Hundred Places,* ed. Richard Wilk and Livia Barbosa (London: Bloomsbury Academic, 2012), 127–130.

40. Dória, "Beyond Rice Neutrality: Beans as Patria, Locus, and Domus in the Brazilian Culinary System,"124–126.

41. W. J. Broughton, G. Hernández, M. Blair, S. Beebe, P. Gepts, and J. Vanderleyden, "Beans (Phaseolus ssp.): Model Food Legumes," *Plant and Soil* 252 (2003): 55–128.

42. Marília Lobo Burle et al., "Integrating Phenotypic Evaluations"; and Fábio de Oliveira Freitas, "Evidências genético-arqueológicas sobre a origem do feijão comum no Brasil," *Pesquisa agropecuária brasileira* 41 (2006), on-line version, accessed 12 June 2017, https://seer.sct.embrapa.br/index.php/pab/article/view/7265.

43. Livia Barbosa, "Rice and Beans, Beans and Rice: The Perfect Couple," in *Rice and Beans: A Unique Dish in a Hundred Places,* ed. Richard Wilk and Livia Barbosa (London: Bloomsbury Academic, 2012), 103.

44. Quoted in Livia Barbosa, "Rice and Beans, Beans and Rice," 104.

45. Francisco Vidal Luna and Herbert S. Klein, *Slavery and the Economy of São Paulo, 1750–1850* (Palo Alto: Stanford University Press, 2003), 97–102; and Elizabeth Anne

Kuznesof, *Household Economy and Urban Development: São Paulo, 1765–1836* (Boulder: Westview Press, 1986), 135.

46. Francisco Vidal Luna, Herbert Klein, and William Summerhill, "The Characteristics of Coffee Production and Agriculture in the State of São Paulo in 1905," *Agricultural History* 90 (2016): 22–50.

47. Afranio Garcia and Moacir Palmeira, "Traces of the Big House and the Slave Quarters: Social Transformation in Rural Brazil during the Twentieth Century," in *Brazil: A Century of Change,* ed. Ignacy Sachs, Jorge Wilheim, and Paulo Sérgio Pinheiro, trans. Robert N. Anderson (Chapel Hill: University of North Carolina Press, 2009), 33–46.

48. Clibas Vieira, *O feijoeiro comum, cultura, doenca e melhoramento* (Vicosa: Universidade Federal de Vicosa, 1967).

49. John O. Browder, "Surviving in Rondonia: The Dynamics of Colonist Farming Strategies in Brazil's Northwest Frontier," *Studies in Comparative International Development* 29 (1994): 45–69.

50. Maurício Borges Lemos and Valdemar Servilha, *Formas de organização de arroz e feijão no Brasil* (Brasilia: BINAGRI, 1979), 40.

51. Instituto Brasileiro de Geografia e Estatística, *Censo agropecuário,* 2006.

52. Lemos and Servilha, *Formas de organização de arroz e feijão no Brasil,* 53.

53. N. L. Johnson, Douglas Pachico, and O. Voysest, "The Distribution of Benefits from Public International Germplasm Banks: The Case of Beans in Latin America," *Agricultural Economics* 29 (2003): 277–286.

54. Marília Lobo Burle, Jaime Fonseca, James A. Kami, and Paul Gepts, "Microsatellite Diversity and Genetic Structure among Common Bean (*Phaseolus vulgaris* L.) Landraces in Brazil, a Secondary Center of Diversity," *Theoretical and Applied Genetics* 121, no. 5 (2010): 801–813; and Marília Lobo Burle et al., "Integrating Phenotypic Evaluations with a Molecular Diversity Assessment of a Brazilian Collection of Common Bean Landraces."

55. Lobo Burle et al., "Microsatellite Diversity and Genetic Structure."

56. Lemos and Servilha, *Formas de organização de arroz e feijão no Brasil,* 53; and Lobo Burle et al., "Integrating Phenotypic Evaluations with a Molecular Diversity," 2678.

57. John Soluri, *Banana Cultures: Agriculture, Consumption and Environmental Change in Honduras and the United States* (Austin: University of Texas Press, 2005); Steve Striffler, *In the Shadows of State and Capital: The United Fruit Company, Popular Struggle, and Agrarian Restructuring in Ecuador, 1900–1995* (Durham: Duke University Press, 2002); Fe Iglesias García, *Del ingenio al central* (Havana: Editorial de Ciencias Sociales, 1999); and Gillian McGillivray, *Blazing Cane: Sugar Communities, Class, and State Formation in Cuba 1868–1959* (Durham, NC: Duke University Press, 2009).

58. Louis A. Pérez, Jr., *Winds of Change: Hurricanes and the Transformation of Nineteenth-Century Cuba* (Chapel Hill: University of North Carolina Press, 2001), 44–46.

59. Stanley J. Stein, *Vassouras: A Brazilian Coffee County, 1850–1900* (Princeton, NJ: Princeton University Press, 1985 [1958]), 33–36 and 173–178.

60. Judith Carney and Nicolas Rosomoff, *In the Shadow of Slavery: Africa's Botanical Legacy in the Atlantic World* (Berkeley: University of California Press, 2009).

61. Thomas Holloway, *Immigrants on the Land: Coffee and Society in São Paulo, 1886–1934* (Chapel Hill: University of North Carolina Press, 1980), 90 and 132.

62. Holloway, *Immigrants on the Land,* 87–94; and Christian Brannstrom, "Coffee Labor Regimes and Deforestation on a Brazilian Frontier, 1915–1965," *Economic Geography* 76 (2000): 326–346.

63. Verena Stolcke, "The Labors of Coffee in Latin America: The Hidden Charm of Family Labor and Self-Provisioning," in *Coffee, Society and Power in Latin America,* ed. William Roseberry, Lowell Gudmundson, and Mario Samper Kutschbach (Baltimore: Johns Hopkins University Press, 1995), 65–93.

64. Wilson Picado Umaña, Rafael Ledezma Díaz, and Roberto Granados Porras, "Territorio de coyotes, agroecosistemas y cambio tecnológico en una región cafetalera de Costa Rica," *Revista Historia* 59–60 (2009): 119–165.

65. Lowell Gudmundson, "Peasant, Farmer, Proletarian: Class Formation in a Smallholder Coffee Economy, 1850–1950," in *Coffee, Society and Power in Latin America,* 136–138.

66. William Roseberry, "The Rise of Yuppie Coffees and the Reimagination of Class in the United States," *American Anthropologist* 98 (1996): 762–775; Daniel Jafee, *Brewing Justice: Fair Trade Coffee, Sustainability and Survival* (Berkeley: University of California, 2007); and Andrés Guhl, *Café y cambio del paisaje en Colombia, 1970–2005* (Medellín, Colombia: Universidad EAFIT, 2008).

67. V. Ernesto Méndez et al., "Agrobiodiversity and Shade Coffee Smallholder Livelihoods: A Review and Synthesis of Ten Years' Research in Central America."

68. Ivette Perfecto and Inge Armbrecht, "The Coffee Agroecosystem in the Neotropics: Combining Ecological and Economic Goals," in *Tropical Agroecosystems,* ed. John H. Vandermeer (Boca Raton, FL: CRC Press, 2002), 157–192.

69. Warren Dean, *With Broadax and Firebrand: The Destruction of Brazil's Atlantic Forest* (Berkeley: University of California Press, 1995); Stefania Gallini, *Una historia ambiental del café en Guatemala: La Costa Cuca entre 1830 y 1902* (Guatemala: Asociación para el Avance de las Ciencias Sociales en Guatemala, 2009); and Brannstrom, "Coffee Labor Regimes and Deforestation on a Brazilian Frontier, 1915–1965."

70. Anapaula Iacovino Davila, *O pequeno produtor de café no Brasil e na Colombia* (São Paulo: Annablume, 2009), 23–24.

71. *Censo agrícola, ganadero y forestal 2007* (Mexico: INEGI).

72. Bertola and Ocampo, *The Economic Development of Latin America since Independence*; and Victor Bulmer-Thomas, *The Economic History of Latin America since Independence,* 3rd ed. (Cambridge: Cambridge University Press, 2014).

73. The centers are members of the Consultative Group on International Agricultural Research (CGIAR); see "Our Research Centers," CGIAR, accessed 4 April 2017, http://www.cgiar.org/about-us/research-centers/.

74. Brookfield and Stocking, "Agrodiversity: Definition, Description and Design," 77–80.

 CHAPTER 8

Hoofprints
Cattle Ranching and Landscape Transformation

Shawn Van Ausdal and Robert W. Wilcox

In 1976, geographer James Parsons warned of an "almost mindless mania for converting forest to pasture."[1] As the din of chainsaws and the crackle of burning underbrush crescendoed, so too did the condemnations of cattle— and hamburgers—as the primary driver of deforestation throughout tropical Latin America. Yet for all the significance of stock raising as a major source of environmental change, this history has been misinterpreted in two ways. First, there is a prevailing notion that the environmental impact of ranching was felt to any serious degree only from the mid-twentieth century. Second, as ranchers pushed into Latin America's forests, many observers claimed they were driven fundamentally by incentives located outside of the cattle economy. Here we qualify these perceptions by showing how the environmental impact of livestock has a much longer (and more varied) history than assumed, and by arguing that rising demand (especially domestic) and the biological advantages of cattle have been key to the phenomenal expansion of ranching. While we recognize the wide variety of livestock that have been raised in Latin America—mules, goats, sheep, pigs, and llamas, to name just a few—we focus primarily on beef cattle.

The population explosion of Old World domesticates introduced to Latin America in the wake of European conquest is well known. In the early sixteenth century, Alonzo de Zuazo claimed that a cattle herd, let loose in the benign environment of the Antilles, would increase tenfold within three or four years. On the great grasslands of Latin America—the Pampas, the Llanos, the Cerrado, and northern Mexico—cattle populations grew quickly, eventually numbering in the millions.[2] European livestock also adapted to high Andean plateaus, temperate hillsides, xeric scrubland, and other environments. These animals were critical to Iberian colonialism, as they provided labor, tallow, hides, and meat in support of a colonial "civilization of leather."[3] Extensive areas—often taken from collapsing (and resettled) indigenous populations— were thus given over to livestock throughout the colonial era.

What were the environmental consequences of this livestock boom? In one of the classic works of Latin American environmental history, Elinor Melville claimed that they were disastrous: by the late sixteenth century, the proliferation of sheep in the Mezquital Valley of north-central Mexico had caused the resource base to collapse, leading to desertification.[4] Other scholars, however, contend that Melville exaggerated the environmental impact of sheep, partly by overestimating their numbers and downplaying their seasonal movements.[5] Undoubtedly, European livestock did help to change the species composition of their forage base through selective grazing or even overgrazing. Though sometimes beneficial, introduced plant species also began to reshape local ecologies and could become invasive, as in the case of citrus trees on Hispaniola or thistles (*Cynara cardunculus*) on the Pampas. Through fire and swidden agriculture, ranchers and farmers also formed new grasslands on forest margins. But the declensionist narratives typical of early environmental history appear overstated when it comes to the early impact of livestock on their New World environments.[6] Herds usually ranged extensively on natural grasslands and their numbers rarely appear excessive relative to the local resource base. Most evidence suggests that the environmental impact of cattle and other livestock was limited through the colonial period.

The (Extra-)Long Nineteenth Century

Starting in the mid- to late-nineteenth century, however, ranching became one of the driving forces of landscape change. Greater integration into the North Atlantic economy spurred the expansion of stock raising. In some cases, the export of hides, wool, and meat provided the incentive. In others, export-led growth stimulated domestic demand, which the doubling of the population between 1850 and 1900 (and again by 1940) propelled yet further. In turn, expanding livestock markets stimulated the settlement of the frontier, where the raising of cattle and sheep often consolidated territorial control. Through such expansion—as well as the modernization of ranching, however slow and uneven—livestock, especially beef cattle, began to reshape Latin American environments.

Between roughly 1850 and 1950, this impact was felt principally in three biomes (see map 8.1). The first is the region's seasonal (dry) tropical forests. Ranging from lush to xeric, these semi-deciduous forests tend to occur in tropical lowlands where an annual rainfall regime of less than two thousand millimeters is marked by a dry season of several months. With moderate levels of biomass and precipitation, which expedited clearing, limited weeds, and reduced soil leaching, these forests became targets for conversion as ranchers began moving beyond the confines of colonial stock raising. Starting in the mid-nineteenth century, some ranchers converted land abandoned by export

Map 8.1. Ranching biomes of Latin America.

Seasonal tropical forests
Temperate savannas
Tropical savannas
Chaco
Caatinga
Pantanal
Lowland rain forests

Sources: Encyclopedia of Earth; Dirzo, Young, and Mooney (2011: 5);
http://www.biodiversidad.gob.mx/ecosistemas/mapas/mapa.html
Base map: https://commons.wikimedia.org/wiki/File:Map-Latin_America.svg

Shawn Van Ausdal

Map by Shawn Van Ausdal.

commodity production, such as sugar (Cuba), tobacco (Colombia), and coffee (Brazil) into pasture; others monopolized village commons and Indian reserves by taking advantage of liberal reforms that undermined communal forms of property. But one of the most significant developments was clearing forests to plant grass.

Expansion of the ranching frontier was aided by the widespread introduction of African grasses such as *pará, guinea,* and *yaraguá* (see illustration 8.1).[7] Their rapid and dense growth helped to keep forests from regenerating, facilitating the creation of "artificial" or planted pastures.[8] An early twentieth-century rancher from Costa Rica claimed that, by choking off the underbrush,

Illustration 8.1. Guatemalan cowboy in a field of guinea grass, 1917.

Harry O. Sandberg, "Central America—Cattle Countries," *Bulletin of the Pan American Union* 44, no. 4 (1917): 450 [449–464].

guinea grass "opens up space for itself."[9] These grasses were also more productive and resilient than most native species, increasing carrying capacity, accelerating the fattening process, and improving animal health.[10] They did not, however, eliminate the work and expense of pasture formation. In Colombia, many ranchers were really grass farmers, employing far more workers to plant and care for pastures than look after livestock. A hectare of grass was also worth ten times an equal area of forest. Such upfront costs underscore the economic imperative of ranching, a point we develop below.

The extent and timing of such pasture-led deforestation is hard to determine, but scattered evidence across Latin America suggests that it was well under way by the end of the nineteenth century and accelerated over the first half of the twentieth century. On Colombia's Caribbean lowlands, "The first trials with artificial grasses for the rainy season [circa 1850] were so successful that all the stock raisers ... cleared the virgin forests around the *ciénagas* and ... planted the grass."[11] By the 1870s, ranchers in Costa Rica were following suit. The integration of Mexico's northern grasslands into the U.S. market during the Porfiriato encouraged ranchers to expand production elsewhere, frequently at the expense of the forest.[12] By the 1940s, ranchers in the state of São Paulo, Brazil, depended on "leys sown down on land reclaimed from forest," and "waving fields of guinea grass" had replaced the forests around Matagalpa, Nicaragua.[13] A decade later, Mexico had over one million hectares of "artificial pastures," much of which was planted on land formerly in forest.[14]

In Colombia, this "relentless destruction of the jungle and forest" contributed to the formation of about ten million hectares of planted pasture.[15] The conversion of seasonal tropical forests to pasture continued beyond mid-century, and by the late 1980s, conservation biologist Daniel Janzen considered them to be "the most endangered major tropical ecosystem."[16]

The second biome is the Pampas of Argentina, Uruguay, and southern Brazil. On these extensive temperate grasslands, huge herds of semi-wild cattle developed over the colonial period, providing the basis of Latin America's best-known ranching region and one of its most celebrated archetypes, the gaucho. While territorial consolidation extended the livestock economy geographically, it was the rapidly expanding export trade that provided the principal dynamic. Hide exports rose from one million to nearly two million annually between the end of the colonial period and the mid-nineteenth century.[17] Dried beef (*tasajo*) exports also expanded quickly from the 1820s. Within three decades, crude processing plants (*saladeros*) annually converted close to half a million head into this cheap form of protein largely destined for slaves in Cuba and Brazil. This export trade encouraged a fourfold extension of ranching lands over the first half of the nineteenth century.[18]

Despite the long-term presence of cattle, it was wool that led to the first extensive ecological transformation of the Pampas. Driven by growing demand in industrializing Europe, the sheep population of Argentina jumped from a couple million at the beginning of the nineteenth century to some forty million in 1865 and seventy-four million by 1895.[19] By the 1860s, the value of wool exports from Buenos Aires surpassed that of hides, and rising land values displaced cattle to south of the capital's hinterland.[20] Given their notoriously efficient feeding habits, further encouraged by the extension of wire fencing (over 180 thousand kilometers by 1888 in the province of Buenos Aires alone), sheep extended the ecological impact of cattle by converting the tall bunchgrasses into "a lawn-like, low, dense sod."[21] The pressure of selective grazing, which reduced the "good" grasses, plus the trampling of hundreds of thousands of cloven hooves, diminished the productive capacity of the land.

By the end of the century, rapidly expanding beef markets in Europe, made accessible by advances in refrigeration, helped the pendulum to shift back in favor of cattle, introducing further changes to the pampa ecosystem. The first refrigerated slaughterhouses (*frigoríficos*) in Argentina date from the 1880s. While these British-financed operations initially exported mutton, it was beef that became the driving force of the industry. The cattle that passed through the *frigoríficos* were no longer the wiry *criollo* animals (locally adapted Iberian breeds) utilized by the hide and *tasajo* trade. Instead, the export of chilled beef required the upgrading of native cattle with "improved" British breeds, most noticeably Angus, Shorthorn, and Hereford, which were capable of producing the "marbled" meat so prized in Britain, Argentina's principle export market. From the 1880s, the nation's beef cattle herd gradually moved from *criollo* to purebred and mestizo (European-*criollo* crossbreeds), permitting Argentina, by World War I, to replace the United States as the world's largest beef exporter.[22]

While the temperate climate of the Pampas facilitated this genetic transformation, success was also dependent on the development of high-quality alfalfa pastures. By subdividing properties and leasing parcels to immigrant grain farmers with the requirement that alfalfa be planted at contract termination, cattle raisers devised an inexpensive means of converting natural grasslands to pasture—almost nine million hectares by 1920.[23] Even where alfalfa pastures were not successful, as in Uruguay and Rio Grande do Sul, the ecological transformation of the Pampas persisted.[24] By the 1930s, Albert Boerger considered that the "deterioration of the sward" in Uruguay was primarily caused by "the introduction of grasses and pasture weeds emanating in the main from Europe."[25] Even in the province of Buenos Aires, one-third of the plant species in natural grasslands were of exotic origin.[26] As with sheep raising earlier, these transformations, along with the falling productivity of alfalfa pastures and an impressive expansion of cereal agriculture linked to increased immigration and the expansion of railways, shifted the locales of ranching, generating concern about the future of the sector, though ranching remained key to Pampas economies for decades.

Outside of northern Mexico, the Pampas, and some high-altitude regions, most cattle grazed on tropical savannas (see map 8.1). While these grasslands vary greatly in topography and ecology, many are characterized by flat or rolling terrain with low-fertility soils and are marked by sharp dry seasons in which the abundant natural grasses become tough and lose much of their nutritive value. In the past, cattle foraged widely, often needing ten or twenty hectares per head and, even then, lost weight during the dry season. On other grasslands, annual flooding was a determining characteristic. In the Pantanal of Brazil and Bolivia, ranchers molded their activities to the annual fluctuations of the Paraguay River and its tributaries. Their animals also adapted to these special conditions, developing longer legs and wider hooves, and spontaneously migrating to higher ground during flooding.[27] While the adversities of tropical savannas delayed substantial population explosion, by the end of the colonial period, two of the most notable regions of tropical savannas—the Llanos of Venezuela and eastern Colombia, and the Brazilian *Cerrado*—had substantial cattle herds. Others included the interior savannas of Central America and Cuba; the intermontane valleys of the Andes; the grasslands of northern Argentina and eastern Paraguay; the arid Caatinga of the Brazilian northeast; and the Chaco of Paraguay, Argentina, and Bolivia.

Raising cattle on these rangelands led to significant environmental changes. While selective grazing and trampling altered species composition, the key cause was fire. Burning was the customary means to remove dried grass and encourage new growth dating from the colonial period, but repeated burnings eventually degraded the resource base. Observers of the Cerrado began to recognize these effects as early as the mid-nineteenth century. By the second quarter of the twentieth century, warnings about rangeland deterioration became increasingly common. In 1944, botanist Agnes Chase wrote that repeated burning "has diminished the better grasses, until much of the eastern llanos . . . are occupied by inferior ones which are more resistant to elimination by fire."[28]

In some cases, ranchers transformed native rangelands by planting more productive "exotic" grasses. The botanist Frances Pennell, traveling through the Savannas of Bolívar (Colombia) in 1918, remarked that "so solidly are these [imported grasses] grown that . . . there appears to be no native flora left."[29] Similarly, by the 1940s, pará and guinea grasses were common on Cuban savannas, and the natural pastures of São Paulo state had "been improved in some parts by planting *Melinis minutiflora* and *Hyparrhenia rufa.*"[30] Nonetheless, this kind of wholesale transformation proceeded slowly, limiting concern about what many perceived to be an inexhaustible frontier.

All of these grasslands played an important role in the territorial consolidation of Latin America's republics. Replacing inhospitable forest with grass was considered the first step in "civilizing" the land. Besides, few other activities were viable in such far-flung regions. Because livestock could walk to market, ranching was critical to the formation of incipient national economies, and cattle drives became commonplace. In Colombia, drovers walked groups of about a hundred head up narrow mountain trails. Where the landscape was more open, such as Brazil, cowboys established five- to twenty-meter wide trails—often the precursors of modern roadways—to move substantially larger herds.[31] While railroads and boats were sometimes used, transporting cattle "on the hoof" was common throughout Latin America until trucks took over in the mid-twentieth century. Driving cattle from the Cerrado to the fattening pastures near Rio de Janeiro and São Paulo could take up to three months. At the end of such drives, cattle had frequently lost 15 to 30 percent of their body weight and required refattening, which necessitated additional pastures near the point of slaughter.[32]

A scarcely researched aspect of Latin American beef markets is the contribution of slaughterhouses to environmental contamination. William Cronon's account of the pollution of the Chicago River with waste from the city's meat-packing industry was an extreme example of a problem likely common to many large Latin American cities.[33] Eyewitnesses described the sights and smells of nineteenth-century *saladeros* in Buenos Aires as noxious.[34] But the ecological impact of their effluent was mitigated by converting most of the animal into a variety of useful products. The scale of these processing plants dwarfed the typical municipal slaughterhouse in Latin America, which persisted late into the twentieth century due to the absence of regional cold chains.[35] By contrast, leather tanning, which tended to be more geographically concentrated, was a more notorious polluter. Additionally, the demand for tanning agents led to the heavy exploitation of some mangroves and forests, most notably *quebracho colorado* in Argentina and Paraguay.

The Post–World War II Boom

The 1950s mark both an intensification of earlier trends and a qualitative shift in the environmental history of ranching in Latin America. Between 1950 and 1990, the cattle population of Latin America more than doubled, driven by new export markets as well as surging domestic demand and government incentives. As cattle numbers multiplied, ranchers intensified their push into the forest, aided by chainsaws, bulldozers, and DDT. Latin America's humid forests, hitherto largely ignored, became a key site of new pasturelands. Similarly, the transformation of the region's natural savannas through the planting of "exotic" grasses became widespread. And cattle themselves experienced a genetic revolution as Indian (and some European) animals subsumed *criollo* breeds.

Central America illustrates how beef exports encouraged the expansion of ranching. In the 1950s, rising beef consumption and the development of the fast food industry in the United States prompted a search for new sources of cheap cattle. With South American beef off-limits because of foot-and-mouth disease, the U.S. Department of Agriculture began certifying Central American meat-packing plants in 1957. Within two decades, there were twenty-eight such plants around the Isthmus exporting almost 120 thousand tons of beef.[36] Such exports, along with foreign aid and road building, helped spark a ranching boom. By the late 1970s, cattle herds had doubled and ranchers had increased the forage base by 150 percent.[37] Some of this expansion occurred by appropriating peasant farms and extending pastures on existing estates, but much of it occurred at the expense of the region's forests. The initial impulse was to continue clearing seasonal forests on the Pacific side of the isthmus, where 60 percent of new pastures were developed through the 1970s. But ranchers also moved into the humid forests of the Caribbean lowlands, which became the main front of the ranching frontier in the 1980s. In just four decades after 1950, ranchers helped clear almost eleven million hectares, or close to 40 percent, of Central America's forests.[38] Between 1961 and 1981, 38 percent of regional production was exported, causing Norman Myers to implicate the U.S. "hamburger connection" in the rapid destruction of the region's forests.[39] Nonetheless, beef consumption in Central America more than doubled over this same period, drawing attention to the home market in the expansion of the ranching frontier.[40]

The Brazilian experience further highlights the mixture of domestic concerns and export markets behind the advance of ranchers into the forest. Beginning in the 1950s and 1960s, the government turned its attention to the country's vast interior. The possibility of expanding beef production with an eye to exports was one consideration. Cheap beef policies to pacify urban frustration from stagnant wages, in the context of a rapidly growing population, was another. Pushing ranchers deep into the Cerrado and Amazon also extended the state's geopolitical control of its frontier regions and better integrated them into the national economy.[41] Generous government subsidies provided a major incentive for the rapid expansion of pasture planting in the Cerrado and in the Amazon basin, even though some contemporary observers questioned the long-term financial viability of cattle production in those regions. The resulting speculative environment intensified the perception of land as a commodity, resulting in an exponential increase in environmental impact. Although partially imagined as an escape valve for landless peasants, ranchers frequently followed (and pushed) small-scale colonizers deep into the forest, concentrating land ownership through purchase and intimidation. The result was a radical spatial redistribution of the national cattle herd, jumping from 17 percent in these regions in 1950 to over 50 percent by 2006.[42] By the early twenty-first century more than 60 million hectares of the Amazon (especially along its southern and eastern edges), or over 10 percent of the entire Amazon Basin, were cleared, mostly for pasture.[43] In the adjacent Cerrado, home to one-third of all Brazilian biodiversity, planted pasture increased by over 520 percent between 1970 and 1995. By 2004 over 41 percent of original vegetation had been replaced by planted pasture, exceeding 65 million hectares.[44]

Critical to these developments was the modernization of tropical ranching. Postwar Brazilian governments expanded technical knowledge through the development of a federal agricultural research institute eventually called EMBRAPA. The same occurred in various other Latin American nations, such as INIFAP in Mexico, and an international center, CIAT, in Colombia.[45] Government agents, private ranchers, and international agencies promoted "improved" ranching methods, including forage and breed development and attention to animal and plant diseases and parasites. The introduction of additional African-origin grasses, most recently *brachiaria* (signal grass), which began first in Brazil in the 1950s, was an important step in the expansion of ranching into that country's interior. By the early 2000s, over one hundred million hectares of pastures, 80 percent in brachiaria, blanketed the country.[46] This so-called rationalization of ranching, which promoted employment of the most up-to-date scientific knowledge, was duplicated in other regions, facilitating deforestation and transforming tropical savannas.

A further contributor to ranching-related environmental change involved the genetic makeup of cattle themselves. At the beginning of the twentieth century, most cattle in tropical Latin America were the descendants of *criollos*. Due to repeated failures to upgrade native cattle with European imports, some far-thinking ranchers, especially from Brazil, began importing Indian-origin Zebu. While fast-maturing European breeds suffered from conditions akin to heat stroke, Zebu, the product of similar ecosystems in the Indian subcontinent, thrived in the American tropics. Spectacular success, specifically in the Cerrado, radically transformed the Brazilian cattle industry by mid-century and eventually spread to all of the tropical and semitropical Americas.[47] Between 1960 and 2012, the total Latin American beef cattle herd increased from an estimated 175 million to a little over 400 million head, the majority Zebu or Zebu crosses.[48] The biological collaboration of Zebu from India and exotic grasses from Africa facilitated the postcolonial occupation of previously inhospitable regions, and directly contributed to widespread environmental transformations.

Cattle Cultures and the Logic of Ranching

None of this history would have been possible without the participation of cowboys and their steeds. Those who handled cattle on a regular basis, whether *vaqueiro*, gaucho, *huaso*, or *llanero*, became central figures to the livestock industry, and sometimes to national culture. Some cowboys were indigenous, others were African slaves or their descendants, but beginning in the nineteenth century they became increasingly *mestizo*. While cogs in the ranching machine, both cowboy and horse represented an independence of spirit that captured the imagination of both urbanites and peasants. J. Frank Dobie considered the Mexican *vaquero* to be a "child of nature," naming "every hill and hallow" and wise to the "virtue of every bush and herb."[49] Their skills and valor became litmus tests of masculinity: racing and breaking horses, corralling and tailing cattle, betting and sometimes fighting. Such practices were institutionalized in the amateur bull fights (*corralejas* in Colombia) and rodeos (*charreadas* in Mexico). The rapid conversion of these horsemen into mounted troops inspired admiration as well as considerable trepidation over the perceived threat they posed to property and politics. To Domingo Saramiento, Facundo Quiroga and his gaucho cavalries were the essence of "barbarity." Often only after the state and ranchers tamed the frontier did its extinct frontiersmen become re-signified as central to national or regional identity.[50] And while this aspect of the rural world was defined largely by elites, mounted cowboy motifs and Brahman bulls stenciled on buses and trucks throughout Latin America today attest to the continuing symbolism of the region's cattle culture. Likewise, while mechanization has reduced the demand for steeds in ranching across the region, horses—themselves often adapted to local environmental conditions—are still key work animals and, through rodeos and cavalcades, symbols of rural life and status.

Despite the strong association between landed elites and ranching, cattle (and other livestock) have been integral to the lives of many peasants.[51] Oxen helped plow the fields and carry goods. Cheese made from the milk of a cow or two—"the pride and happiness of the household"—improved the family's diet or income without much added labor or additional resources.[52] Money pooled for common cause, called a *vaca* (literally cow) in much of Latin America, underscores the role of cattle as informal savings accounts. With a little more land, a peasant might breed cattle as well, selling yearlings to ranchers to raise and fatten. On the agrarian frontier, such enterprises reinforced property claims and captured the labor of colonizing the forest in a saleable commodity: grass.[53] While the skewed ownership of cattle, not just land, highlights Latin America's enduring inequalities, the share of peasants and small ranchers has often been larger than generally imagined. The tendency of small producers to breed animals that were traditionally slaughtered at four to five years has meant that snapshots of ownership patterns downplay the percentage of animals that originated on peasant farms.[54]

For ranchers, raising cattle was a lifestyle. Even if absentee, they often felt an affinity for the countryside and valued physical work under open skies, although they might have done little of it themselves. They took pride in cinching deals with a handshake, competing in livestock shows, and being the object of respect. As James Parsons and many others have pointed out, "raising cattle is a prestigious activity in Latin America."[55] Sometimes this status is generalized too quickly. On the frontier, there might be little in the way of material comforts or education for an urbanite to differentiate rancher and peasant. However, ranchers moved with swagger and assumed their place at the top of the social structure. In Mexico's Huasteca region, Claudio Lomnitz found that they felt "superior to [their] peones not because they [were] essentially different from them, but because they [were] better, more distilled, versions of the same."[56] This shared culture (at times), as well as personal relations, entitlements, and debt tied ranch workers to their *patrón*. The neighboring peasantry, noting the rancher's spare land and friendly relations with mayor and magistrate, deferred to that authority, if often reluctantly.

But it was primarily the biological advantages of cattle that attracted Latin American elites. Before the spread of rail and roads, transportation costs greatly limited the areas where field crops could be profitably grown. Cattle and other livestock could walk to market, making them especially important in colonizing the frontier. They also acted as organic machines, harvesting the grass crop on the hoof and reducing labor requirements. Where there was a scarcity of labor but an abundance of land, raising livestock made sense. Similarly, where drainage and irrigation projects were impractical, cattle's mobility enabled stockowners to adapt to climatic fluctuations and lower the risks typically associated with agricultural production. Comparatively easy to manage from a distance, ranching also enjoyed economies of scale that crop farming typically did not until the diffusion of farm machinery. Where access to credit remained limited, the relative liquidity of livestock made them a desirable investment and a way for ranchers to tap domestic capital. In fact, before postwar industrialization and economic expansion generated more investment options, considerable amounts of Latin American capital were tied up in cattle and grass. In some rural areas today, ranching remains one of the few viable ventures. While the profits may have been modest, they tended to be secure. Over the long run, ranchers could also profit from rising land values as frontier regions became integrated into the national economy. Jesse Knight, who speculated in land in Colombia during the 1940s and 1950s, stated that raising cattle was "one of the best means I know of to show dominion and the exercise of possession over one's property."[57]

It is a mistake to believe, however, that "land, not profits from beef production" motivated ranchers.[58] The economic viability of ranching (and the display of mobile capital) made its other attributes possible, not vice versa. Susanna Hecht has shown how the incentives offered by the Brazilian government to settle the Amazon created a speculative bubble in which rapidly rising land prices made the operating costs of ranching an afterthought despite the fact that "pastureland created from forests is expensive to implant and to maintain."[59] This situation was atypical in the larger history of ranching. Ranchers did speculate in land, but the high cost of developing pastures, or the potential losses from a poorly run ranch, forced them, in the interim, to keep their eye on the bottom line. Likewise, the privilege and power derived from ranching was usually insufficient to compensate for a money-losing operation. In the Amazon, ranching continued to expand even after the speculative bubble burst. Today, low land values, falling transportation costs, and improved pasture management has tended to make raising cattle in the Amazon more profitable than elsewhere in Brazil.[60]

Although not the only factor, growing demand for beef has been a driving force of ranching expansion. On the Pampas, where cattle long outnumbered the human population, as well as Central America (between the 1960s and 1980s) and northern Mexico (which has sent over one million head per year to the United States since the mid-1980s), export markets provided the incentive. Brazil started exporting beef during World War I, but the percentage of cattle destined for foreign markets remained limited until a sharp uptick starting in the late 1970s. By 2010, 20 percent of domestic production was exported, with the Amazon supplying 20 percent of that total.[61] While exports have usually been imagined in terms of British roast beef and American hamburgers, inter-regional trade in live animals, and more recently semen and embryos, has also been important.

Yet the bulk of cattle production has been absorbed by domestic consumption. While domestic demand was partly governed by the fluctuations of the export sector, it was tied primarily to growing populations who sought meat as an essential part of their diet. The availability of meat, especially beef, from the colonial period helped make it a significant part of numerous local diets. Alexander von Humboldt famously remarked that Caracas, with a population one-tenth that of Paris, "consumed more than one-half the quantity of beef annually used in the capital of France."[62] In Mexico City a few years earlier, the price of beef—considered the "meat of the poor" (the rich preferred mutton)—was a little more than double that of tortillas.[63] Meat consumption, however, has varied greatly by region and class. While few statistical studies have been attempted, beef consumption appears to have increased over the second half of the nineteenth century as growing economies and populations encouraged ranchers to increase their herds and expand into new lands. Consumption patterns varied widely between nations. By the late nineteenth century, residents of Buenos Aires ate, on average, 100 to 120 kilos of beef per year.[64] This was extreme; in Colombia, per capita consumption may have been only fifteen kilos, rising to twenty-five kilos by the 1920s.[65] But the social significance of meat, especially beef, was pervasive. In the early twentieth century, the government regularly provided prisoners in Cartagena (Colombia) with half pound of meat daily, and middle class residents of Santiago (Chile) rioted after tariffs imposed on imported Argentine cattle caused beef prices to rise.[66] At times the politics of beef stemmed from an elite desire to improve the "race" by raising their protein intake.[67] At other times they sought to appease urban discontent. Overall, Latin Americans consumed more and more beef as the population tripled between 1940 and 2000 and per capita incomes rose over 60 percent. By the 1980s, Lovell Jarvis remarked that the "importance of beef among meats and among all foods is a striking" characteristic of Latin America.[68] While the region's beef exports doubled between 1961 and 2000, so too did domestic consumption. More significantly, since 1961 beef exports have averaged only 9 percent of total Latin American production, underscoring the significance of domestic markets.[69]

Recent Developments

Over the last twenty-five or so years, ranching expansion in Latin America has slowed. While the cattle population has continued to rise—by about seventy-seven million head or almost 25 percent since 1990—the average annual growth rate has fallen from almost 3 percent, to a little over 1 percent.[70] Similarly, the impulse to expand pastureland has diminished. The total area of grazing land in South America increased by 35 percent, or more than eighty-four million hectares, between 1961 and 1989 (outside of Argentina and Uruguay), but since then it has only grown by 4 percent.[71]

Three factors help to explain this slowdown: declining demand, rising productivity, and the elimination of subsidies. Starting in the 1980s, Central American ranching became less profitable as the U.S. market shrank due to falling consumption and new import restrictions.[72] For the most part, however, the tapering of demand has originated within Latin America, as the slowing growth rate of cattle herds parallels a drop in human birth rates. Chicken, consumption of which has skyrocketed since the 1980s as its price fell, has also become a viable substitute for beef. The result has been rising overall meat consumption, while preference for beef has either dropped in absolute terms (as in Argentina, Colombia, Mexico, and Costa Rica), or its rate of growth has slowed significantly. Contributing to this slowdown has been the gradual but steady intensification of ranching. While beef production grew largely through pasture expansion to about 1980, since then the amount of meat produced per animal has grown, on average by slightly more than 1 percent annually. Additionally, subsidized credit for ranching has dried up throughout the region, along with much of government-sponsored colonization of the humid tropics.

The environmental effects of this have been uneven. New pastures continue to emerge along Brazil's "arc of deforestation," in Central America's Caribbean lowlands, and in northwestern Ecuador, despite the reduced rates of forest loss.[73] But the slowing expansion of ranching, along with rural outmigration and better conservation efforts, has contributed to a relatively new phenomenon: forest regeneration in abandoned pastures.[74] In Central America, for instance, the total area of pastureland has contracted by 2 percent since 1990.[75] This spontaneous reforestation also has its own uneven geography, with more than 40 percent of the regrowth in xeric landscapes, such as northeastern Brazil, while reforestation rates in moist and dry forests have been outpaced by continued forest loss.[76] Aggregated statistics also hide important shifts at smaller scales: while pastures are abandoned in some locations, new ones are opened elsewhere.

The reforestation of abandoned pastures is part of a larger effort toward green ranching. Growing interest in reducing the environmental footprint of cattle has started to make headway. In Brazil, stricter enforcement of environmental legislation contributed to a 70 percent reduction in the rate of forest clearing in the Amazon since 2004.[77] In addition, there have been efforts to encourage environmentally responsible grass-fed, low-input beef production to appeal to the growing global demand for natural and organic beef. For example, Brazilian Zebu breeders developed a campaign in the early 2000s to promote Zebu as a "*boi ecológico.*" Limited international demand has diminished the campaign's zeal, but organic beef carved out a niche in the region. By 2012, almost 4.7 million hectares of Latin America's grazing land (mostly in Argentina and Uruguay) was certified organic. This is over 50 percent more than all the land in organic production in the United States and Canada combined.[78]

But other environmental issues remain. In the second decade of the century, Argentine meat production had declined so significantly that exports and prices were restricted by government mandate in order to satisfy local consumption demands, causing considerable conflict between the government and producers. One response has been to transform cattle production to feedlots. A 2012 study estimated that up to 50 percent of all Argentine cattle were finished in feedlots, though informal observations have suggested more. As consumer preferences change, Mexican producers, especially in the north, have also turned to feedlots to finish cattle for the domestic market—in part trying to recapture the market from the five million pounds of grain-fed beef imported annually from the United States since 1997. The same trend is beginning to grow in Brazil to supply export demand, though a 2014 study indicated total feedlot percentage of production cycle was low, at 7 percent.[79] In large part these production shifts can be attributed to cheap access to grains for feed (corn, soybeans); in 2017, the rest of the region had little to no feedlot presence. The environmental impacts of these transformations largely have been apparent in pollution through increased concentrations of methane and ammonium emissions and waste runoff affecting local water quality, as well as the expansion of soy and corn agriculture. From a human health standpoint, the rapid expansion of feedlots in Argentina have contributed to the highest incidence of E. coli infection in the world. Further studies undoubtedly will reveal additional impacts from this expanding cattle sector.[80]

An additional concern has been the role of livestock in global climate change. Since the beginning of the twenty-first century, increasing attention has been paid to the contribution of millions of ruminants to methane levels through belching and flatulence, as well as carbon dioxide emissions through clearing and burning of forests and savannas for the planting of pasture. Some studies argue that, globally, cattle have contributed more to greenhouse gases than automobiles, though this includes all ancillary activities involved in raising livestock. Overwhelming evidence demonstrates that ranching furthers land degradation, reduces and pollutes water resources, and decreases biodiversity, though the magnitude of these impacts has been strongly debated.[81] Taking into consideration the increasing vertical integration of modern livestock raising, there is little doubt that ranching is a significant factor in calculating overall global environmental transformation. Ultimately, ranching not only led Latin American landscape change but also has become an important contributor to global environmental transformation.

Conclusion

An environmental history of ranching requires an understanding of a variety of geographical, societal, and biological influences. Beginning in the mid-nineteenth century, and accelerating from the mid-twentieth century, ranchers and cattle have been one of the major drivers of landscape change across the region as they sought to stake out a place for themselves in the countryside and to profit from growing demand for beef and land. The diverse geography of Latin America often determined various approaches to ranching and uneven impacts on landscapes. Sometimes demand was external, but more often than not it emanated from the desire of Latin Americans themselves to consume more meat. That beef has traditionally been the most ubiquitous meat in Latin American diets has much to do with the biological advantages of cattle as well as the culture and status of ranching. These benefits, and the imagined ability of cattle and grass to "civilize" the frontier, further encouraged the simplification (and monopolization) of vast areas. And innovations such as new breeds and introduced grasses strongly influenced how cattle are raised and, ultimately, consumption patterns. By addressing these influences, we come to understand more deeply how the material, ecological, and social foundations of the beef cattle industry collectively determined its spectacular historical trajectory.

Shawn Van Ausdal is an associate professor of history at the Universidad de los Andes in Bogotá (Colombia). With a Ph.D. in geography from the University of California, Berkeley, he has also been a visiting scholar at the Rachel Carson Center for Environment and Society in Munich. He has published widely on the history of cattle ranching in Colombia and is currently working on a manuscript on the subject that spans much of the nineteenth and twentieth centuries. He has also published articles and book chapters on the history of food and development discourse in Latin America.

Robert W. Wilcox is an associate professor of history at Northern Kentucky University, Highland Heights, Kentucky. He also has been a visiting professor at the Universidade Federal de Grande Dourados, Dourados, Mato Grosso do Sul, and the Universidade Federal de Mato Grosso, Cuiabá, Mato Grosso, Brazil. He has published several articles and book chapters on the economic and environmental history of cattle ranching in Latin America, most specifically central Brazil. His recent book, *Cattle in the Backlands: Mato Grosso and the Evolution of Ranching in the Brazilian Tropics* (University of Texas Press, 2017) is a detailed study of ranching development in Mato Grosso from 1870 to 1950.

Notes

1. James Parsons, "Forest to Pasture: Development or Destruction?," in *Hispanic Lands and Peoples: Selected Writings of James J. Parsons,* ed. William Denevan (Boulder: Westview Press, 1989), 278.

2. Alfred Crosby, *The Columbian Exchange: Biological and Cultural Consequences of 1492* (Westport, CT: Greenwood Press, 1972), 76, 85–92.

3. Domingo Faustino Saramiento, quoted in Horacio Giberti, *Historia económica de la ganadería argentina* (Buenos Aires: Ediciones Solar, 1961), 73.

4. Elinor Melville, *A Plague of Sheep: Environmental Consequences of the Conquest of Mexico* (Cambridge: Cambridge University Press, 1994).

5. Karl Butzer and Elizabeth Butzer, "The Sixteenth-Century Environment of the Central Mexican Bajío: Archival Reconstruction from Colonial Land Grants and the Question of Spanish Ecological Impact," in *Culture, Form and Place,* ed. Kent Mathewson (Baton Rouge: Louisiana State University, 1993), 89–124; Andrew Sluyter, *Colonialism and Landscape: Postcolonial Theory and Applications* (Lanham: Rowman & Littlefield Publishers, 2002).

6. The effect of the new ungulates on indigenous populations, however, mostly through crop predation and trampling, was often devastating. See León García Garagarza, "The Year the People Turned to Cattle: The End of the World in New Spain, 1558," in *Centering Animals in Latin American History,* ed. Martha Few and Zeb Tortorici (Durham: Duke University Press, 2013), 31–61.

7. Pará (*Brachiaria mutica*); guinea (*Panicum maximum*); yaraguá, or *jaraguá* in Portuguese (*Hyparrhenia rufa*). For the introduction and diffusion of these grasses around Latin America, see James J. Parsons, "Spread of African Pasture Grasses to the American Tropics," *Journal of Range Management* 25, no. 1 (1972): 12–17.

8. Shawn Van Ausdal, "Pasture, Power, and Profit: An Environmental History of Cattle Ranching in Colombia, 1850–1950," *Geoforum* 40, no. 5 (2009): 712.

9. Quoted in Marc Edelman, *The Logic of the Latifundio: The Large Estates of Northwestern Costa Rica since the Late Nineteenth Century* (Palo Alto: Stanford University Press, 1992), 75.

10. Shawn Van Ausdal, "Productivity Gains and the Limits of Tropical Ranching in Colombia, 1850–1950," *Agricultural History* 86, no. 3 (2012): 10.

11. Louis Striffler, *El río San Jorge* (Barranquilla: Gobernación del Atlántico, 1994 [1886]), 103.

12. Éric Léonard, "Ganadería y construcción de la propriedad territorial en el trópico seco mexicano. Raíces y fracasos de una reforma agraria," in *Historia ambiental de la ganadería en México,* ed. Lucina Hernández (Xalapa, México: Instituto de Ecologia, A.C., and L'Institut de Recherche pour le Développement, 2001), 199.

13. P. de Lima Corrêa, cited in G. M. Roseveare, *The Grasslands of Latin America* (Aberystwyth, UK: Imperial Bureau of Pastures and Field Crops, 1948), 115; Samuel H. Work and Leo R. Smith, "The Livestock Industry of Nicaragua," *Foreign Agricultural Report* 12 (1946): 10.

14. United Nations, *Livestock in Latin America: Status, Problems and Prospects,* vol. 1, *Colombia, Mexico, Uruguay and Venezuela* (New York: United Nations, 1962), 36.

15. Dimas Badel, *Diccionario histórico-geográfico de Bolívar* (Bogotá: Gobernación de Bolívar, Instituto Internacional de Estudios del Caribe, Carlos Valencia Editores, 1999 [1943]), 304; United Nations, *Livestock,* 14.

16. Daniel Janzen, "Dry Tropical Forests: The Most Endangered Major Tropical Ecosystem," in *Biodiversity,* ed. E. O. Wilson (Washington, DC: National Academy Press, 1988), 130.

17. Osvaldo Barsky and Julio Djenderedjian, *Historia del capitalismo agrario pampeano,* vol. 1, *La expansión ganadera hasta 1895* (Buenos Aires: Siglo XXI Editores Argentina, 2003), 144.

18. Barsky and Djenderedjian, *La expansión ganadera,* 338–339; Andrew Sluyter, "The Hispanic Atlantic's Tasajo Trail," *Latin American Research Review* 45, no. 1 (2010): 100; Samuel Amaral, *The Rise of Capitalism on the Pampas: The Estancias of Buenos Aires, 1785–1870* (Cambridge: Cambridge University Press, 1998), 123, 126.

19. Hilda Sabato, *Agrarian Capitalism and the World Market: Buenos Aires in the Pastoral Age, 1840–1890* (Albuquerque: University of New Mexico Press, 1990), 26.

20. Amaral, *The Rise of Capitalism,* 240–241.

21. Barsky and Djenderedjian, *La expansión ganadera,* 330; Roseveare, *Grasslands,* 18.

22. Carmen Sesto, *Historia del capitalismo agrario pampeano,* vol. 2, *La vanguardia ganadera bonaerense (1856–1900)* (Buenos Aires: Siglo XXI Editores Argentina, 2005), chapter 5.

23. Roseveare, *Grasslands,* 22.

24. Stephen Bell, *Campanha Gaúcha: A Brazilian Ranching System, 1850–1920* (Palo Alto: Stanford University Press, 1998), 128–132.

25. Quoted in Roseveare, *Grasslands,* 36.

26. Roseveare, *Grasslands,* 28.

27. Robert W. Wilcox, "Cattle and Environment in the Pantanal of Mato Grosso, Brazil, 1870–1970," *Agricultural History* 66, no. 2 (1992): 243–244.

28. Robert W. Wilcox, "'The Law of the Least Effort': Cattle Ranching and the Environment in the Savanna of Mato Grosso, Brazil, 1900–1980," *Environmental History* 4, no. 3 (1999): 352; Chase quoted in Roseveare, *Grasslands,* 123.

29. Francis W. Pennell, "A Botanical Expedition to Colombia," *Journal of the New York Botanical Garden* 19, no. 222 (1918): 134.

30. Roseveare, *Grasslands,* 114, 135.

31. Wilcox, "'Law of Least Effort,'" 357–358.

32. Robert W. Wilcox, *Cattle in the Backlands: Mato Grosso and the Evolution of Ranching in the Brazilian Tropics* (Austin: University of Texas Press, 2017), 152–153.

33. William Cronon, *Nature's Metropolis: Chicago and the Great West* (New York: W.W. Norton, 1991), 249–250.

34. Andrew Sluyter, *Black Ranching Frontiers: African Cattle Herders of the Atlantic World, 1500–1900* (New Haven: Yale University Press, 2012), 183.

35. Little has been written about slaughterhouses in Latin America. Two important exceptions are Jeffrey Pilcher, *The Sausage Revolution: Public Health, Private Enterprise, and Meat in Mexico City, 1890–1917* (Albuquerque: University of New Mexico Press, 2006), and Maria-Aparecida Lopes, "Struggles over an 'Old, Nasty, and Inconvenient Monopoly': Municipal Slaughterhouses and the Meat Industry in Rio de Janeiro, 1880–1920s," *Journal of Latin American Studies* 47, no. 2 (May 2015): 349–376.

36. Robert C. Williams, *Export Agriculture and the Crisis in Central America* (Chapel Hill: University of North Carolina Press, 1986), 204.

37. David Kaimowitz, *Livestock and Deforestation in Central America in the 1980s and 1990s: A Policy Perspective* (Jakarta: Center for International Forestry Research, 1996), 11–12.

38. Kaimowitz, *Livestock and Deforestation,* 6, 12.

39. Norman Myers, "The Hamburger Connection: How Central America's Forests Become North America's Hamburgers," *Ambio* 10, no. 1 (1981): 2–8.

40. FAOSTAT (Food and Agriculture Organization of the United Nations), "Food Balance–Food Supply–Livestock and Fish Primary Equivalents," accessed 21 May 2014, http://faostat3.fao.org/faostat-gateway/go/to/download/FB/CL/E.

41. Susanna B. Hecht, "Cattle Ranching in Amazonia: Political and Ecological Considerations," in *Frontier Expansion in Amazonia,* ed. Marianne Schmink and Charles H. Wood (Gainesville: University Press of Florida, 1984), 368–371.

42. Instituto Brasileira de Geografia e Estatística, Conselho Nacional de Estatística, Serviço Nacional de Recenseamento, *Série nacional,* vol. 2, *Brasil, censo agrícola* (Rio de Janeiro: IBGE, 1956), 49.

43. Sergio Margulis, *Causes of Deforestation of the Brazilian Amazon,* World Bank Working Paper No. 22 (Washington, DC: The World Bank, 2004), 6, 10.

44. Mauro Augusto dos Santos, Alisson Barbieri, José Alberto Magno de Carvalho, and Carla Jorge Machado, *O cerrado brasileiro: notas para estudo* (Belo Horizonte: UFMG/ Cedeplar, 2010), 6, 8; Carlos A. Klink and Ricardo B. Machado, "A conservação do Cerrado brasileiro," *Magadiversidade* 1, no. 1 (2005): 149.

45. Respectively, Empresa Brasileira de Pesquisa Agropecuária; Instituto Nacional de Investigaciones Forestales, Agrícolas y Pecuarias; Centro Internacional de Agricultura Tropical.

46. A. da S. Mariante, M. do S. M. Albuquerque, A. A. do Egito, and C. McManus, "Advances in the Brazilian Animal Genetic Resources Conservation Programme," *Animal Genetic Resources Information* 25 (1999): 107–121.

47. Robert W. Wilcox, "Zebu's Elbows: Cattle Breeding and the Environment in Central Brazil, 1890–1960," in *Territories, Commodities and Knowledges: Latin American Environmental History in the Nineteenth and Twentieth Centuries,* ed. Christian Brannstrom (London: Institute for the Study of the Americas, 2004), 218–246.

48. Hugo H. Montaldo, Eduardo Casas, José Bento Sterman Ferraz, Vicente E. Vega-Murillo, and Sergio Iván Román-Ponce, "Opportunities and Challenges from the Use of Genomic Selection for Beef Cattle Breeding in Latin America," *Animal Frontiers* 2, no. 1 (2012): 23. The authors note that this is 29 percent of the entire global herd.

49. Quoted in Richard Slatta, *Cowboys of the Americas* (New Haven: Yale University Press, 1990), 40.

50. Domingo Faustino Sarmiento, *Facundo: Civilization and Barbarism,* trans. Kathleen Ross (Berkeley: University of California Press, 2003).

51. The importance of draft oxen to plantation economies should not be overlooked. In Honduras, one-fifth of (nonforested) United Fruit Company banana plantation lands were in pasture, while on large Colombian coffee estates the area in grass could equal or double that of coffee. Oxen were also central to Cuban sugar production even after the extension of railway lines. See John Soluri, *Banana Cultures: Agriculture, Con-*

sumption, and Environmental Change in Honduras and the United States (Austin: University of Texas Press, 2006), 50; Marco Palacios, *Coffee in Colombia, 1850–1970: An Economic, Social, and Political History* (Cambridge: Cambridge University Press; 1980), 35, 94; Reinaldo Funes, "Animal Labor and Protection in Cuba," in *Centering Animals in Latin American History,* ed. Martha Few and Zeb Tortorici (Durham: Duke University Press, 2013), 212–219.

52. Rafael Ospina Pérez, "Elección del ganado vacuno seleccionado para mejorar el antioqueño," *Boletín Agrícola* 1, no. 9 (1918): 345.

53. Sally Humphries, "Milk Cows, Migrants, and Land Markets: Unraveling the Complexities of Forest-to-Pasture Conversion in Northern Honduras," *Economic Development and Cultural Change* 47, no. 1 (1998): 95–124.

54. Shawn Van Ausdal, "Un mosaico cambiante: notas sobre una geografía histórica de la ganadería en Colombia, 1850–1950," in *El poder de la carne. Historias de ganaderías en la primera mitad del siglo XX,* ed. Alberto Flórez (Bogotá: Universidad Javeriana, 2008), 81–94.

55. Alejandro Reyes, "Entrevista con James Parsons," *Estudios Sociales* 1 (1986): 210.

56. Claudio Lomnitz-Adler, *Exits From the Labyrinth: Culture and Ideology in the Mexican National Space* (Berkeley: University of California Press, 1992), 170. See also Peter Rivière, *The Forgotten Frontier: Ranchers of North Brazil* (New York: Holt, Rinehart and Winston, 1972), 89–92.

57. Rockefeller Archives (Sleepy Hollow, NY), Record Group 2, Series C, Box 113, Folder 855, Jesse Knight to Lawrence Rockefeller, May 3, 1943.

58. James Nations, "Terrestrial Impacts in Mexico and Central America," in *Development or Destruction: The Conversion of Tropical Forest to Pasture in Latin America,* ed. Theodore Downing, Susanna B. Hecht, H. A. Pearson, and Carmen Garcia Downing (Boulder: Westview Press, 1992), 194.

59. Susanna Hecht, "Environment, Development, and Politics: Capital Accumulation and the Livestock Sector in Eastern Amazonia," *World Development* 13, no. 6 (1985): 678.

60. Robert Walker et al., "Ranching and the New Global Range: Amazônia in the 21st century," *Geoforum* 40, no. 5 (2009): 737.

61. Walker et. al., "New Global Range," 238.

62. Alexander von Humboldt and Aimé Bonpland, *Personal Narrative of Travels to the Equinoctial Regions of America during the Years 1799–1804,* vol. 3 (London: George Bell & Sons, 1908), accessed 6 April 2015, http://www.gutenberg.org/files/7254/old/qnct310.txt.

63. Roger Horowitz, Jeffery M Pilcher, and Sydney Watts, "Meat for the Multitudes: Market Culture in Paris, New York City, and Mexico City over the Long Nineteenth Century," *The American Historical Review* 109, no. 4 (2004): 1066; Enriqueta Quiroz, "El consumo de carne en la ciudad de México, siglo XVIII," Online paper, Instituto de Investigaciones Dr. José María Luis Mora, Mexico, D.F. (no date), 13, accessed 22 September 2017, http://www.economia.unam.mx/amhe/memoria/simposio08/Enriqueta%20QUIROZ.pdf.

64. Barsky and Djenderedjian, *La expansión ganadera,* 357.

65. Shawn Van Ausdal, "When Beef was King: Or Why Do Colombians Eat So Little Pork?" *Revista de Estudios Sociales* 29 (2008): 97.

66. Benjamin Orlove, "Meat and Strength: The Moral Economy of a Chilean Food Riot," *Cultural Anthropology* 12, no. 2 (1997): 1–35.

67. Shawn Van Ausdal, "Reimagining the Tropical Beef Frontier and the Nation in Early Twentieth Century Colombia," in *Trading Environments: Frontiers, Commercial Knowledge and Environmental Transformation, 1820–1990*, ed. Gordon Winder and Andreas Dix (London: Routledge, 2016), 166–192.

68. Lovell Jarvis, *Livestock Development in Latin America* (Washington, DC: The World Bank, 1987), 3.

69. FAOSTAT, "Production—Live Animals" and "Trade—Crops and Livestock products," accessed 11 June 2014, http://faostat3.fao.org/faostat-gateway/go/to/download/T/*/E.

70. FAOSTAT, "Production—Live Animals," accessed 23 May 2014, http://faostat3.fao.org/faostat-gateway/go/to/download/Q/QA/E.

71. FAOSTAT, "Inputs—Land," accessed 23 May 2014, http://faostat3.fao.org/faostat-gateway/go/to/download/R/RL/E.

72. Thomas Rudel, *Tropical Forests: Paths of Destruction and Regeneration* (New York: Columbia University Press, 2005), 46.

73. T. Mitchell Aide et al., "Deforestation and Reforestation of Latin America and the Caribbean (2001–2010)," *Biotropica* 45, no. 2 (2013): 5.

74. Susanna B. Hecht, "The New Rurality: Globalization, Peasants and the Paradoxes of Landscapes," *Land Use Policy* 27, no. 2 (2010): 161–169.

75. FAOSTAT, Inputs—Land," accessed 23 May 2014, http://faostat3.fao.org/faostat-gateway/go/to/download/R/R"/E.

76. Aide et al., "Deforestation and Reforestation," 5.

77. Daniel Nepstad et al., "Slowing Amazon Deforestation through Public Policy and Interventions in Beef and Soy Supply Chains," *Science* 344, no. 6188 (2014): 1118–1123.

78. FiBL and IFOAM, *The World of Organic Agriculture: Statistics and Emerging Trends*, accessed 5 March 2015, https://www.fibl.org/fileadmin/documents/shop/1636-organic-world-2014.pdf.

79. Claus Deblitz, "Feedlots: A New Tendency in Global Beef Production?" Agri Benchmark, Beef and Sheep Network, working paper 2/2011, updated July 2012, p. 2, PDF, accessed 17 February 2017, http://www.agribenchmark.org/beef-and-sheep/publications-and-projects/working-paper-series.html; Derrell S. Peel, Kenneth H. Mathews, Jr., and Rachel J. Johnson, "Trade, the Expanding Mexican Beef Industry, and Feedlot and Stocker Cattle Production in Mexico," *Economic Research Service*, U.S. Department of Agriculture, August 2011, 1–24, PDF, accessed 17 February 2017, https://www.ers.usda.gov/webdocs/publications/ldpm20601/6818_ldpm20601.pdf; Danilo Domingues Millen, Rodrigo Dias Lauritano Pacheco, Paula M. Meyer, Paulo H. Mazza Rodrigues, and Mario De Beni Arrigoni, "Current Outlook and Future Perspectives of Beef Production in Brazil," *Animal Frontiers* 1, no. 2 (2011), 46, 47.

80. A. R. García, S. N. Fleite, D. Vazquez Pugliese, and A. F. de Iorio, "Feedlots and Pollution: A Growing Threat to Water Resources of Agro-Production Zone in Argentina," *Environmental Science and Technology* 47, no. 21 (2013): 11932–11933; "Is Feedlot Beef Bad for the Environment?," *The Wall Street Journal*, 12 July 2015, accessed 17 February 2017, https://www.wsj.com/articles/is-feedlot-beef-bad-for-the-environment-1436757037; Natalia Amigo, Elsa Mercado, Adriana Bentancor, Pallavi Singh, Daniel Vilte, Elisabeth Gerhardt, Elsa Zotta, Cristina Ibarra, Shannon D. Manning,

Mariano Larzábal, and Angel Cataldi, "Clade 8 and Clade 6 Strains of Escherichia coli O157:H7 from Cattle in Argentina have Hypervirulent-Like Phenotypes," *PloS ONE* 10, no. 6 (June 1, 2015): 1–17, PDF, accessed 27 February 2017, http://journals.plos .org/plosone/article/file?id=10.1371/journal.pone.0127710&type=printable.

81. Henning Steinfeld, Pierre Gerber, Tom Wassenaar, Vincent Castel, Mauricio Rosales, and Cees de Haan, *Livestock's Long Shadow: Environmental Issues and Options* (Rome: Food and Agriculture Organization of the United Nations, 2006): xxi–xxiii.

CHAPTER 9

Extraction Stories

Workers, Nature, and Communities
in the Mining and Oil Industries

Myrna I. Santiago

In 1969, labor activist Sergio Almaraz described the desolation he witnessed in a Bolivian mining camp:

> The poverty in the mines has its own attendants; permanently wrapped in wind and cold, curiously ignoring man. It has no color. Nature has dressed itself in gray. The mineral, having contaminated the earth's belly, has turned it into a wasteland. At an altitude of four or five thousand meters, where not even wild grass grows, the mining camp is located. The mountain, irritated by man, wants to expel him. From that mineralized belly, poisoned water flows. In the mining tunnels, the continuous dripping of a yellowish and stinky liquid called *copajira*, burns the miners' clothes. Hundreds of kilometers below, where there are rivers and fish, death follows in the shape of liquid poison from the mines' detritus. The minerals are extracted and cleaned, but the earth gets spoiled. Wealth becomes misery. And right there, in that freezing cold, seeking protection in the mountain's lap, where not even weeds dare to stray, are the miners. Camps lined up with the symmetry of prisons, short huts, walls of mud and rock, zinc roofs, dirt floors.... The *pampa* winds sneak in through the cracks and the family, squeezed in improvised beds— generally no more than crude leather—if it doesn't freeze, it risks asphyxia. Hidden behind those walls are the people of hunger and of sick lungs.[1]

Almaraz captured a stark reality: the inhospitable ecology of the high mountains where miners and nature became entangled in an intimate and punishing relationship with disastrous consequences for both. Nature and human beings—who are part of nature but also separate from it—experienced the biological, social, and environmental sequelae of extraction under capitalist regimes typical of the nineteenth and twentieth centuries. Notions of progress and modernization espoused by metropolitan elites dissolved in the thin

Andean air like rock under mercury, replaced by the dual exploitation of labor and nature that disfigured both. As a result, extraction sites doubled as locations of social struggle. The miners, shivering from the combination of nature's mountain climate and the poverty created by capitalist labor relations, their lungs scarred by inhaled particles loosened from the earth they excavated, acted. They identified the mining companies as the agents who brought them to the arid sierras and caused their misery. They organized unions and strikes, making possible the nationalization of Bolivia's tin mines in 1952. Later they resisted the dictatorship in power when Almaraz visited the camps.[2] They embodied the complicated reality of communities that shared geographies with extractive industries and rebelled as a result.

The history of mining and petroleum in Latin America in the nineteenth and twentieth centuries demonstrates that extraction follows nature, and conflict follows extraction. Nature determined the geographical territories where mining and oil companies dug or drilled, often localities with challenging ecosystems lacking the population necessary to mine successfully. Those landscapes required immigrant labor. Upon their arrival, workers found harsh environments that exposed them to nature in addition to the dangers inherent in digging into the earth. The communities that formed around extraction sites endured the hardships of inhabiting unfamiliar landscapes in circumstances of poverty: the high altitudes of the mountains, the tropical humidity of the rainforests, the searing heat of the deserts. All shared the environmental effects of extraction: soil, air, and water pollution; rampant disease; ingestion of toxic chemicals; contamination of the food chain; and, in the case of petroleum, extreme vulnerability to fire. By design, however, foreign and domestic capitalist enterprises prioritized investment in technology and infrastructure. Like their Spanish colonial mining forefathers,[3] nineteenth- and twentieth-century extractive companies minimized the resources destined for workers and local communities. That behavior translated into the scene Almaraz encountered in 1969. Rarely did those living in the shadow of extractive industries reap the benefits of their labor, their bodily sacrifice, and the radically transformed landscapes of extraction. The subterranean wealth that the miners and oil workers ripped from the earth never stayed on site. The affluence accrued elsewhere, in the glittering foreign metropolises of London and New York or national capitals—Lima, São Paulo, Caracas.

In time, workers or community members tired of the grueling social and environmental milieus that extractive companies subjected them to and mobilized in protest. When those movements occurred, what forms they took, how they unfolded, and what they accomplished differed by country and time period. Over the nineteenth and twentieth centuries, extraction in Latin America comprised every nation and the diversity of nature: nitrates, copper, tin, silver, gold, bauxite, coal, emeralds, pearls, guano, oil, gas, and, in the twenty-first century, lithium. The extractors were similarly varied: giant multinational conglomerates, national capitalists, state-owned companies, or small miners combing through abandoned mines[4] or infiltrating ecosystems previously undisturbed by mining in search of their own mythical El Dorado.[5]

Not every case can be addressed here. Space limits the choices, but the historiography also narrows the recounting of extraction stories. While some miners and petroleum workers have received scholarly attention precisely because they represented vanguard movements,[6] the environmental history of extraction is in its infancy.[7] The cases explored here—Peruvian guano, Chilean copper, Mexican, Venezuelan, and Ecuadorean oil, and "megamining"—are not meant to be paradigmatic. Rather, they illustrate how social and ecological conditions engendered socioenvironmental conflict, including "resource wars,"[8] involving different social actors as the centuries passed. What extractive communities share is a particularly conflictive intersection between nature and society. Their stories link environmental and social history and illuminate the local, hidden, and often hurtful costs of globalization and modernity over the last two hundred years.

To examine the local dynamics of extraction and conflict, four cases suffice. Peru's nineteenth-century guano extraction represents conflict without successful mobilization. Chilean mining, by contrast, became the crèche of labor radicalism leading to nationalization. The third case compares petroleum in Mexico, Venezuela, and Ecuador in the twentieth century—different contexts with conflict aplenty, heterogeneous actors, and distinct social and political results. Lastly, the chapter reviews the "super cycle" of extraction engulfing the continent today and the internationalization of antimining activism in its wake.[9]

The Brief and Violent Episode of Guano Extraction

The depletion of Peruvian guano in four decades (1840–1880) represents an intensely exploitative, conflict-ridden, and violent instance of extractive economic activity of dubious long-term benefit for the country. While the Inca utilized bird droppings as fertilizer, they avoided the Chincha islands that cormorants, boobies, and pelicans occupied off the Peruvian coast for millennia. The terrain was forbidding for humans: whipping winds, scorching temperatures, and extraordinary dryness, the perfect landscape for birds to excrete digestive waste until, as the colonial chronicler Garcilaso de la Vega observed, the "heaps of manure look like the peaks of snowy mountains."[10] As the British spied an agricultural commodity in the birds' refuse, the Peruvian military government claimed ownership over nature and declared guano property of the nation in 1841. British merchants quickly negotiated the purchase of that very nature, offering the caudillos loan advances on future profits from bird droppings.[11] However, since slavery was in its death throes and the islands were uninhabited, the procurement of labor posed a problem. The Peruvian government tackled it by shipping convicts, army deserters, and debt peons offshore. By 1849, the British resolved the matter with Chinese "coolie" labor. Contracted or kidnapped by British merchants, more than ninety-two thousand Chinese had crossed the ocean to reach the Chincha islands by 1874,[12] a voyage determined by nature as much as by capital.

The encounter between man and nature in the Chincha Islands was cruel. Nature seemed unforgiving, accosting workers with blistering sun and thrashing winds without respite or local sources of drinking water. Ammonia saturated the air and burned laborers' lungs with every breath. Working conditions were likewise brutal. Without protective equipment, workers inhaled excreta for as many hours as they took to shovel their quota of four tons of guano per day. They slipped in excrement powder, fell into the ocean, and drowned. Guano landslides buried them alive.[13]

Occupied by men subject to a near-slavery labor regime, the islands metamorphosed into violent landscapes. Workers' camps "gained a notorious reputation for fights, robbery, murder, gambling, prostitution, alcohol, and opium abuse," writes historian Gregory Cushman. Labor protests erupted. Mutinies aboard ships and unrest amid the dung erupted regularly. Overseers responded to defiance of authority with "floggings," the hallmark of labor relations under conditions of bondage. Physically ill, suffering from cultural, linguistic, psychological, and ecological dislocation, and isolated by the natural barriers islands represent, Chinese workers despaired. Suicide became prevalent, an option more effective than mobilization to escape a grim environment and the inherent violence of bonded labor.[14]

The value of fertilizer, moreover, was so high that it provoked two international conflicts—resource wars that were in essence violent struggles over nature. As Europeans modernized their agriculture, they roamed the world seeking natural additives. Thus in 1864 Spain invaded the Chinchas to take the birds' feces home. Peru repelled them by 1866, but at great expense—necessitating further exports to repay the debts incurred over the control of nature. By the 1870s, the heaps of guano had dwindled to naught, scraped bare by thousands of unfree men. However eternal the process of bird excretion is, perceived as "a natural resource," guano was exhaustible, and Peruvians knew it. Desperate for revenue, the government turned its gaze to another product of nature useful as fertilizer, the nitrates on its southern border deserts. In 1875, hurt by dwindling guano earnings and a global depression, Peru expropriated the Chilean-owned nitrate mines in Tarapacá. That decision contributed to the Pacific War (1879–1883), a devastating resource war that engulfed Peru, Chile, and Bolivia. Victorious, Chile reconfigured the region's geographical territory.

It took Peru's southern coast and Bolivia's access to the ocean, increasing its territory by one third and gaining access to "the wealth of the desert."[15] Without guano or nitrates at the end of the nineteenth century, what did Peruvians gain from exporting bird excreta?

While historians debate the details, the consensus is that Peru failed to develop despite the guano boom and the state's monopoly on nature. Between 1840 and 1878, Peru sold 12.6 metric tons of guano worth 750 million dollars.[16] European merchants and farmers, the Limeño elite, and the Peruvian government reaped the benefits from the biological rubbish that birds dumped on the islands, but none fostered national development. Guano credit bought luxury goods and financed railroad construction and the growth of a state bureaucracy and the military. But it also indebted Peru to European lenders for decades. The majority of the population, meanwhile, saw not a penny of investment in their future.[17] The islands, cleansed of guano and people, returned to the birds.

Chile: Copper and Mineral and "Gente de Lucha"

Nature had led Chileans to the desert since the 1830s, but it was the Pacific War that turned the economy toward mining. The Argentinian singer Mercedes Sosa signaled that identity in a 1969 song describing, "*Un verde Brasil besa a mi Chile, cobre y mineral*" (a green Brazil kisses my Chile, copper and mineral).[18] Simultaneously, the workers who migrated to desert landscapes made a name for themselves as "gente de lucha" (belligerent), deeply influencing Chilean politics throughout the twentieth century.[19] The northern environment was inimical for migrant workers mining nitrates (1890–1930) and copper (since 1830). At three thousand to nine thousand feet (914 to 2743 meters), the Atacama desert that Chileans took from Bolivia and Peru is one of the driest places on earth, flanked by the towering Andes cordillera, windswept and subject to extreme temperatures. There, nature had accumulated nitrate two meters deep, the treasure British companies sought when they purchased the desert from a cash-strapped Chilean government in 1882.[20] Miners blasted the desert floor with dynamite, a high-risk maneuver they followed with dangerous hand sifting of ore and the hauling of 140-kilo (300-pound) sacks on their backs to carts headed for smelters. In the smelters, workers were safe from explosions and falling boulders, but they risked bodily injury from steam, dust, boiling liquids, and grinding machinery.[21] Burns and mutilations were so frequent that "amputees and otherwise disabled nitrate workers became a common sight in Iquique, Antofagasta, and other nitrate towns."[22] Nevertheless, some fifty thousand Chileans inhabited the desert by 1920, as European fertilizer needs captured Latin American landscapes for a global capitalist economy. Fluctuations in world prices, however, translated into periodic unemployment, out-migration, and dislocation.[23]

As did guano, labor and environmental conditions in the nitrate fields fed conflict. Chilean miners, however, built a militant movement. The first strike in Atacama on record occurred in 1865. But the confrontation that left a deep imprint on the national psyche happened at Iquique in December 1907, "when thousands of striking nitrate workers (and their families) converged on the port in great order and dignity to seek redress of grievances."[24] General Roberto Silva Renard's retort was machine-gun fire that killed hundreds.[25] The Iquique massacre did not end activism in the British-owned nitrate fields, however. World War I did. As Europeans decimated each other and their landscapes, the demand for fertilizer plummeted. The Great Depression that followed permanently eclipsed it. By then, another of nature's products had emerged as the driver of Chile's economy: copper. Those miners, in turn, faced another bleak experience in nature and extraction. But they also inherited a history of combativeness.

The history of copper mining began in the 1850s. Dug into in the semiarid Norte Chico below the Atacama desert by Chilean-owned small companies, the richest veins were exhausted by the 1880s. Nomadic miners left behind pitted landscapes strewn with contaminated rock and dirt from Maipo to Copiapó. In forty years, the companies deforested nearby steppes to fuel some nine hundred smelters, polluting the air and degrading soils in the process.[26] The miners received little for their labor, toiling in environments that Bolivians would endure a century later. The Polish metallurgist Ignacio Domeyko visited the opulent Carrizal mine in the 1870s and observed that "the outside view is something sad: immense mounds of stones, large hills of rock fragments . . . here and there miserable hovels tied with reeds . . . incessantly entering and exiting [are] seminude naked men . . . with immense baskets over their shoulders."[27]

Charles Darwin, who witnessed miners at work in Coquimbo and Jahuel in the same period, noted that the companies treated the men like "truly beasts of burden." He wrote, "It was quite revolting to see the state in which they reached the mouth of the mine; their bodies bent forward, leaning with their arms on the steps, their legs bowed, their muscles quivering, the perspiration streaming from their faces over their breasts, their nostrils distended, the corners of their mouths forcibly drawn back, and the expulsion of their breath most laborious."[28]

To prevent protest, the companies enforced a disciplinary regime that consisted of surrounding the mining camps with armed guards licensed to beat workers accused of stealing ore. Into this desolate and tortuous ground, following nature's subterranean riches, Big Copper (*la gran minería*) arrived. It deepened socioecological conflict and shaped Chile's natural and political landscape in the twentieth century.

Spurred by the electricity revolution and the exploitation gap left by Euro-peans during World War I, U.S. giants Anaconda and Kennecott opened the largest mines in the world in Atacama (Chuquicamata and Potrerillos) and in el Norte Chico (El Teniente). Not intimidated by nature, the North Amer-icans unleashed massive environmental change. At El Teniente, for instance, Kennecott built mills, a smelter, an aerial tramway, offices, a mining camp for nine thousand people, and a railroad that "was literally carved out of the can-yon wall." New technology affected the mountains profoundly. An innovative method used at El Teniente, for example, entailed dynamiting the roof of the mine's tunnels until they collapsed and the mountain itself caved in.[29] At the smelters, novel technologies that submerged crushed ore in acidic water ex-tracted even minute amounts of copper, promoting mountain consumption further. Whereas in the nineteenth century miners dug four to six tons of rock per ton of copper, Big Copper tore up fifty to sixty tons of mountain per ton of metal.[30] Smelters also secreted highly toxic waste over wide territories. The Atacama smelters dumped their "tailings," the pulverized chemical paste left after leaching the copper, directly into the ocean. El Teniente deposited its detritus into two local rivers until farmers objected. Thereafter, Kennecott engineered the *tranque,* an embankment or dam-like wall that contained the tailings. As nature would have it, tranques were vulnerable to the tectonic movements that created the Andes in the first place. In 1928, for example, an earthquake cracked El Teniente's tranques, dispatching three million tons of slag down the Cachapoal River, burying two mining camps and a village along the way. The story repeated itself in 1966, when a 7.6 earthquake buck-led the tranques in four communities in Valparaíso. The aptly named village of El Cobre was buried by an avalanche of mud and rock, killing its 350 to 400 inhabitants, half of whom were children.[31]

The scarring of the landscape was inscribed in and on workers' bodies too. In the 1940s, conditions at El Teniente were "inhuman and brutish," the min-ister of labor declared. Men spent ten or more hours pounding the earth in claustrophobic spaces without proper protective equipment or ventilation, breathing air impregnated with toxic dust and smoke from repeated dynamite blasts, their bodies painfully contorted and subjected to extreme heat. Miners' muscles cramped painfully. Their hearing disappeared in months. Bronchial infections, asthma, tuberculosis, and silicosis corroded their lungs. Adding insult to injury, Kennecott established a policy of firing and blacklisting men with silicosis as being "physically unfit." Accidents stalked the miners: chemical burns, bone-breaking falls, limb-tearing explosions, toxic gas poisonings, elec-tric shocks, and tunnel fires and collapses maimed and killed them routinely.[32]

The strenuous nature of mining combined with exploitative working condi-tions created conflict of major consequences for Chile. Instead of sinking into despondency as countless guano workers once did, copper miners adopted

militant nationalist and anti-imperialist identities with roots in the desert's nitrates. Nitrate miners dismissed during the Depression migrated to the copper
mines, sharing stories of "epic struggles" in the tough natural environment of
"the northern sea, the desert, the nitrate fields." Fearless in their masculinity,
the men became a leading segment among Chile's working class. Envisioning
themselves as "gente de lucha" and protectors of national honor, their families,
and their communities, the miners called for an alternative to transnational
and private ownership of the mines.[33] By the 1930s, they demanded the nationalization of copper.[34] El Teniente workers maintained that demand in the
hard-fought strikes of 1946–1947. In 1954, the copper miners organized an
industry-wide strike that ended with the government arresting the leadership
and militarizing the desert through a stage of siege. Finally, in 1965, a general
strike shut down the country and decades of militant activism began to pay
off. President Eduardo Frei started the "Chileanization" of Big Copper in 1967,
not via expropriation but by buying 51 percent of the companies. The miners
kept up the pressure, becoming key actors in the election of socialist candidate
Salvador Allende to the presidency in 1970.[35] When Allende finalized the nationalization of the copper, coal, and nitrate mines in December 1970, Chilean
miners could feel satisfied about their contribution to national control over the
country's economic heart—in effect, the nationalization of nature.

The Chilean socialist experiment however was short-lived. The challenge
Chilean socialism posed to capitalism in the Americas ended on 11 September 1973, when General Augusto Pinochet overthrew Allende. Pinochet kept
mining under state control. He militarized extraction, however, with generals
running the mines. The regime's attitude toward workers was diaphanously
explained when General Oscar Bonilla told labor leaders, "Stop using the word
'demand'; don't forget that this is a dictatorship."[36]

After a decade of ruthless repression, El Teniente's union led the first protest against the dictator. Alongside textile workers and public employees, the
miners staged a successful one-day general strike on 11 May 1983. They remained active in oppositional politics until Pinochet was forced to leave office
in 1990.[37]

Guano and copper share similarities and differences across the nineteenth
and twentieth centuries. Both were instrumental for Western capitalist development and produced intensive local environmental destruction. But while
island guano was exhausted quickly, Chile's vast northern desert sustained
extraction for over a century. Workers felt the full impact of a bitter environment and abusive treatment in both cases. The social conflicts characteristic of
extractive industries soon followed. Guano workers faced the supervisor's lash
and Chilean miners the soldiers' guns, but the effects were different. Unlike
immigrant Chinese laborers, migrant Chilean miners forged a vibrant union
movement, changed the ownership of the mines, and influenced the fate of the

nation, even if dictatorship cut their victories short. Further north, meanwhile, another extractive activity, petroleum, was chasing nature and generating its own frictions.

Petroleum, Conflict, and Activism in Tropical Environments

Petroleum extraction changed the world in ways that are obvious compared to guano or copper. From its humble beginnings in the mid-nineteenth century as an illumination fuel, oil revolutionized the globe in one hundred years as transportation fuel, fertilizer, pesticide, and more. No surprise, then, that oil moves armies across oceans and over mountains and into deserts in wars over its control. Following the pattern of extraction, oil spurred deep social and environmental changes that led to strife. Throughout the twentieth century, petroleum companies exploited environment and people as wantonly as any mining company, with the additional characteristic that the material nature of oil makes it extraordinarily dangerous. It is easy to spill and flammable, which means hydrocarbons expose larger territories to hazards than mining. Mobilization, therefore, included social sectors beyond labor. Local communities, indigenous peoples, and, by century's end, environmental organizations also raised their voices, using new discourses to challenge industry practices. The cases of Mexico, Venezuela, and Ecuador exemplify those shifting patterns.

Mexico was the first country in Latin America to extract the exceedingly ignitable product of millennia of underground fossil decomposition that is petroleum. Mexico was also one of the first countries in the world to nationalize oil. The Mexican Revolution (1910–1940) provided the context for such a bold move, but it was the companies' exploitation of nature and labor that provoked workers into revolutionary politics and expropriation. As was the case with guano and copper, nature called the shots on the location of hydrocarbon deposits. In Mexico, such place was the northernmost continental tropical rainforest, the Huasteca, northern Veracruz, on the Gulf coast. Oil had seeped through the earth's crust onto the surface since pre-Columbian days. As the Inca used guano, the Huastec also utilized naturally emanating pitch. They sealed canoes, painted pottery, and burned *chapopote* as incense.[38] Similarly, hydrocarbons were not commodified until European and U.S. entrepreneurs invented uses and created a market for them. In Mexico, an ecological disaster advertised the existence of petroleum when a well owned by the Englishman Weetman Pearson exploded in July 1908. It drenched marshes, mangroves, and streams and burned for fifty-six days, expelling clouds of sulfur dioxide that killed humans and animals. The amount of petroleum that blanketed the landscape, including the waters of the Mexican Gulf, made it the worst oil spill in the Western hemisphere to date.[39] The damage to nature was without precedent and permanent. In 1929, two decades after the explosion, a newspaper reported that on location, "in a broad extension, everything is charred, ashen in color, shuddering the soul. The light itself has livid reflections and there are no vestiges of life. No foliage on the trees, no birds in the sky, no living being in the solitude of the place. . . . There is nothing left but an awful lagoon. . . . But the grayish waters seem to boil like lead, emanating gases that close up, for an enormous radius, fill the air. All the trees have been robbed of their greenery, burned and seem to raise up to the heavens, with anguishing contortions, their bare and gray branches."[40]

In 2005, the lagoon was still tinted with rainbow oil swirls and provoked headaches among passers-by. In 2017, the "Dos Bocas" pool, perfectly round like an open mouth, was easily identified on Google Earth, testimony to the destructive power of crude oil and nature.[41] But in 1908, European and U.S. oil companies celebrated the devastation as proof of imminent prosperity. Mexico entered a frenetic petroleum export period that lasted until 1921, when the easily accessible oil had been sucked dry—sufficient time to shape a radicalized labor force.

Mexican oil workers emerged as political radicals in the context of exploitative labor conditions, an inhospitable environment, and the Mexican Revolution. Starting in the late 1910s, the workers organized unions to demand better wages and an eight-hour workday. In response to hazardous and insalubrious work and natural environments, they also demanded safety protocols and equipment, health care, sanitary working and living conditions, compensation for accidents on the job, and a stop to discriminatory treatment by foreign managers[42] (see illustration 9.1). Infused with the spirit of contemporary anarchist, socialist, and communist critiques of capitalism, Mexican oil workers forged confrontational masculine identities and positioned themselves on the left wing of the revolution. Their reality demonstrated that class position affected the experience humans had in and with nature. Those at the bottom of the social hierarchy were systemically more exposed to the natural and man-made dangers of extraction, such as burns and asphyxia, than those higher up the corporate ladder.[43] In the aftermath of the bust that took place in 1921, those men still employed consolidated unions while the Great Depression gave impetus to alternatives to foreign and private ownership. In 1937 union men proposed that the government expropriate the oil field at Poza Rica, Veracruz, for them to run as a cooperative.[44]

The workers accomplished some of their goals when President Lázaro Cárdenas expropriated the oil companies in 1938. Militantly nationalist, the government nationalized oil more than a decade before the Bolivian Revolution of 1952 expropriated the mines and three decades before the beginning of the Chileanization of Big Copper. Tied to a nationalist development project that included responsiveness to labor, the acquisition of the oil industry led

Illustration 9.1. Mexican workers at an earthen oil dam at Potrero del Llano, northern Veracruz, circa 1914.

Courtesy of DeGolyer Library, Southern Methodist University.

to gains in workers' health, safety, and social benefits.[45] In time, the state incorporated the oil workers' union into its apparatus and blunted its militant force, while the national oil company, *Petróleos Mexicanos*, financed much of the country's diversified economic development. What state ownership of hydrocarbons would mean for environmental policy, however, was not obvious in 1938.

As the new war technologies of World War I (airplanes, tanks, and battleships) converted petroleum into a military commodity of strategic value, Venezuela became the second Latin American nation to extract hydrocarbons and undergo socioecological turmoil in the process. In the 1920s, British and American companies migrated from Mexico to Venezuela, transforming Lake Maracaibo and the country's economy for the rest of the century. Utilizing the same practices they did in Mexico, the companies caused comparable ecological calamities in Lake Maracaibo. By 1926, as the United States Consul Alexander Sloan wrote, "The waters . . . are covered with oil which is carried up to the shore by the waves and blackens all vegetation which it touches. . . . Along the shore are rows of palm trees whose leaves are so covered with oil that they droop to the ground, or rows of what used to be palm trees but which are now stumps denuded of all foliage by fire. . . . Oil is spattered everywhere on the vegetation and the houses, it is carried into the offices and dwellings on the shoes or the clothes of those who enter."[46]

Fishermen quickly blamed fish deaths on the pollution, while farmers decried the loss of cacao trees to spills and fires, and ranchers complained that their cattle could no longer drink lake water.[47]

The oil companies, likewise, introduced the devastating force of oil fires to Venezuela. Gas leaked, wells blew up, and workers burned alive. The principal lakeside town, Lagunillas, burned to the ground three times between 1928 and 1929 when spills caught fire and devoured the wooden structures.[48] Thus panicked, nearby Cabinas residents banded together in 1928 to protest how the Venezuelan Oil Concessions Corp was drilling within town limits, "63 meters away from the . . . plaza, 87 meters from the church, 24 meters from the [store] and 97 meters from a Chinese restaurant," putting everyone at risk of fire.[49]

Protest in Venezuela, however, happened in a context of dictatorship. That meant repression accompanied oil extraction. Learning from the Mexican experience, the dictator Juan Vicente Gómez decided against building refineries on Venezuelan soil "in order to avoid large concentrations of workers with their attendant labor problems."[50] Nevertheless, the oil field workers led the labor movement several times during the first half of the twentieth century.

Despite Gómez's ban on unions, the oil workers organized the first major strike in Venezuelan history in July 1925. Their demands included the eighthour workday and more than the dollar-a-day the companies paid. Like the Mexican oil workers, Venezuelan men decried the food shortages in the fields, the racism and discrimination their foreign bosses practiced, the lack of compensation for accidents, the high rates of malaria, and the lack of health care and housing.[51] Goméz's response was to use troops to break the strike, blaming the workers' activism on "Mexican agitators."[52] A decade would pass before the workers rebuilt their organizations and protested again.

In December 1936, the oil workers led a national movement for the second time. The political demands focused on the end of autocratic rule, while the labor demands included health care, sanitation, safety, and potable water. Venezuelan workers across the country sympathized and organized a general strike in repudiation of the companies and "militarism." The government called out the army once again and opened fire on strikers at the Mene Grande oil field, killing five. In January 1937, the government arrested the union leadership and exiled them, destroying the movement.[53]

The state then modified policy and some tactics. In 1937, the government made the momentous ecological decision "to sow petroleum," that is, to make oil the core of the economy. To guarantee a successful "harvest," however, labor peace was imperative. Thus the government became paternalistic toward

workers, mediating between the companies and the unions, seeking to neu-
tralize labor militancy. Between 1946 and 1948, the government brokered the
first collective bargaining contracts. But when the workers organized their
third strike in 1950, the government crushed the unions mercilessly, effectively
depoliticizing them.[54]

Sowing petroleum became "the excrement of the devil" by 1976.[55] The
"petrolization" of the economy meant not only rampant environmental degradation, but also the queasy feeling among Venezuelans that agriculture had
been sacrificed on the altar of oil, in addition to other social and economic
distortions. Food insecurity plagued the nation, leading to a food crisis in 1977
and riots in the 1980s.[56] The nationalization of oil in 1976 thus owed little to
labor activism in comparison to Mexico, Bolivia, or Chile. Subsequently, the
Venezuelan state demobilized oil workers as a social force.

Mexicans learned that nationalization had real ecological limits too, in the
second half of the century. PEMEX showed no more regard for nature and
communities than foreign private enterprise. The company deforested and
contaminated soil, water, and air wherever it found hydrocarbons. Chasing
nature into highly challenging terrain, moreover, PEMEX began drilling offshore in the 1970s. Catastrophe ensued when on 3 June 1979, the well "Ixtoc I"
exploded and caught fire. The earth hemorrhaged petroleum into the Gulf for
290 days, polluting some fifteen thousand square kilometers of water and soil-
ing beaches from Yucatán to Texas.[57] Thus the destructive power of extraction
spread to fishing communities and marine life across the entire Gulf.

In Ecuador, oil followed similar patterns but brought different local historical actors to the forefront. Texaco and Gulf found oil in 1967 in Amazonia,
in territory occupied by the Quijos, Canelos, Quichua, Huaorani, Siona-
Secoya, and Cofán indigenous groups. A boom ensued, reenacting the historical script of extractive industries colonizing relatively isolated and forbidding
environments: immigration, haphazard urbanization, deforestation, and the
fragmentation and loss of habitats leading to wildlife reduction. The unique
contribution of petroleum to ecological destruction, the oil spill, appeared
as well. Spurting from wells, pipelines, tanks, and oil "waste" pits, petroleum
blanketed the forest floor and water supplies.[58]

Ecuador also showed the ancillary effects of road construction in rainforest ecosystems. A flood of landless subsistence farmers followed the oil trail,
adding pressure to nature and indigenous communities already under siege.
Loggers arrived and accelerated deforestation. Ranchers filed in, replacing
trees with grass and livestock. Violence between indigenous peoples and immigrants exploded, while Texaco built a 315-mile pipeline that climbed over
the Andes to the coast to export the crude oil.[59] In 1992 the company left, but
its waste resurfaced embodied in the communities' health as rashes, miscar-
riages, birth defects, and cancer.[60]

Just as workers had done elsewhere, the indigenous peoples of *Oriente* organized. They did not seek nationalization, however. They focused instead on recovering their health and ecologies, using innovative mobilization tactics. One was to ally with Ecuadorean and North American environmentalists and human rights lawyers to file suit against Texaco in U.S. court. In 1993, representing thirty thousand Amazonians, the plaintiffs claimed that Texaco had poured more than eighteen billion gallons of petroleum waste into waterways, discarded more rubbish in over nine hundred pits, flared natural gas that contaminated the air and killed vegetation (see illustration 9.2), and spilled over seventeen million gallons of oil along the pipeline.[61] The plaintiffs demanded compensation for damages to nature and the communities' health, in addition to effective clean-up and environmental restoration. Multiple legal maneuvers later, Chevron, having merged with Texaco in 2001, moved the case to Ecuador in 2003, hoping to win handily. However, the court ruled that Chevron owed indigenous groups nineteen billion dollars. In 2013, the company appealed in a New York court, accusing the Amazonian legal team of extortion and "racketeering," like a mafia.[62] In March 2014, judge Lewis A. Kaplan dismissed Ecuador's jurisdiction and award, freeing Chevron from all responsibilities.[63] Whether the communities can afford to appeal remains a question.

Thus Ecuador shows that the effects of oil extraction on local ecosystems and human bodies linger. Two decades after Texaco's departure and dismissal of all its workers, Amazonia has not recovered from the ecological damage the company caused. Indigenous people, likewise, continue falling ill from the pollution the company left behind. Nevertheless, the lawsuit against Texaco points to changes in the extraction complex at the turn of the twenty-first century. New actors mobilized in struggles over extraction, using new tactics and new tools, forging international alliances, and making explicit environmental demands. They scored victories, suffered defeats, and learned to be even more vigilant—significant developments as a "super cycle"[64] of extraction intensified the exploitation of Latin American nature.

"Megamining": New Challenges and New Actors

At the turn of the twenty-first century, mining and oil projects of unprecedented scale accosted the Latin American landscape. Local resource wars flared across the continent in tandem, as diverse communities felt or feared the ecological impacts of mining and oil extraction. Indebted in the 1980s, in the 1990s Latin American governments forsook extractive nationalism in favor of neoliberal policies. Under pressure from international lending agencies, governments jettisoned national ownership of nature, legislating investor protections and granting truly extraordinary access to nature to extraction com-

Illustration 9.2. Ermenegildo Criollo, Cofán leader, in Amazonia, Ecuador.

Photo by Lou Dematteis (2008).

panies. Spurred by booming high-tech industries and the appearance of China on the superpower stage, new extractive enterprises rushed in to meet the demand. Brazilian, Argentinian, and Australian corporations competed against British and U.S. companies for Latin America's landscape. And although indigenous peoples practiced mining and Spanish colonial authorities expanded on the practice, the sheer size of Canadian mines and their disfiguring open pits inspired the term "megamining" to distinguish it from all previous historical experience.[65]

The current extraction cycle intensifies all aspects of mining. The amount of territory under concession has no historical precedent. By 2011, Peru, for example, had granted 13.6 percent of its national territory to 410 mining companies, some 17.4 million hectares.[66] Argentina, similarly, had conceded ten square kilometers for a single project, the Cordón Esquel gold mine in Patagonia.[67] Chile, meanwhile, topped the list with three hundred million hectares of mining lands in private hands in 2000.[68] Conflict over land, therefore, skyrocketed throughout the continent.[69]

"Waterscapes," the term scholars have coined to analyze social conflict over water, are at the center of conflict today as well. Technologies such as open pit mining are water-intensive and highly polluting. Heavy metals, mercury, cyanide, and arsenic measured in tons leach into the water table and contaminate surface water. Megamining, moreover, casts a long ecological shadow on populations distant from the mines, as the companies extract the water those communities utilize for their survival.[70] In the petroleum industry, likewise, complex technologies such as ultra-deep offshore drilling, hydraulic fracturing ("fracking"), and acidizing are shaping waterscapes and reshaping nature in Mexico, just as they are doing in the United States. The process of injecting water, sand, and chemicals miles underground to crack the rocks that trap oil or gas means competition for scarce water, contamination of water tables, and even earthquake swarms.[71] Offshore drilling, as the 2010 British Petroleum spill in the Mexican Gulf demonstrated, can have an exceptionally high environmental cost.

Extraction-related mobilization and conflict, therefore, is at a historic high. The leading actors are no longer labor unions, however, as they lost much of their power in the second half of the twentieth century under repression and cooptation. Instead, community groups, environmentalists, indigenous peoples, women, and religious organizations head the resistance against megamining. The language they use is also new. While nationalist appeals have not disappeared altogether, contemporary social actors have elaborated novel environmental discourses to analyze extractive praxis. Notions such as the rights of nature, respect for Pachamama (Mother Earth), human rights as inclusive of nature protection and indigenous identities, and environmental justice are part of the continent's lexicon.[72] Modes of activism, too, have evolved.

Strikes and mass demonstrations have broadened to include road blockades and transnational networking. Organizing now occurs on a global scale, with grassroots activists joining efforts to protect nature and all its creatures worldwide.[73] The magnitude of the challenges demands no less.

Conclusion

An environmental history of extraction in Latin America in the nineteenth and twentieth centuries reveals a landscape forged in social struggle. Ore and oil led companies to places where nature abounded but humans did not: islands, mountains, deserts, and rainforests. Labor history illuminates how workers experienced hostile natural environments, exploitative capitalist labor relations, and exceedingly dangerous work. The intersection of environmental and social history shows that neither communities nor landscapes were spared socioecological transformations and that extraction sites became landscapes of conflict, often landscapes of violence, which, in turn, could give birth to movements that influenced the history of their nations. A joint environmental and social history approach to extractive industry leads to histories that expose the ecological roots of social conflict in this sector of the Latin American economy. The result is a set of stories that illustrate some of the complexity of the human experience in nature across time and geographical space. Yet they represent but a sample. The environmental and social history of extraction in Latin America still needs to be written. And as global capitalism demands more of Latin American landscapes, the human stories continue to accumulate—challenging historians to dig and recover the ecological memory of five hundred years of mining.

Myrna I. Santiago received her Ph.D. in history from the University of California at Berkeley. She teaches Latin American history at Saint Mary's College of California. Her research focuses on environment, labor, and social history. Her book, *The Ecology of Oil: Labor, Environment, and the Mexican Revolution, 1900–1938,* won the 2007 Elinor Melville Prize for the Best Book in Latin American Environmental History and the 2007 Bryce Wood Book Award for the Outstanding Book on Latin America in the Social Sciences and the Humanities. She has published in the *Journal of American History, Environmental History,* and *Journal of Women's History.*

Notes

1. Quoted in Gregorio Iriarte, *Los mineros: Sus luchas, frustraciones y esperanzas* (La Paz, Bolivia: Ediciones Puerta del Sol, 1982), 30–31.

2. June Nash, *We Eat the Mines and the Mines Eat Us: Dependency and Exploitation in Bolivian Tin Mines* (New York: Columbia University Press, 1979), 259, 280–281.

3. See Kendall W. Brown, *A History of Mining in Latin America from the Colonial Era to the Present* (Albuquerque: University of New Mexico Press, 2012).

4. See José Miguel Sánchez C. and Sara María Enríquez B., "Impacto ambiental de la pequeña y mediana minería en Chile," Paper prepared for the World Bank, December 1996; Leonardo Güiza, "La pequeña minería en Colombia: Una actividad no tan pequeña," *Dyna* 80, no. 181 (October 2013): 109–117.

5. See Sebastião Salgado, *Workers: An Archaeology of the Industrial Age* (New York: Aperture Foundation and Eastman Kodak Company, 1993), 300–319.

6. See Alberto Flores Galindo, *Los mineros de la Cerro de Pasco, 1900–1930* (Lima: Pontificia Universidad Católica del Perú, 1974); Héctor Lucena R., *Las relaciones laborales en Venezuela: el movimiento obrero petrolero: proceso de formación y desarrollo* (Caracas: Ediciones Centauro, 1982); Thomas Miller Klubock, *Contested Communities: Class, Gender, and Politics in Chile's El Teniente Copper Mine 1904–1951* (Durham: Duke University Press, 1998); Lief S. Adleson, "Historia social de los obreros industriales de Tampico, 1906–1919" (Ph.D. diss., El Colegio de México, 1982); Thomas C. Greaves and William Culver, eds., *Miners and Mining in the Americas* (Dover, NH: Manchester University Press, 1985).

7. See Nikolas Kozloff, "Maracaibo Black Gold: Venezuelan Oil and Environment during the Juan Vicente Gómez Period, 1908–1935" (Ph.D. diss., Oxford University, 2002); Mauricio Folchi, "Historia ambiental de las labores de beneficio en la minería del cobre en Chile, siglos XIX y XX" (Ph.D. diss., Universidad Autónoma de Barcelona, 2006); Daviken Studnicki-Gizbert and David Schecter, "The Environmental Dynamics of a Colonial Fuel-Rush: Silver Mining and Deforestation in New Spain, 1522 to 1810," *Environmental History* 15, no. 1 (2010): 94–119; Micheline Cariño and Mario Monteforte, "De la sobreexplotación a la sustentabilidad: Nácar y perlas en la historia mundial," *El Periplo Sustentable* 12 (May 2007): 81–131; Myrna I. Santiago, *The Ecology of Oil: Environment, Labor, and the Mexican Revolution* (Cambridge University Press, 2006).

8. See Michael T. Klare, *Resource Wars: The New Landscape of Global Conflict* (New York: Henry Holt and Company, 2002).

9. See Anthony Bebbington and Jeffrey Bury, eds., *Subterranean Struggles: New Dynamics of Mining, Oil, and Gas in Latin America* (Austin: University of Texas Press, 2013).

10. Quoted in Edward D. Melillo, "The First Green Revolution: Debt Peonage and the Making of the Nitrogen Fertilizer Trade, 1840–1930," *American Historical Review* 117, no. 4 (October 2012), 1037–1038.

11. Peter Findell Klarén, *Peru: Society and Nationhood in the Andes* (Oxford: Oxford University Press, 2000), 160.

12. Melillo, "The First Green Revolution," 1041; Gregory T. Cushman, *Guano and the Opening of the Pacific World: A Global Ecological History* (Cambridge: Cambridge University Press, 2013), 55.

13. Melillo, "The First Green Revolution," 1039; Cushman, *Guano and the Opening of the Pacific World*, 55.

14. Cushman, *Guano and the Opening of the Pacific World*, 55; Melillo, "The First Green Revolution," 1039, 1042.

15. Klarén, *Peru: Society and Nationhood in the Andes,* 175, 180–181; Brian Loveman, *Chile: The Legacy of Hispanic Capitalism* (New York: Oxford University Press, 1979), 197.
16. Melillo, "The First Green Revolution," 1042.
17. Klarén, *Peru: Society and Nationhood in the Andes,* 159, 162, 172, 178.
18. Armando Tejeda Gómez, "Canción con todos," 1969, accessed 31 May 2015, http://www.cancioneros.com/nc/173/0/cancion-con-todos-armando-tejada-gomez-cesar-isella.
19. Quoted in Klubock, *Contested Communities,* 86.
20. Simon Collier and William F. Sater, *A History of Chile, 1908–1994* (Cambridge: Cambridge University Press, 1996), 144.
21. Collier and Sater, *A History of Chile,* 163.
22. Loveman, *Chile,* 198.
23. Loveman, *Chile,* 209.
24. Collier and Sater, *A History of Chile,* 196.
25. Hobart A. Spalding, Jr., *Organized Labor in Latin America: Historical Case Studies of Urban Workers in Dependent Societies* (New York: Harper & Row, 1977), 32.
26. Folchi, "Historia Ambiental de las labores," 311, 320, 330, 341.
27. Quoted in Folchi, "Historia ambiental de las labores," 262.
28. Quoted in Klubock, *Contested Communities,* 21–22.
29. Klubock, *Contested Communities,* 28, 30.
30. Folchi, "Historia ambiental de las labores," 292.
31. Folchi, "Historia ambiental de las labores," 364–366, 383–386.
32. Klubock, *Contested Communities,* 134; Joanna Swanger, "Defending the Nation's Interest: Chilean Miners and the Copper Nationalization," in *Workers' Control in Latin America, 1930–1979,* ed. Jonathan C. Brown (Chapel Hill: University of North Carolina Press, 1997), 274.
33. Klubock, *Contested Communities,* 88.
34. Swanger, "Defending the Nation's Interest," 276.
35. Collier and Sater, *A History of Chile,* 254, 268.
36. Quoted in Collier and Sater, *A History of Chile,* 359–360.
37. Collier and Sater, *A History of Chile,* 376.
38. Román Piña Chan, "El desarrollo de la tradición huasteca," in *Huaxtecos y Totonacos: Una antología histórico-cultural,* ed. Lorenzo Ochoa (Mexico City: Consejo Nacional para la Cultura y las Artes, 1990), 172.
39. Santiago, *The Ecology of Oil,* 129–130.
40. *El Mundo,* 2 June 1929.
41. Google Earth Imagery, accessed 15 December 2010, 21°31'59.37"N 97°37'13.30"W.
42. Santiago, *The Ecology of Oil,* 231–232.
43. Santiago, *The Ecology of Oil,* 197–199.
44. Mario Román del Valle and Rosario Segura Portillo, "La huelga de 57 días en Poza Rica," *Anuario* 5 (1988): 77–83.
45. Angelina Alonso Palacios and Roberto López, *El sindicato de trabajadores petroleros y sus relaciones con PEMEX y el estado, 1970–1985* (Mexico City: El Colegio de México, 1986), 30.
46. Quoted in Edwin Lieuwen, *Petroleum in Venezuela: A History* (New York: Russell & Russell, 1967), 52.

47. Kozloff, "Maracaibo Black Gold," 242, 251, 282.

48. Kozloff, "Maracaibo Black Gold," 77, 92–93, 121–122, 126, 147, 236–237, 243.

49. Quoted in Kozloff, "Maracaibo Black Gold," 195.

50. Fernando Coronil, *The Magical State: Nature, Money, and Modernity in Venezuela* (Chicago: University of Chicago Press, 1997), 107.

51. Edwin Liewen, "The Politics of Energy in Venezuela," *Latin American Oil Companies and the Politics of Energy,* ed. John D. Wirth (Lincoln: University of Nebraska Press, 1985), 50.

52. Miguel Tinker Salas, *The Enduring Legacy: Oil, Culture, and Society in Venezuela* (Durham: Duke University Press, 2009), 122–123.

53. Wolfgang Hein, "Oil and the Venezuelan State," in *Oil and Class Struggle,* ed. Petter Nore and Terisa Turner (London: Zed Press, 1980), 231, 233; Jesús Prieto Soto, *Luchas proletarias populares y petroleras: ¿Ocaso de PDVSA? Quinta edición* (Maracaibo: Imprenta Internacional, 2004), 31, 37, 45, 61, 72.

54. Coronil, *The Magical State,* 183, 630.

55. Tinker Salas, *The Enduring Legacy,* 243.

56. Lieuwen, "The Politics of Oil in Venezuela," 73; Hein, "Oil and the Venezuelan State," 244.

57. Arne Jernelöv and Olof Lindén, "Ixtoc I: A Case Study of the World's Largest Oil Spill," *Ambio* 10, no. 6, The Caribbean (1981): 299–306.

58. Lou Dematteis and Kayana Szymczak, *Crude Reflections / Cruda Realidad: Oil, Ruin and Resistance in the Amazon Rainforest / Petróleo, devastación y resistencia en la Amazonía* (San Francisco: City Lights, 2008), 18–26, 46–47, 60–61, 84–85.

59. Paul Sabin, "Searching for a Middle Ground: Native Communities and Oil Extraction in the Northern and Central Ecuadorian Amazon, 1967–1993," *Environmental History* 3, no. 2 (1998): 144–168; Judith Kimerling, "Oil, Lawlessness, and Indigenous Struggles in Ecuador's Oriente," in *Green Guerrillas: Environmental Conflicts and Initiatives in Latin America and the Caribbean,* ed. Helen Collinson (London: Latin American Bureau, 1996), 61–73.

60. Dematteis and Szymczak, *Crude Reflections / Cruda Realidad,* 30–45, 49–59.

61. "Supreme Court Won't Consider Blocking $18B Judgment against Chevron," CNN, 24 October 2012, accessed 3 June 2013, http://www.cnn.com/2012/10/10/world/amricas/chevron-ecuador-lawsuit.

62. Paul M. Barrett, "Chevron's Lawyers at Gibson Dunn Get Tough in Ecuador Pollution Case," *BloombergBusinessweek,* 18 April 2013, accessed 3 June 2013, http://www.businessweek.com/articles/2013-04-18/chevrons-lawyers-at-gibson-dunn-get-tough-in-ecuador-pollution-case.

63. Clifford Krauss, "Big Victory for Chevron over Claims in Ecuador," *The New York Times,* 4 March 2014, accessed 12 April 2014, http://www.nytimes.com/2014/03/05/business/federal-judge-rules-for-chevron-in-ecuadorean-pollution-case.html?_r=0.

64. Bebbington and Bury, *Subterranean Struggles,* passim.

65. See Liisa North, Timothy David Clark, and Viviana Patroni, eds., *Community Rights and Corporate Responsibility: Canadian Mining and Oil Companies in Latin America* (Toronto: Between the Lines, 2006).

66. Oxford Business Group, *The Report: Peru 2012* (Oxford: Oxford Business Group, 2012), 111.

67. Mariana Walter and Joan Martínez-Alier, "How to be Heard When Nobody Wants to Listen: Community Action against Mining in Argentina," *Canadian Journal of Development Studies* 30, no. 1–2, Special Issue, Rethinking Extractive Industry: Regulation, Dispossession, and Emerging Claims (2010), 290.

68. Timothy David Clark, "Canadian Mining in Neo-Liberal Chile: Of Private Virtues and Public Vices," in *Community Rights*, 94.

69. José de Echave, "Mining and Communities in Peru: Constructing a Framework for Decision-Making," in *Community Rights*, 26.

70. J. Budds and L. Hinojosa, "Restructuring and Rescaling Water Governance in Mining Contexts: The Co-Production of Waterscapes in Peru," *Water Alternatives* 5, no. 1 (2012): 127, 131.

71. "Fracking Fights Loom Large in Mexico," 9 July 2014, accessed 29 June 2015, fnsnews .nmsu.edu/fracking-fights-loom-large-in-mexico/; Richard A. Oppel, Jr., and Michael Wines, "Industry Blamed as Earthquakes Jolt Oklahoma," *New York Times*, 4 April 2015.

72. Rights for Nature in Ecuador's Constitution, accessed 29 June 2015, http://therightsof nature.org/wp-content/uploads/pdfs/Rights-for-Nature-Articles-in-Ecuadors-Consti tution.pdf; CNN Mexico, "Bolivia crea una ley que considera a la madre tierra un sistema viviente," 15 October 2012, accessed 29 June 2015, http://mexico.cnn.com/ planetacnn/2012/10/15/bolivia-crea-una-ley-que-considera-a-la-madre-tierra-un-sistema-viviente; Dinah Shelton, "Human Rights, Environmental Rights, and the Right to Environment," *Stanford Journal of International Law* 28 (1991): 103.

73. Evidence and Lessons from Latin America (ELLA), ella.practicalaction.org, accessed 29 June 2015; Environmental Justice Organizations, Liabilities and Trade, "Mapping Environmental Justice," accessed 29 June 2015, www.ejolt.org; Environmental Justice Atlas, accessed 29 June 2015, https://ejatlas.org/.

 CHAPTER 10

Prodigality and Sustainability
The Environmental Sciences and the Quest for Development

Stuart McCook

Since the early nineteenth century, environmental change in Latin America has been decisively driven by developmentalist projects. The historian Kenneth Pomeranz argues compellingly that "developmentalism and resistance to it frame much of the environmental history of the last several centuries."[1] Nations across the globe embraced developmentalist projects, which Pomeranz characterizes as a broad set of practices and ideologies that emphasized "state building, sedentarization, and the exploitation of resources."[2] States across the political spectrum—from imperial Great Britain to Communist China—all embraced this developmentalist paradigm. Similarly, in Latin America, diverse states such as imperial Brazil, Porfirian Mexico, socialist Cuba, and neoliberal Chile also pursued developmentalist projects. Developmentalist projects were highly diverse and never fully coherent, either within countries or between countries. Nonetheless, for nineteenth- and twentieth-century Latin America, it is possible to discern three overarching phases of developmentalism. During the first phase, which lasted from the early nineteenth century through to the Great Depression, liberal developmentalism—which emphasized export-led development—prevailed. During the second phase, which lasted from the Great Depression through the early 1990s, developmentalism in Latin America tended to be more inward looking, emphasizing industrialization and an increased role for the state. The third phase, neoliberal developmentalism, began in the 1980s; this model sought to curtail the power of the state and increase the power of private enterprise—especially large corporations—in developmentalist projects.[3]

Developmentalism in Latin America unleashed a process of constant environmental change, and managing that change became a central challenge for developmentalism. Since the early nineteenth century, then, the environmental sciences have become an essential part of these developmentalist projects. In Latin America, such projects usually involved managing the natural world—from rural environments, such as the cane fields of Cuba, to urban environments of twentieth-century megalopolises like Mexico City. For much of the nineteenth century, scientists worked on projects to accelerate economic growth—such as conducting inventories of natural resources and acclimatizing exotic plants with potential economic value. Beginning in the late nineteenth century, scientists became involved in managing the environmental challenges produced by these same developmentalist projects, such as catastrophic soil erosion, urban air pollution, and species loss. These concerns shaped the nature of scientific research in Latin America as scientists aligned their research with the developmentalist priorities of their patrons.

Scientists played an increasingly important role in shaping public discourses about development and the environment. The proliferation of environmental problems created a space for science in Latin America, especially after World War II. Some scientists argued that their expert knowledge could provide the best answers to the most pressing environmental problems of the day. This was a technocratic, top-down worldview, in which scientists exercised social power through institutions—usually, but not exclusively, public institutions. Some scientists were often aghast at the environmental consequences of developmentalism. But most of them worked for states that wanted scientists to help make developmentalist projects work effectively, not to question them altogether. This sometimes brought scientists into conflict with their patrons and with the public in whose name they presumed to operate. So scientists often found themselves trying to strike a balance between producing and conserving. By the last quarter of the twentieth century, public groups, fearing the abuse of authority by experts, also began to question technocratic environmental decision-making, although they did not necessarily question the science.

The new nations of Latin America rebuilt their scientific institutions and communities at a moment when the old discipline of natural history gradually branched into an ever-wider range of specialized disciplines and subdisciplines, which we can characterize as the environmental sciences. Broadly understood, the environmental sciences are those that study our organic and physical environments.[4] Some of these emergent disciplines—botany, zoology, geography, ecology—were mainly directed at understanding natural processes in their own right. Others—such as plant pathology, forestry, wildlife management, public health, and atmospheric chemistry—were more utilitarian, primarily focused on understanding nature with respect to human activities. These could be divided into the pure and applied environmental sciences, but this distinction is misleading. The pure sciences often produced research of practical value, and the applied sciences often led to fundamental new understandings of environmental processes.[5] This diversity matters; environmental scientists rarely spoke with a single voice, and their research rarely pointed in a single policy direction.

Progress and Prodigality:
The Long Nineteenth Century, 1780–1930

During the long nineteenth century (roughly from the Haitian Revolution to the Great Depression), Latin America's new nations enlisted science to promote liberal programs of export-led development. The wars of independence marked the almost complete decolonization of Spanish and Portuguese America; only Cuba and Puerto Rico remained in Spanish hands. Most nations (and even the remaining colonies) pursued economic development through commodity exports, producing tropical goods for industrializing markets in the Global North, which had seemingly insatiable appetites for Latin American products. Latin America's elites shared in the myth of "prodigal nature," which understood the region's natural resources as being, for all practical purposes, infinitely abundant and inexhaustible. These resources were to be used to promote national economic development: "Sin azúcar," went the Cuban saying, "no hay país" (Without sugar, there is no nation). By the end of the nineteenth century, Brazil produced four times as much coffee as the rest of the world combined, and Cuba was the world's largest producer of sugar. This ideology of prodigality was perhaps understandable when looking at the vast landscapes of Brazil. But even the elites in comparatively small places like Cuba shared the same vision.

In the early nineteenth century, as in the colonial period, the field of natural history encompassed most of what we now call the environmental sciences.[6] Naturalists could be found in almost every corner of Latin America. In some places, such as Colombia, the communities of naturalists suffered greatly during the wars of independence. But a number of important natural history institutions, such as Mexico's botanical garden, survived into the nineteenth century.[7] Between 1812 and 1840, new national museums were established in Argentina, Brazil, Chile, Colombia, Mexico, Peru, and Uruguay.[8] Elites in many countries organized economic societies, which promoted science as a tool for economic development. Latin America's universities offered rudimentary instruction in science as part of a general curriculum, but (like their counterparts in Europe and North America), did little to promote scientific research. Rather, universities focused on training lawyers, physicians, teachers, and engineers.[9]

During these years, most of Latin America's naturalists either learned research in local museums, academies, institutes, or government agencies, or they pursued advanced education abroad. "Indispensable aliens"—foreign scientists, engineers, and physicians—also provided scientific and technical knowledge. As Leida Fernández Prieto has observed, however, their knowledge was "not diffused in a simple, linear way." Rather, "the construction and dissemination of knowledge must be analyzed both as a process of learning and negotiation, in which many actors became involved," including both foreign and local scientists.[10] Many of Latin America's leading naturalists in the long nineteenth century were emigrants from Continental Europe. They were part of a small wave of middle-class migrants to the New World, attracted by the tremendous opportunities for research and discovery. Leading figures in this group included the Italian Antonio Raimondi (Peru), the Germans Teodor Wolff (Ecuador) and Adolf Ernst (Venezuela), and the Swiss Emil Goeldi (Belém, Brazil) and Henri Pittier (Costa Rica and Venezuela). Many of these foreign scientists established themselves permanently in Latin America and, as immigrants, contributed to the process of nation-building.

The new nations of Latin America were particularly interested in demarcating their national territory and surveying their wild landscapes. Such surveys had both symbolic and practical ends. Symbolically, the surveys and inventories defined the national space, and the state's symbolic ownership of it. Practically, the surveys provided the nation's elites with inventories of natural resources that could be used to fuel development.[11] In 1830, Venezuela's newly independent government hired the Italian engineer Agustin Codazzi to conduct a corographic survey of Venezuela, which he completed in 1839. In Cuba, the naturalist Ramón de la Sagra—working with naturalists from across the island—produced a comprehensive *Historia física, política, y natural de la isla de Cuba* in 1845. Between about 1880 and 1930, almost every nation in Latin America produced an inventory of the nation's plants. These floras were intended to be comprehensive, so naturalists explored their national territories, making arduous journeys to collect plants and to map territories. They attempted to synthesize all of the botanical knowledge about the nation's nature held in museums and botanical gardens abroad. These naturalists also selectively (and often silently) appropriated local environmental knowledge from indigenous groups, farmers, amateur naturalists, and others.[12] While most naturalists embraced the prevailing liberal, developmentalist paradigm, some—such as the Brazilian José Bonifacio de Andrada e Silva (1763–1838)—did express concern about the environmental and social impact of this model. Their critiques of environmental destruction were rooted in utilitarian concerns about the wasteful, destructive use of natural resources essential for national economic and social development, rather than any intrinsic concern about "nature" itself.[13] This position can, in some respects, be seen as a precursor to later twentieth-century discussions of sustainability.

The agricultural sciences gained importance as ecological problems began to present serious obstacles to continued growth in agricultural production. In many parts of Latin America, farmers practiced what the agronomist Franz Dafert described as ecological "robbery," exhausting soils and degrading landscapes, to accumulate capital as quickly as possible.[14] Problems such as this led planters to start taking more interest in chemical fertilizers and other means to restore fertility to the soil. The rapid spread of monocultures also produced an unprecedented wave of crop diseases and pests—what John Soluri has aptly characterized as "commodity diseases." The South American Leaf Blight prevented Brazil from developing rubber plantations; the Panama disease devastated Central America's banana plantations; the mosaic disease of sugar caused significant losses in Cuba and Puerto Rico. These diseases were not random accidents; they were a consequence of intensified production, and also of the accelerated circulation of organisms across the Global South.[15]

Agricultural science had emerged as a research discipline in early nineteenth century Germany. The chemical work of Justus von Liebig at the University of Giessen showed how chemical fertilizers could produce tremendous gains in agricultural productivity. Ironically, perhaps, this global interest in chemical fertilizers produced an environmentally destructive resource boom in Peru. Foreign companies mined Peru's coastal Chincha islands for bird guano, rich in nitrogen, phosphate, and potassium. The guano was exported to enrich the fields of North America and Europe—and also sugar plantations in Cuba, Puerto Rico, and elsewhere in the Caribbean.[16] Over the century, the agricultural sciences expanded to include a wide range of scientific disciplines, including agricultural chemistry, economic entomology, plant pathology, and plant acclimatization and breeding.

The earliest state-sponsored agricultural experiment station in Latin America was the Instituto Agronômico de Campinas, in São Paulo, Brazil. The state government hired an Austrian agronomist, Franz Dafert, to explore the problems of coffee production.[17] Cuba's Círculo de Hacendados founded a chemistry laboratory in 1880, particularly devoted to exploring the problems of sugar production. Public agricultural experiment stations were also established under U.S. auspices in Cuba and Puerto Rico after 1898.[18] Research into commodity diseases did produce some spectacular successes. The Puerto Rican naturalist Carlos Chardón, for example, conducted pioneering work on the mosaic disease of sugarcane, prompting a varietal revolution in sugarcane cultivation in Puerto Rico and Cuba that made it possible to sustain and even increase production in spite of diseases and pests.[19] But in other cases, such as the blights that destroyed cacao plantations in Ecuador and rubber plantations in Brazil, scientists could offer explanations but no solutions.[20] Much of their research was less glamorous but essential "maintenance" research, aimed at sustaining agricultural productivity in the face of declining yields.[21]

Epidemic diseases presented another major obstacle to development in the long nineteenth century. Catastrophic epidemics had, of course, plagued Latin America since the Conquest, and some diseases (such as yellow fever and malaria) were endemic across large areas. These diseases altered the dynamics of independence, as foreign soldiers without immunological experience to these diseases died at a much higher rate than did locals.[22] These were compounded by the arrival of new epidemic diseases—cholera and the bubonic plague, for example—that accompanied the new waves of immigrants from Europe and Asia. Later in the nineteenth century, however, these same diseases arguably acted as an impediment to elite plans for development, discouraging immigration and making large infrastructure projects such as railroads and canals costly and difficult, both in terms of human lives and money. Scientific discoveries in the 1870s and 1880s gave doctors a new understanding of these epidemics. The germ theory of disease, articulated by the German research Robert Koch and the French researcher Louis Pasteur, argued that diseases were caused by microbes (bacteria) rather than by broad environmental conditions. Latin American physicians quickly embraced the germ theory. Building their own research, as well as research from scientists elsewhere in the tropics, they discovered that microbes associated with the epidemics were often spread by animal vectors, especially mosquitoes. These new theories, in turn, gave medicine new social power. Physicians promoted sanitation campaigns to control the ecological niches of the insect and animal vectors that spread the microbes.[23]

Sanitation campaigns produced the first modern alignment of political elites and scientists in Latin America. These campaigns focused on Latin America's cities, which had been growing rapidly during the later years of the export boom (1870–1930). These cities were connected to the global economy as ports and entrepôts; they also received immigrants from around the world. This rapid, unconstrained immigration and urbanization left many of the urban poor living in squalid conditions, vulnerable to disease. Governments provided biomedical scientists (bacteriologists, public health officials) with resources and institutions to conduct research into epidemic diseases, and also with the financial and political backing to run sanitation campaigns aimed at eliminating the foci of disease. In some cases, these campaigns involved the construction of sewers, the provision of clean drinking water, and cleaning away garbage. These campaigns also had a strong coercive element, often directed at the working and lower classes; in some cases this included forced vaccinations and the demolition of housing deemed "unsanitary." Carlos Finlay of Cuba conducted pioneering work on the insect vector for yellow fever, although that work was largely elided by the American Walter Reed.[24] The brilliant Brazilian physician Oswaldo Cruz and his colleagues conducted a wide-ranging national survey of sanitation and disease in Brazil. Beyond this top-down approach, public officials also tried to teach proper sanitary behavior to the masses. These efforts met with mixed results, often because the public health officials failed to appreciate the economic and social challenges faced by the people they were trying to help. In 1904, a forced smallpox vaccination campaign in Rio de Janeiro—directed by Oswaldo Cruz—triggered a popular revolt against the government, known to history as the *Revolta contra vacina*.[25]

The Great Depression ended the age of liberal developmentalism in Latin America, although the model had—in places—faced severe economic and ecological challenges since the late nineteenth century. From the 1880s through to the 1930s, cycles of boom and bust had become commonplace. The economic—and by extension environmental—costs of this model were becoming more apparent. This developmentalist model had left parts of Latin America with severely degraded rural and urban landscapes and a host of new environmental problems. The environmental sciences helped sustain this model of developmentalism in the face of these obstacles. It's difficult imagine how the course of Latin American development would look, for example, if yellow fever and malaria had remained unchecked, or if banana and sugarcane had succumbed to crop epidemics the way that Brazilian rubber had. In key areas, then, science had become a key instrument of national development, and scientific institutions a key part of the national landscape. Still, few states had more than a rhetorical interest in science more generally.

Crises of Abundance and Growth:
The "Short Twentieth Century," 1930–1990

During the short twentieth century, states across Latin America began to explore alternative models of development. The global economy suffered sharp shocks with the Great Depression and World War II; many of Latin America's export markets in Europe were disrupted, and the region became more oriented toward the U.S. market. Politically, the postwar decades were unstable; they were shaped by competing populist and authoritarian visions of development, and also by the larger politics of the Cold War. Regardless of their politics, however, most states pursued more inward-looking programs of development, and the larger ones experimented with industrialization. States played a greater role in planning and directing economic development. They expanded university and government research for economic development. Much of this research was focused on technology and engineering to support industrialization, but states also sponsored research institutions to help deal with the ever wider range of environmental challenges they faced.[26]

Beginning in the 1930s, the environmental sciences began to find a firmer institutional footing across Latin America. This was partly driven by the rapid expansion of postsecondary education, especially after World War II. States established new universities, perhaps most famously the University of São Paulo (1934), or expanded existing universities, opening higher education to a much broader swath of society. For the first time, a small but steady stream of Latin Americans received a scientific education in Latin America itself, even if many still had to go abroad for graduate school. States also began to support scientific research on a larger scale, both within the universities and within state institutions. The Brazilian government established the National Council for Scientific Research (CNPq) in 1951 to plan and fund scientific and technical research and help develop the community of experts necessary to promote national development. Over the next decades, other Latin American countries founded research councils of their own: for example Argentina's CONICET and Venezuela's CONICIT were founded in 1958, and Mexico's CONACYT in 1970.[27]

Major scientific research centers were also established or expanded, some of which conducted world-class research. For example, the Instituto Agronômico de Campinas (São Paulo, Brazil) and the Colombian Centre for Coffee Research (CENICAFE), founded in 1938, conducted globally innovative research in the genetics, breeding, and cultivation of coffee. Between the 1940s and 1970s, a number of international agricultural research centers were also established in Latin America. These included CIMMYT (International Maize and Wheat Improvement Center), founded in Mexico in 1943; CIAT (the International Center for Tropical Agriculture) in Colombia in 1967; CATIE (Tropical Agronomical Research and Higher Education Center) in Costa Rica, founded in 1972 (but with roots in the 1940s); and the CIP (International Potato Center), founded in Lima in 1973. Other civil society groups established new organizations to support the scientific study of the environment and conservation, including the Sociedade Brasileira Protetora dos Animais, Colombia's Sociedad de Ciencias Naturales and the Fundación La Salle, and Mexico's Committee for the Protection of Wild Birds, and the Mexican Forestry Society.[28] Collectively, universities and research institutes such as these helped greatly expand research in environmental sciences in Latin America, and also created new employment opportunities for Latin American scientists.

The natural history inventories of the nineteenth and twentieth century provided valuable data for the new sciences of plant geography, ecology, and others. Early in the twentieth century, scientists had recognized that some landscapes were home to an unexpectedly rich number of plant and animal species. For example, phytogeographers such as Karl Werklé, an Alsatian naturalist and long-time resident in Costa Rica, discovered that the country had an unusually large and diverse flora, and argued that this was because of the country's "extraordinary variation of atmospheric and climatic conditions" as well as its position as a land bridge between several different floristic provinces.[29] Similarly, ornithologists working in nineteenth-century Colombia were astounded at the unmatched diversity of bird species they discovered. As Camilo Quintero has argued, "the work of Charles Darwin and Alfred Russel Wallace helped to focus attention on the importance of geographical distribution in reaching new conclusions about the natural world."[30] Although the word "biodiversity" was not coined until much later, the fundamental insight about Latin America's biological wealth has its roots in this period.[31]

These inventories and ecological studies generated a growing sense that key species of wild plants and animals were under threat from indiscriminate hunting, collecting, and habitat loss. The vogue for chinchilla coats in Europe and North America led to the virtual extinction of these animals in the Southern Cone. Likewise, the demand for feathers in women's fashion produced a large-scale trade (both legal and illegal) in bird feathers during the first decades of the twentieth century. Naturalists in Brazil, Colombia, and elsewhere called for the state to enact laws to protect endangered species and their habitats. Across Latin America, scientists, foresters, and engineers—such as the Mexican forester Miguel Ángel Quevedo—were instrumental in promoting conservation policies and the formation of national parks in Latin America, as Emily Wakild's chapter on conservation explores more fully (see chapter 11). In the 1930s, some Latin American states seemed to accept this view. The Brazilian constitution of 1937 recognized "natural objects" as part of the national patrimony. External forces also helped promote conservation policies in Latin America. Several Latin American states joined the Convention on Nature Protection and Wildlife Preservation in the Western Hemisphere, which was signed in 1940.[32] Still, as Emily Wakild argues, many Latin American states failed to ratify this convention. Between the 1940s and the 1970s, developmentalist policies trumped conservation efforts. More effective conservation programs did not begin until several decades later.

Attitudes and policies toward wild nature began to change during the 1970s and 1980s. This was partly driven by the new domestic and international conservation movements concerned with environmental destruction and the loss of biodiversity, especially in the face of the pressures of urbanization, industrialization, internal colonization, and agricultural expansion and modernization. National and international scientists working in Latin America had long called attention to this destruction. Through technocratic government bureaucracies and conservation-minded NGOs, scientists pressured national governments to enact and enforce conservation legislation. In contrast with earlier periods, this legislation enjoyed a measure of success. Between 1970 and 1980, for example, the Costa Rican state established a series of national parks and conservation areas that covered more than a quarter of the national territory. As elsewhere, there were many social, environmental, and economic rationales for doing so. Government officials in Costa Rica cited the "scientific value" of these landscapes, both for education and for the value of the "genetic material" that these landscapes contained. Civil society organizations also established conservation areas. In Costa Rica, the world-famous Monteverde Cloud Forest Reserve was established by a group of American Quakers, while the Caribbean Conservation Corporation was established by the American naturalist Archie Carr.[33]

Agriculture continued to play an important economic role in much of Latin America, and states made important investments in agricultural research. This

Illustration 10.1. A coffee farm near Sasaima, Colombia.

This seemingly mundane photo of a coffee farm near Sasaima, Colombia, captures larger trends the "technification" of agriculture in the twentieth century, a process that included scientific breeding and the intensive use of agricultural chemicals. It shows dwarf hybrid coffee plants densely cultivated in neat rows. The cultivar shown here has been developed since the 1970s. It is the product of national and global scientific research and exchanges between institutions in Colombia, Brazil, and Portugal. It includes coffee germplasm collected as far away as Timor. The cultivar was developed for its productivity, its suitability for Colombia's landscapes, and its resistance to coffee rust—a catastrophic global disease of coffee. Emergent environmental challenges such as the rust, compounded by climate change, mean that many of Latin America's agricultural landscapes now depend upon constant biological innovation to remain economically and ecologically viable.

Photo by author.

was partly driven by the need to sustain agricultural production in the face of commodity diseases. It was also driven by the pressing need to feed the region's burgeoning populations, especially in the industrializing cities. Beginning in the 1950s, Latin American scientists (along with agricultural scientists across the globe) promoted the "technification" of agriculture, using hybrid plants and chemicals (among other technologies) to dramatically increase agricultural productivity, particularly to feed the region's burgeoning cities. Significantly, these efforts at improving productivity focused on food crops as well as cash crops. These new Green Revolution technologies included packages of hybrid seeds, chemical fertilizers, fungicides, and pesticides. Producers of traditional export crops also technified their farms, often under the guidance of scientists and extension agents from national experiment stations and government agencies, such as Brazil's EMBRAPA (Brazilian Agricultural Research Corporation), founded in 1973. For example, coffee farmers eliminated shade trees and replaced their old varieties of coffee with new hybrid "sun" coffees (many of which had been developed in Brazil). In the short term, technification could produce tremendous spikes in productivity. While prodigal science could, it appeared, replace prodigal nature, these increases in productivity often had tremendous economic and ecological costs.[34] In Mexico, as Chris Boyer and Micheline Cariño Alvero's chapter in this volume explains (see chapter 1), the Green Revolution technologies penetrated almost the entire country and triggered conflicts between peasant producers and large farmers, and also intensified the commodification of the natural world.

Environmental researchers also turned their focus to research on problems associated with urbanization and industrialization. In the decades following World War II, Latin American cities began growing explosively. This growth was not limited to the capital cities alone: for example, the population of Belo Horizonte, Brazil, grew from 350 thousand to more than a million between 1950 and 1964. This rapid growth was accompanied by the growing use of motor vehicles, power plants, and other sources of pollution. These processes, in turn, produced a host of new problems—many of which were fundamentally environmental. Water pollution, which had long been an urban environmental problem, worsened as poorly regulated factories dumped their effluent into the water systems. Air pollution joined water pollution as a major new issue; heavy smog became a daily fact of life for urban Latin Americans. Chronic pollution in turn caused chronic health problems and premature mortality, particularly in infants. Urban mortality levels climbed rapidly. Nonetheless, the science and policy response to these problems was at first slow and uneven. The earliest studies of air pollution in Mexico City date to the early 1960s. They were conducted by Humberto Bravo, of the Air Pollution Department (Sección de Contaminación Ambiental) of the National University of Mexico (UNAM). Nonetheless, systematic studies of Mexico City's air quality only began in the 1980s, several decades after the problem was first identified.[35]

Beginning in the 1960s, ecology (as a discipline) gained renewed global importance, partly because of the emergence of local and global environmental movements—which were triggered in part by growing popular concern about the environmental and health impacts of rapid urbanization and agricultural modernization. In the Global North, Rachel Carson's *Silent Spring* (1962) mobilized an environmental movement concerned about the impact of agricultural chemicals on people, animals, and landscapes. In the following decades, the ties between ecologists and environmental movements grew stronger and expanded to include concerns such as industrial pollution and deforestation. In 1968, as Wakild's chapter describes in greater detail, the IUCN organized a major conference on conservation in Latin America, held in Argentina, and scientists, foresters and wildlife specialists figured prominently in the discussions. Reflecting this emergent global concern for the environment, in 1972 the United Nations convened a global Conference on the Human Environment in Stockholm. One of the conference's key outcomes was the creation of the United Nations Environment Program (UNEP), which promoted the scientific study of the environment and helping member states develop policies and practices aimed at preserving the environment. The Stockholm conference even had a deep impact on some of Latin America's most conservative governments. After the Stockholm conference, for example, Brazil's military government created a national Secretariat of the Environment (best known by its Portuguese acronym SEMA), and placed a scientist, Paulo Nogueira Neto, at its head. The Brazilian government also began to develop meaningful environmental legislation.[36]

During the 1970s and 1980s, biodiversity and conservation became major political concerns, both nationally and internationally. Biologists and ecologists had developed the concept of biodiversity—a word first coined in English—in 1985. But the concept had earlier roots in the inventory sciences, biogeography, and ecology, as scientists cataloged and mapped species and studied their interconnections. Since the early twentieth century, naturalists in Latin America had already recognized that the region was home to an unusually large number of species. Over the twentieth century, further studies had measured genetic diversity and ecosystem diversity. At a global level, the concept of biodiversity—and growing alarm at environmental destruction—was often most vocally articulated by foreign scientists working in Latin America, such as Thomas Lovejoy (who worked in Brazil) and Daniel Janzen (who worked in Costa Rica). They raised a global concern about habitat loss and species extinction in the tropics. Scientists formed a new discipline—conservation biology, which focused on the study of ecological systems that were "perturbed, either directly or indirectly, by human activities or other agents." It was meant to be an applied discipline, a "crisis discipline" in the words of one scientist, which prioritized action above theory.[37] At the local level, Latin American scientists also used the language of biodiversity to make a case for the preservation or conservation of wild areas and parks. This new scientific concept had economic and political traction.[38] During the 1980s and 1990s, scientists identified the Brazilian Amazon as a global biodiversity hotspot.

By the 1980s, the environmental sciences became embedded in state institutions, and in the growing university sector. Most Latin American nations could train and employ scientists locally, and research into local environments and local development issues flourished. Driven by internal and global trends, the environmental sciences had expanded and diversified. Still, the environmental sciences did not always all point in the same direction. For example, the growing technification of agriculture drew criticism from scientists concerned about the conservation of natural environments and about the impact of agricultural chemicals on rural peoples and landscapes. And while scientists enjoyed unprecedented institutional power, some civil society groups criticized their power, arguing that technocratic decision-making was inconsistent with democratic decision-making. These concerns were to rise to the surface in the 1980s and 1990s, as Latin American states and their development programs faced fundamental challenges in the face of changing domestic and global politics.

Global Challenges and Sustainable Futures since 1990

The end of the Cold War and the advent of the debt crisis and structural reforms (beginning in the late 1980s and continuing into the 1990s) marked the beginning of a new period in Latin American history. Most Latin American countries gradually made the transition to democracy. At the same time, structural reforms imposed by international lending agencies forced states to slash public spending. Scientific research institutions had their budgets cut or were closed altogether. All of this took place just at a moment when global attention was focused on Latin American environments. In 1992, Rio de Janeiro hosted the United Nations Conference on Environment and Development, which included representatives from 170 states and several thousand NGOs. This convention marked a major shift in emphasis: partly as a result of the conference, the theme of sustainability became a central focus for research in the environmental sciences, both pure and applied. Another shift, related to this, is that the environment—particularly Latin America's vast tropical rainforests—itself became an object of concern and debate. Claudia Leal's chapter in this volume eloquently describes the growing political mobilization around Latin America's forests (see chapter 5).[39]

Environmental scientists also became much more directly involved in addressing environmental problems in Latin America's cities. As Lise Sedrez and Regina Horta Duarte argue in this volume, Latin America's cities are complex environments, with distinctive and often acute problems of their own. These problems are becoming increasingly urgent in the wake of rapid urbanization; now, 80 percent of Latin Americans live in urban areas. One problem, among many, is air pollution. Since the 1960s, university-based scientists had monitored air pollution and developed sophisticated assessments of its impacts. Along with NGOs, they advocated for a climate agenda. NGOs and different

Illustration 10.2. Sign from the Araucárias Municipal Natural Park near Guarapuava, Brazil.

PARQUE NATURAL MUNICIPAL DAS ARAUCÁRIAS

O Parque Natural Municipal das Araucárias é uma unidade de conservação que foi criado em 1991, enquadrando-se no grupo de Unidades de Proteção Integral.
Essa área tem por objetivo básico a preservação de ecossistemas naturais de grande relevância ecológica e beleza cênica, possibilitando a realização de pesquisas científicas e o desenvolvimento de atividades de educação e Interpretação ambiental, de recreação em contato com a natureza e de turismo ecológico.

SNUC/ Lei N° 9.985/2000

Sign from the Araucárias Municipal Natural Park near Guarapuava, Brazil: "This area's objective is to preserve the natural ecosystems of great ecological relevance and scenic beauty, making possible scientific research and the development of environmental education and interpretation, recreation in contact with nature, and ecotourism."
Photo by author.

levels of government—especially municipal governments—developed policies aimed at addressing the worst effects of this pollution. The metropolitan government of Mexico City, for example, launched its first program to improve air quality—named Proaire—in 1996. Transnational organizations such as the World Health Organization and the UNEP have sponsored a study on urban air pollution in the world's megacities. Public-private partnerships also played an increasingly important role. The Science & Technology Alliance for Global Sustainability, for example, sponsored a pioneering study of air pollution in Mexico City during the early 2000s. Through this combination of scientific, policy, and government initiatives, Mexico City has made some headway toward addressing pollution, although significant problems remain.[40]

The discourse of sustainability also made its way into the agricultural sciences. In agriculture, sustainability became just as important as productivity, which had dominated agricultural research. In some small niches, producers experimented with fully organic agriculture. This organic agriculture is not a return to before the Green Revolution. In fact, modern organic agriculture often draws upon a sophisticated range of sciences—albeit a different range of sciences than those that drove the Green Revolution."This includes botany and newer disciplines such as agroecology and agroforestry. Latin American scientists, such as the Chilean Miguel Altieri (who works at the University of California at Berkeley) and the Mexican Efraím Hernández Xolocotzi, (who spent most of his career at Mexico's Escuela Nacional de Agricultura en Chapingo), were global leaders in establishing these new, more holistic, disciplines. Among other things, these disciplines seek to control diseases and pests without using chemicals, using techniques such as integrated pest management and biological control. Agroecology seeks to integrate ecological knowledge into agricultural practice, and is open to engaging with local agricultural knowledge as well (see chapter 7 by Soluri in this volume).[41] Certified organic crops—coffee, bananas, and cacao, among others—are most often export crops destined for foreign consumers. So far, many of these alternative forms of agriculture are principally funded by NGOs and other civil organizations; they have received less attention from the state. In the 1990s, Cuba turned to organic agriculture in the face of catastrophic shortages of petroleum and chemicals after the fall of the Soviet bloc.[42] While organic agriculture has gained popularity under certain conditions, conventional agriculture is still dominant.

The paradigm of sustainability has not supplanted the productivist and utilitarian paradigms of the nineteenth and twentieth centuries. While agricultural researchers pay ever more attention to sustainability, many practices of high modernist agriculture persist. The rapid expansion of soybean agriculture across southern South America involves many of the hallmarks of modernist agriculture, including the use of genetically modified organisms, agricultural chemicals, and large-scale landscape change. This modern "United Republic of Soy" (in the words of an advertising campaign by Syngenta) encompasses parts of Brazil, Bolivia, Paraguay, Uruguay, and Argentina. EMBRAPA, Brazil's public agricultural research division, has, with the support of private companies, conducted vital research allowing for soybean cultivation to expand into new regions.[43] Scientists are experimenting with the introduction of genetically modified organisms (GMOs) into some crops. They continue to work in the struggle against the commodity diseases that plague some of Latin America's traditional crops, as well as some of its newer exports, such as Chile's salmon industry.[44] In other cases, agricultural intensification has sometimes even permitted economic diversification and the return of forests—a sharp distinction from earlier patterns of intensification. In Colombia, for example, the intensification of coffee farming allowed for coffee production to increase while the total amount of land under coffee cultivation has actually *decreased*. Coffee farmers could then cultivate other crops (cacao, beans, sugarcane, corn), and in some cases allow secondary forest to recover.[45]

In the early twenty-first century, many of Latin America's most pressing problems are now connected to climate change. For decades, scientists had conducted pioneering work on global weather patterns on the Guano Islands off the coast of Peru. Scientists connected unusual patterns in Pacific Ocean currents—El Niño and La Niña—with abnormal weather patterns across the tropics.[46] Unusual weather patterns have contributed to natural disasters, such as Hurricane Mitch, which devastated Central America in 1998. Warming global temperatures have caused glaciers in the high Andes and the Antarctic to melt. In the Andes, these melting glaciers threaten supplies of drinking water, and power for hydroelectric projects—to name just two potential consequences. Scientists of all kinds—climate scientists, geologists, agronomists, and physicians—are being mobilized to conduct research to predict and mitigate the myriad impacts of climate change on Latin American communities. For example, the World Bank has funded a project on the "Adaptation to Rapid Glacier Retreat in the Tropical Andes."[47] In Colombia and Central America, coffee agronomists are working to breed varieties of Arabica that will be adaptable to the changing coffee ecosystems. The threat of climate change has also suggested a potentially new global role for Latin America's tropical rain forests, as providing environmental services for sequestering greenhouse gases.[48]

Conclusions

In the early twenty-first century, the ecological sciences have become one of the most important—and politically powerful—ways of framing nature and developmentalist projects in Latin America. Scientific ideas such as "biodiversity" have entered the social, economic, and political realms. Scientific ideas have offered new ways of thinking about nature and of valuing it. Over the previous two centuries, the ecological sciences had helped Latin America's nation-states claim, develop, and ultimately conserve their natural resources. Scientists helped identify, frame, and fight some of the region's worst ecological problems. The environmental sciences gained cultural authority—particularly in the second half of the twentieth century—precisely because they offered concrete tools for addressing these problems. Finally, the environmental sciences gained authority as they allied themselves with a broader range of actors. The state remained a major supporter of scientific research, a major employer of research scientists, and a major consumer of scientific knowledge. But in the latter half of the twentieth century, the environmental sciences also aligned themselves with other groups, such as universities, national and transnational corporations, national and international environmental NGOs, multilateral organizations, and research institutes. Returning to Pomeranz's definition of the developmentalist project, we can see how the sciences in Latin America have contributed—in many ways—to state-building and the exploitation of natural resources. At the same time, our story raises the possibility that the developmentalist era *may* be coming to an end. It is too early to tell whether the emergent scientific and political discourses on sustainability represent the beginning of a postdevelopmentalist phase in human history, or whether the developmentalist project will continue largely unabated.

Stuart McCook is professor of history at the University of Guelph, Canada. He received his Ph.D. from Princeton University in 1996. He is the author of *States of Nature: Science, Agriculture, and Environment in the Spanish Caribbean, 1760–1940* (University of Texas Press, 2002). His current project is a global environmental history of the coffee leaf rust. He has published several articles and chapters on this topic and is currently completing a monograph. His broader research interests are in the environmental history of tropical crops, and the history of science in Latin America.

Notes

1. Kenneth Pomeranz, "Introduction: World History and Environmental History," in *The Environment and World History*, ed. Edmund Burke III and Kenneth Pomeranz (Berkeley: University of California Press, 2009), 7.
2. Pomeranz, "World History and Environmental History," 7.
3. For an overview of the economic history of Latin America in the nineteenth and twentieth centuries, see V. Bulmer-Thomas, *The Economic History of Latin America since Independence* (Cambridge: Cambridge University Press, 1994).
4. Peter J. Bowler, *The Norton History of the Environmental Sciences* (New York: W.W. Norton, 1993), 1–7. Another good introduction to the history of the environmental sciences (albeit with a European and North American focus) are Frank N. Egerton's two classic articles "The History of Ecology: Achievements and Opportunities, Part One," *Journal of the History of Biology* 16, no. 2 (1983): 259–310, and "The History of Ecology: Achievements and Opportunities, Part Two," *Journal of the History of Biology* 18, no. 1 (1985): 103–143. The first focuses on the history of pure ecology, and the second on applied ecology.
5. Bowler, *History of the Environmental Sciences*, 204–211.
6. Regina Horta Duarte, "Between the National and the Universal: Natural History Networks in Latin America in the Nineteenth and Twentieth Centuries," *Isis* 104, no. 4 (December 2013): 777–787.
7. Rick A. López, "Nature as Subject and Citizen in the Mexican Botanical Garden, 1787–1829," in *A Land between Waters: Environmental Histories of Modern Mexico*, ed. Christopher R. Boyer (Tucson: University of Arizona Press, 2012), 73–99.
8. Maria Margaret Lopes and Irina Podgorny, "The Shaping of Latin American Museums of Natural History, 1850–1990," *Osiris* 15 (1 January 2000): 108–118; Maria Margaret Lopes, "A formação de museus nacionais na América Latina independente," *Anais Museu Histórico Nacional* 30 (1998): 121–145.

9. Frank Safford, *The Ideal of the Practical: Colombia's Struggle to Form a Technical Elite* (Austin: University of Texas Press, 1976); Ana Waleska P. C. Mendonça, "A universidade no Brasil," *Revista brasileira de educação* 14 (August 2000): 131–150.

10. Leida Fernández Prieto, "Islands of Knowledge: Science and Agriculture in the History of Latin America and the Caribbean," *Isis* 104, no. 4 (1 December 2013): 790–791.

11. Stuart McCook, *States of Nature: Science, Agriculture, and Environment in the Spanish Caribbean, 1760–1940*, 1st ed. (Austin: University of Texas Press, 2002), 11–20; Lopes, "A formação de museus nacionais."

12. McCook, *States of Nature*, ch. 2.

13. José Augusto Pádua, *Um sopro de destruição: pensamento político e crítica ambiental no Brasil escravista, 1786–1888* (Rio de Janeiro: Jorge Zahar Editor, 2002); José Augusto Pádua, "A profecia dos desertos da líbia: conservação da natureza e construção nacional no pensamento de José Bonifácio," *Revista Brasileira de Ciências Sociais* 15, no. 44 (October 2000): 119–142.

14. Warren Dean, "The Green Wave of Coffee: Beginnings of Tropical Agricultural Research in Brazil (1885–1900)," *Hispanic American Historical Review* 69, no. 1 (1989): 91–115.

15. Warren Dean, *Brazil and the Struggle for Rubber: A Study in Environmental History* (Cambridge: Cambridge University Press, 1987); John Soluri, *Banana Cultures: Agriculture, Consumption, and Environmental Change in Honduras and the United States* (Austin: University of Texas Press, 2005); Stuart McCook, "The Neo-Columbian Exchange: The Second Conquest of the Greater Caribbean, 1720–1930," *Latin American Research Review* 46, special issue (2011): 11–31; on commodity diseases, see John Soluri, "Something Fishy: Chile's Blue Revolution, Commodity Diseases, and the Problem of Sustainability," *Latin American Research Review* 46 (2011): 55–81.

16. Gregory T. Cushman, *Guano and the Opening of the Pacific World: A Global Ecological History* (Cambridge: Cambridge University Press, 2012), 47.

17. Dean, "Green Wave of Coffee."

18. Leida Fernández Prieto, *Espacio de poder, ciencia y agricultura en Cuba: El Círculo de Hacendados, 1878–1917* (Madrid: CSIC, 2009), ch. 3.

19. McCook, *States of Nature*, 77–104.

20. Stuart McCook, "Las epidemias liberales: Agricultura, ambiente y globalización en Ecuador (1790–1930)," in *Estudios sobre historia y ambiente en América*, vol. 2, ed. Bernardo García Martínez and María del Rosario Prieto (Mexico, DF: Instituto Panamericano de Geografía e Historia, 2002), 223–246; Dean, *Brazil and the Struggle for Rubber*.

21. A. L. Olmstead and P. Rhode, *Creating Abundance: Biological Innovation and American Agricultural Development* (Cambridge: Cambridge University Press, 2008), 62.

22. J. R. McNeill, *Mosquito Empires: Ecology, Epidemics, Empires and Revolution in the Greater Caribbean, 1640–1914* (Cambridge: Cambridge University Press, 2010).

23. Mariola Espinosa, *Epidemic Invasions: Yellow Fever and the Limits of Cuban Independence, 1878–1930* (Chicago: University of Chicago Press, 2009); Eric D. Carter, *Enemy in the Blood: Malaria, Environment, and Development in Argentina* (Tuscaloosa: University of Alabama Press, 2012).

24. Mariola Espinosa, "Globalizing the History of Disease, Medicine, and Public Health in Latin America," *Isis* 104, no. 4 (December 2013): 798–806.

25. David S. Parker, "Civilizing the City of Kings: Hygiene and Housing in Lima, Peru," in *Cities of Hope: People, Protests, and Progress in Urbanizing Latin America, 1870–1930*, ed. Ronn F. Pineo and James A. Baer (Boulder, CO: Westview Press, 1998), 153–178; Espinosa, *Epidemic Invasions*; Nancy Stepan, *Beginnings of Brazilian Science: Oswaldo Cruz, Medical Research and Policy, 1890–1920* (New York: Science History Publications, 1976); Jeffrey D. Needell, "The Revolta Contra Vacina of 1904: The Revolt against 'Modernization' in Belle-Époque Rio de Janeiro," *Hispanic American Historical Review* 67, no. 2 (1987): 233–269.

26. Hebe M. C. Vessuri, "Academic Science in Twentieth-Century Latin America," in *Science in Latin America: A History*, ed. Juan José Saldaña (Austin: University of Texas Press, 2006), 197–230.

27. Adriana Feld, "Planificar, gestionar, investigar. Debates y conflictos en la creación del CONACYT y la SCONACYT (1966–1969)," *Eä-Journal of Medical Humanities & Social Studies of Science and Technology*, 2, no. 2 (December 2010), accessed 04 October 2017, http://www.ea-journal.com/en/numeros-anteriores/62-vol-2-n-2-diciem bre-2010/570-planificar-gestionar-investigar-debates-y-conflictos-en-la-creacion-del-conacyt-y-la-seconacyt-1966-1970; Mendonça, "A universidade no Brasil"; S. Motoyama, "A gênese do CNPq," *Revista da Sociedade Brasileira de História da Ciência* 3 (1985): 27–46; Vessuri, "Academic Science in Twentieth-Century Latin America."

28. Regina Horta Duarte, "Passaros e cientistas no Brasil: Em busca de proteção, 1894–1938," *Latin American Research Review* 41, no. 1 (2006): 3–26; Regina Horta Duarte, *A biologia militante: o Museu Nacional, especialização científica, divulgação do conhecimento e práticas políticas no Brasil, 1926–1945* (Belo Horizonte: Editora UFMG, 2010); Camilo Quintero, *Birds of Empire, Birds of Nation: A Place for Science and Nature in U.S.-Colombia Relations* (Bogotá: Universidad Nacional de Colombia, 2012); Lane Simonian, *Defending the Land of the Jaguar: A History of Conservation in Mexico* (Austin: University of Texas Press, 1995).

29. McCook, *States of Nature*, 30–31.

30. Quintero, *Birds of Empire, Birds of Nation*, 17.

31. Megan Raby argues that biodiversity science developed in tandem with the expansion of U.S. hegemony in the Caribbean. See Megan Raby, *American Tropics: The Caribbean Roots of Biodiversity Science* (Chapel Hill: University of North Carolina Press, 2017).

32. Duarte, "Passaros e cientistas no Brasil"; Duarte, *A biologia militante*; Quintero, *Birds of Empire, Birds of Nation*.

33. Sterling Evans, *The Green Republic: A Conservation History of Costa Rica*, 1st ed. (Austin: University of Texas Press, 1999), 7–11.

34. For a thoughtful case study of the Green Revolution, see Wilson Picado's study of rice improvement in Costa Rica, "Las buenas semillas. Plantas, capital genético y Revolución Verde en Costa Rica," *Historia Ambiental Latinoamericana y Caribeña* 2, no. 2 (20 February 2013), accessed 04 April 2014, http://www.fafich.ufmg.br/halac/index.php/periodico/article/view/61.

35. Humberto Bravo A., "Variation of Different Pollutants in the Atmosphere of Mexico City," *Journal of the Air Pollution Control Association* 10, no. 6 (1960): 447–449; Talli Nauman, "Mexico City's Battle for Cleaner Air," *Contemporary Review* 281, no. 1642 (November 2002): 279ff.

36. Ronald A. Foresta, *Amazon Conservation in the Age of Development the Limits of Providence* (Gainesville: University of Florida Press, Center for Latin American Studies, 1991), 14; José Drummond and José de Andrade Franco, "Nature Protection: The FBCN and Conservation Initiatives in Brazil, 1958–1992," *Historia Ambiental Latinoamericana y Caribeña* 2, no. 2 (2013): 351.

37. Michael E. Soulé, "What Is Conservation Biology?," *BioScience* 35, no. 11 (December 1985): 727–734; Raby, *American Tropics*.

38. Emily Wakild, "Parables of Chapultepec: Urban Parks, National Landscapes, and Contradictory Conservation in Modern Mexico," in *A Land between Waters: Environmental Histories of Modern Mexico*, ed. Christopher R. Boyer (Tucson: University of Arizona Press, 2012), 201; Quintero, *Birds of Empire, Birds of Nation*.

39. Earth Council, *The Earth Summit, Eco 92: Different Visions* (San Jose, Cost Rica: Earth Council, Inter-American Institute for Cooperation on Agriculture, 1994).

40. Patricia Romero Lankao, "How Do Local Governments in Mexico City Manage Global Warming?," *Local Environment* 12, no. 5 (2007): 519–535; Luisa T. Molina and Mario J. Molina, *Air Quality in the Mexico Megacity: An Integrated Assessment* (Dordrecht: Kluwer Academic Publishers, 2002); OECD, *OECD Environmental Performance Reviews: Mexico 2013* (Paris: Organisation for Economic Co-operation and Development, 2013), accessed 09 June 2014, http://www.oecd-ilibrary.org/content/book/9789264180109-en.

41. Miguel A. Altieri, *Agroecology: The Science of Sustainable Agriculture* (Boulder, CO: Westview Press, 1995); Efraím Hernández Xolocotzi, *Agroecosistemas de México: contribución a la enseñanza, la investigación y la divulgación agrícola* (Chapingo: Colegio de Postgraduados, 1977).

42. Peter Rosset and Medea Benjamin, *The Greening of Cuba: A National Experiment in Organic Farming* (San Francisco: Global Exchange, 1994).

43. Margarida Cassia Campos, "Modernização da agricultura, expansão da soja no Brasil e as transformações socioespaciais no Paraná," *Revista Geografar* 6, no. 1 (30 June 2011): 161–191.

44. Soluri, "Something Fishy."

45. Andrés Guhl, *Café y cambio de paisaje en Colombia, 1970–2005* (Medellín: Banco de la República and Fondo Editorial Universidad EAFIT, 2008).

46. Cushman, *Guano and the Opening of the Pacific World*.

47. Mark Carey, *In the Shadow of Melting Glaciers: Climate Change and Andean Society* (New York: Oxford University Press, 2010).

48. Philip M. Fearnside, "Deforestation in Brazilian Amazonia: History, Rates, and Consequences," *Conservation Biology* 19, no. 3 (1 June 2005): 680–688.

 CHAPTER 11

A Panorama of Parks

Deep Nature, Depopulation, and the Cadence of Conserving Nature

Emily Wakild

During the twentieth and early twenty-first centuries, Latin American countries set aside nature for conservation in thousands of parks and reserves. Today, such designations cover more than one-fifth of Latin America's territory (see map 11.1). Parks range from Costa Rica's tiny coastal Manuel Antonio National Park, consisting of a mere sixteen square kilometers, to those as large as Brazil's Tumucumaque National Park, at nearly fifteen thousand square kilometers, a size larger than Belgium. Conservation areas famously include some of the region's and even the world's largest tropical forests, but also its driest deserts, highest mountain ranges, biggest waterfalls, deepest marine reserves, and even parts of urban landscapes. Political and economic circumstances, social pressures, cultural preferences, scientific theories, development mandates, and individual personalities shaped how, when, and why Latin American nations have conserved nature.

Precisely what conservation meant, to whom it appealed, and why it proved politically palatable depended on the time and the place. In the 1920s, Chilean conservationists acted on their desires to restore and protect temperate forests, while at the same time Argentina's National Park Service placed tourism and border security high on their agenda. Many of Mexico's national parks have their origins during the social revolution of the 1930s, while Costa Rica's owe their existence to peaceful and deliberate state-building in the 1970s. The conceptual simplicity of the idea of national parks has driven their ubiquity, and sometimes parks have been only symbolic, with few actual changes on the ground. For instance, in 1956, Guatemala simply declared every volcano a national park with little lasting effect. Conservation is nearly always contested, reflecting internal hierarchies of power, including those grafted along colonial lines, but in Latin America there is little reason to believe that parks were external creations imposed by a "global conservation aristocracy."[1] More accurately, conservation worked like an echo that refracted society onto na-

Map 11.1. Designated conservation areas in Latin America, 2015.

National Parks

Designated Nature Protection Areas

Map includes all conservation areas in Latin America within the World Database of Protected Areas (WDPA).

ture, furthering the mutually intertwined relationship between the two. Parks did not merely enclose a piece of nature away from society; by designating a particular space as part of a national or international protected area, the landscape was set on a different ecological trajectory through a societal action. To choose one place and not another altered the evolutionary path of human and nonhuman communities along the way. This means conservation areas form artifacts of the past, present, and future.

To see conservation as simply a phenomenon of the twentieth century is to misunderstand the effects of history on both landscape and culture. In some ways, the deep past, especially the Pleistocene extinctions of more than ten thousand years ago, shaped nature's dynamic course and determined what remained to be conserved. Mass extinctions of large animals coincided with the arrival of humans and transformed prairies into forests, among other effects. Other absences also mattered to the landscapes and the cultures that conserved them. Five hundred years ago, when another group of humans arrived, the disease and violence they brought decimated native populations and created a cascade of changes socially and environmentally. Forests resurged in fields that went unplanted, and new species emerged where fires were no longer set. These transformations set nature on new paths. Nature alone does not determine conservation; conservation's contours are unleashed by human conceptions of whether a forest or a desert has special value, or if some places retain sacred attributes that deserve to exist beyond the reach of commerce. To explain how this played out, this chapter explores some ways deep histories of extinction and shared experiences of colonialism shaped nature. The chapter then offers four possible origin stories for national parks before providing a rough timeline summarizing twentieth-century conservation efforts.

Deep Nature, 16,500 to 10,000 Years Ago

The contours of the peopling of the Americas are commonly known, yet the specifics remain controversial and dynamic, changing with new techniques and discoveries. Most likely, bands of people crossed the Bering Strait during periods of low sea levels caused by glacier expansion. People then either followed animals into and across North America, hunting and leaving clues such as the Clovis point, or they followed the bountiful Pacific coastlines as far south as Chile.[2] What people did as they spread across the continents remains debated but that these migrations transformed the landscape is obvious: by the Late Pleistocene, mammals exceeding a hundred pounds were reduced by more than seventy percent and only relatively small species survived.[3] It is likely that climate change, disease, and other factors played a role but the Americas were already relatively depauperate in large mammals, and these extinctions further depleted the stocks. Glaciation, human distribution, and large animal extinction combined over thousands of years to bequeath a particular configuration of nature.

The consequences of these processes for twentieth-century conservation may seem far removed, but the cascading effects of mass extinction remade nature. Grasslands became forests without large grazers to browse back seedlings and fertilize fields. Medium-sized animals such as white-lipped peccaries, anteaters, and capybaras took refuge in tropical forests and changed the forests' composition as they foraged and spread seeds. Because evolution is a process that takes place over generations through species interactions with each other and with their environment, Pleistocene extinctions altered the course of plant and animal evolution. As a result, parks protect birds and monkeys rather than enormous animals such as giant ground sloths (Mylodons) or automobile-sized armadillos (Glyptodonts). Similarly, big game hunting has had a minimal influence on the creation of conservation areas—quite unlike the trajectory of conservation in sub-Saharan Africa.[4]

The Pleistocene matters for more than the past interactions between humans and fauna; it has shaped the scientific theories that drove conservation programs. Pleistocene temperatures ranged much cooler than today, and in the tropics, rainfall likely decreased. Many areas that are today tropical forests were then savannahs due to lower precipitation. Some areas remained forested, and in the 1970s, scientists began to think of these sites as "Pleistocene Refuges," theorizing that the refuges help account for the high species diversity of ecosystems in the Amazon. First published by the German petrochemical engineer Jürgen Haffer in 1969, but also developed by Brazilian zoologist Paulo Vanzolini, this theory posited that the refuges were places species (specifically certain plants, birds, butterflies, and lizards) maintained their populations during the period of precipitation decline.[5] When temperatures warmed and rainfall returned, a new wave of speciation took place as flora and fauna moved outward from the refuges into new niches. A lovely, if all too simple, theory that has today largely been disproved, the notion of Pleistocene Refuges nevertheless drove conservation planning in Brazil in the late 1970s as a way of determining sites of high diversity and endemism. Millions of hectares of parks were designated because certain territories appeared to harbor these natural refuges. The Pleistocene has mattered to how nature in the Americas unfolded and to how societies used their understanding of the ancient past to conserve that nature.

Colonial Canvas

Just as there is no consensus on exactly how and when peoples arrived in the Americas, there is no certainty to the numbers of people who lived in the hemisphere in 1492. It is clear that human societies found and created modes of subsistence, survival, and sophistication in nearly every biome and habitat.[6] From ancient civilizations in Aspero, Peru that plucked protein out of the Humboldt Current to Mexica warriors that feasted on grasshoppers and lizards in their lakebed metropolis of Tenochtitlán, indigenous peoples employed complex strategies of using nature. Agricultural prowess created maize, squash, and potato varieties, and the productivity of cultivation supported tens of millions of people, especially in Mesoamerica and the Andes. Plants stood out as the basis of indigenous farming practices as few animals were domesticated. The development of agriculture and sedentary societies shaped nature, as it did in fertile valleys around the world, but it was the interruption of this process by colonialism—rather than the process itself—that set the Americas apart.

Europeans' arrival was ecologically cataclysmic. It devastated native peoples through disease and warfare and set cultural and spiritual systems into chaos. This reverberated in the landscape. Massive ecosystems, vast beyond the size of European nations, were abandoned to their own processes. The cessation of farming and reduction of fire use led to the regeneration of over fifty million hectares of forest, woody savanna, and grassland by the mid-eighteenth century.[7] Today's "old growth" trees are, at least in part, products of how nature resurged when humans were removed. Colonial depopulation further shaped nature's exuberance by ensuring that many species hardly interacted with humans for centuries. What does it mean, then, to say that wild nature, abundant animals, and prolific forests are products of colonialism? This makes resplendent nature an artifact of human calamities.

Imagining conservation as a colonial concept, as much recent literature on conservation tends to do, oversimplifies the fact that colonialism and conservation did not coincide in Latin America. While colonizers sometimes engaged in conservation—including restrictive but ineffective Portuguese forestry policies—formal colonial powers did not enact the first nature reserves in South America as they did in Southern Africa or South Asia.[8] This discrepancy matters to questions about the sovereignty or intentions of conservationists. Seeking a global synthesis, scholars have argued that "the rapid growth of protected areas prior to 1960 owes something to the anxiety of colonial rulers to set aside land before they lost power."[9] Colonial forces shaped nature and society, but in this region they did not define or even begin conservation. To overlook or dismiss the actions of independent Latin American nations in the realm of nature conservation is to discount the nationalist struggles waged for political and economic autonomy. Rather than critique it, such an argument resurrects a colonial mentality.

In a global context, the colonial experience in Latin America was earlier, longer, deeper, and more systematic, thus leaving shards scattered everywhere. But independence was also sooner, fiercer, and shot through with mixed-caste, cross-class, and popular or intellectual appeals for authority.[10] The temporal distance—from decades to more than a century—between formal colonial rule and national park programs meant that nations had decades to contend with the development of a state capable of managing nature and to struggle through the articulation of national identities. International ideas and norms, and especially those tied to the United States, came to shape and inspire many parks. And yet, rather than a colonial imposition or a direct importation of North American ideas, nature protection in the region proceeded at a disjointed pace and left an uneven patchwork of parks with no progression toward a singular goal. In many instances, parks were tied to the charisma and commitment of specific individuals or small groups who came to have a great influence over official designations.[11]

Diverse Cosmopolitan Roots

While many cultures engaged in conservation at diverse moments, there are at least four places to start the history of national parks in Latin America: Argentina, Brazil, Chile, and Mexico. The first known suggestion for a national park occurred in 1876, when the Brazilian military engineer and abolitionist, André Rebouças, recommended the creation of two national parks, one on the fluvial Bananal Island in the Araguaia River and another of the Seven Falls cataracts on the Paraná River near Iguaçu Falls.[12] While traveling in the United States, Rebouças heard about Yellowstone National Park and imagined similar parks for Brazil, mainly to justify tourism.[13] Brazil created its first official park, Itatiáia, in 1937, and Rebouças' idea was heeded in 1959 with the creation of Araguaia National Park.

By then, Mexico had created many national parks. During the late nineteenth century, concerns over deforestation drove scientifically oriented political advisors, or *científicos,* toward conservation. Miguel Ángel de Quevedo was the most prolific of these, and his European training and elite upbringing earned him the nickname "Apostle of the Tree." Although Quevedo left the country during the first decade of the Mexican Revolution (1910–1940), his championing of conservation influenced the nation's first national park, Desierto de los Leones, which was declared around a popular urban forest in the same year as the new Constitution of 1917. By the mid-1930s, when the constitution's social aims reached their fullest implementation, Quevedo would oversee the creation of a series of parks linked to revolutionary ideas of social justice.[14]

Another beginning is 1903, when Francisco P. Moreno donated three leagues of land to the Argentine government with the explicit demand they be incorporated as the National Park of the South. His donation was immediately accepted, but a formal national park, now called Nahuel Huapi, was not established on the site until 1922, and actual administration developed in the mid-1930s, when several other parks in the region were proposed and created.[15] A fourth origin story is the request of the forester Federico Albert, who insisted that the Chilean government create its first national park, named after the intellectual Benjamín Vicuña Mackenna, to protect hillsides from logging and erosion. The park was created on paper in 1925 but subsequently annulled in 1929 because of conflicts over land tenure. In 1926, the government established Vicente Pérez Rosales National Park, named after the state agent in charge of colonization in the region, directly adjacent to Moreno's donated land in Argentina.[16]

Who, then, merits recognition as the founder of conservation in Latin America? Brazil's black son of a slave, who traveled insatiably within and beyond his native land valuing nature along the way? Mexico's elite *científico*, who worked for four decades and through various regimes until Mexico had more national parks than any country in the world? Argentina's trilingual scientist, frontier explorer, and museum director, who donated private land back to the state for

Illustration 11.1. Families enjoying the central plaza of Huaraz, Peru, during the International Glacier Forum in June 2013.

Large billboards such as the one pictured explained the role of glaciers in the national system of protected areas.

Photo by author.

a park? Chile's first official forester, who emigrated from Germany and spent his life guiding conservation decrees and laws in his adopted country? Each of these individuals point to conservation's diverse and cosmopolitan roots. One thing is certain: there was no single hero and no native "Yellowstone" on which to model other parks. These men, and later several women, point to the ubiquity and commonality—but also the diversity—of conservation's history. Parks in the region were initially driven by charismatic individuals who themselves represent distinct experiences and orientations. As the origin stories indicate, the chronology of conservation is hardly straightforward. National pulses were irregular, yet their combined cadence reveals a possible regional timeline, articulated next. I hope other scholars will later expand, refine, and contest this chronology as more is learned.

A Suggestive Sequence

Given the contours of deep nature, the features of colonial rule, and the idiosyncrasies of individual champions, a rough chronology of conservation actions in Latin America during the twentieth century can provide a useful tool for understanding regional trends. The general timeline consists of two surges with a lull in between. Some caveats about the quantitative data used to sketch out this timeline are in order. Conservation's statistics are notoriously difficult to compile, but the World Database of Protected Areas (WDPA), a joint project of the International Union for the Conservation of Nature and the United Nations Environment Program, collects and updates worldwide information monthly.[17] Although it ultimately relies on reporting by individual countries, which can lag, this is the most reliable way to gather basic dates, places, and sizes. The WDPA includes any area considered by a reporting government to be a conservation unit (with 182 different designations)—including national parks, forest reserves, wetland sites, faunal reserves, and much more. The WDPA is prone to counting areas more than once if they have dual designations, for example a Ramsar wetland might be inside a National Park, so it has two listings. Furthermore, this data does not include most state or local conservation areas. However, it shows general trends clearly.

In a timeline, three eras stand out. A trickle of conservation areas emerged scattered around the region by the early twentieth century. Conservation surged forward in the 1930s, when several countries began putting parks into place alongside administrative bureaucracies such as ministries of agriculture or forestry departments. This initial era might be thought of as an intensely nationalist period, when countries experimented with conservation. Advocates heard about conservation programs elsewhere or observed their necessity and had the autonomy to try them out. This slowed around 1940, and park

Figure 11.1. Total conservation areas created per year in Latin America.

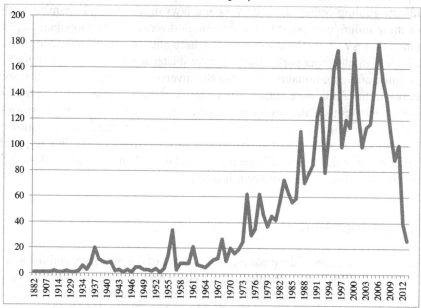

Compiled from data from the World Database of Protected Areas (WDPA) and used with permission, 2015. Data available at www.protectedplanet.net. Chart by author. Please note the decline from 2010 is due to reporting lags, not to a cessation of conservation areas.

creation remained minimal—with the exception Guatemala's 1956 volcano frenzy—until the late 1960s. A second era can be characterized as a hiatus, when conservation hovered with little support and developmentalism forged ahead, accelerating the destruction of nature. The third era began in 1968 with a rapid swell tied to the increased sophistication of international institutions and new interest in tropical environments. This era marked a global turn and the consolidation of standardized conservation principles. This rapid pace accelerated every few years through the end of the century. Future studies may differentiate more recent trends of the twenty-first century.

Nationalism's Nature: Conservation's Arrival, 1920–1940

The first significant era for conservation occurred from roughly 1920 to 1940 and was largely a response to concerns over degradation combined with a curiosity about the notion of protecting land. Justifications for action included largely elite ideas about scenery and tourism potential, intellectual suggestions of the need for scientific study, or strategies to secure frontier lands with national claims. That is, scarcity, scenery, and security shaped early approaches toward conservation. In general, temperate and alpine landscapes of conifers, mountains, and glaciers were preferred for protection, and countries with these landscapes were the first to create conservation areas. As a rule, countries with both tropical and temperate landscapes protected the higher, cooler, more forested areas first. In some countries, such as Mexico or Brazil, these areas hosted the largest populations of people. For example, the pine and fir forests surrounding Mexico City—the nation's undisputed political and economic center—were seen by early conservationists as worthy and in need of protection.

In Chilean and Argentine Patagonia, mountainous and forested landscapes were, and still are, a distant frontier with few settlements and sparse populations. Indeed, only about five percent of Argentina's population lives in the region that takes up more than a third of the country's territory. Exequiel Bustillo, director of Argentina's first Park Service, believed in the authority of the national government to "conserve nature in its virgin state, preserve the beauty of the landscape, and create rapid public access."[18] He prioritized Patagonia, focusing his efforts on roads and amenities for tourism. Parks were also largely stacked along the international borders to fortify territorial claims. If Argentines were preoccupied with security, Chileans had different motives. Chilean parks in the 1920s similarly occupied a frontier region, but one that had experienced disastrous colonization schemes to grow wheat. The resulting deforestation and erosion raised concern in the national government, which turned to forestry as an industry. Concern with erosion led to parks as a strategy to guarantee forests rather than showcase scenery.[19] Tree plantations, state-run nurseries, forest reserves, and national parks formed a portfolio of possible arrangements designed by and managed for state purposes under the umbrella of conservation. These motivations were obviously influenced by connections to the global economy, but were also shaped by domestic agendas. The Valley of Mexico and the Patagonian Andes received some of the first park dedications, again reflecting regional preferences for alpine settings.

Unique, scenic, and economically unviable lands were usually the first to earn conservation status. The creation of Iguazú/Iguaçu Falls National Parks are paradigmatic cases that solidified the argument for monumental scenery. Brazil and Argentina had different reasons for protecting the falls and creating a park, but for both, the falls' immensity as a symbolic landmark gave obvious cause. Argentina moved first. On 9 October 1934, President Augustín Justo signed a law creating the Dirección de Parques Nacionales within the Ministry of Agriculture. The law spelled out the duties, finances, and jurisdiction of the new agency and officially created Parque Nacional Iguazú (alongside formalizing Moreno's donation as Parque Nacional Nahuel Huapi).[20] After the creation of the park on the Argentine side, members of the Brazilian government feared an influx of Argentine settlers, and viewed economic development and territorial control as a way to incorporate a peripheral region. Brazil created Parque Nacional do Iguaçu in 1939 as a culminating change after more than fifty years of measures to nationalize the area that bordered both Argentina and Paraguay. In both countries, the military played a strategic and important role in the creation of the park, as did the idea of state-led development through the colonization of a hinterland.[21] No small bit of posturing and competition drove the policies that shaped the dual park, as both countries went about figuring out what a park was and what it might do.

Park creation appears related to the expansion of stable state institutions because official conservation schemes relied on such institutions for management and legitimacy.[22] Parks echoed important social and economic changes, especially the emergence of political nationalism and the effects of the Great Depression. This marked a shift among the largest economies away from an export-driven model of production and toward policies of import-substitution industrialization (ISI). With the collapse and withdrawal of U.S. and British capital, a space for state growth opened and more robust bureaucracies developed. Personnel, budgets, and tangible structures (such as roads, signs, and maps) turned an official decree into an actual park. Increased nationalist and populist political strategies in Brazil, Argentina, Chile, and Mexico meant advocates latched on to the idea of a national park and adapted it to their own state systems. The more the state had developed, the easier it was to accommodate national parks. Some nations housed their park bureaucracies within their ministries of agriculture or grouped them with forestry departments, while others saw them belonging within secretariats of tourism. Often dismissed or ignored, these early parks and the visions they represent provide a useful historical record that confirms diverse roots.

The Hiatus: Developmentalism above Conservation, 1942–1968

The articulation of nationalist politics and ISI policies opened the door for conservation, but the implementation of these policies created competing demands. A second period, dominated by promises of economic development, started around 1942. Since parks contradicted the message of industrialization, few parks were created in this era, and administration generally languished. There were two years that saw over twenty protected areas designated, but most saw fewer than five. In most countries, politicians focused on industrialization, infrastructure, and integrating newly organized and powerful groups (urban workers, middle-class professionals, etc.) into political systems. While conservation had some popular appeal, it was more associated with intellectuals strengthening the institutional role of science in the region. Parks were not created in meaningful numbers during this period but some investments were made in institutions that would later steward conservation. For example, two internal research institutes that blended scientific and social research, Brazil's National Institute of Amazonian Research and Chile's Institute of Patagonia, were created during the hiatus and signified the continual evolution of formal scientific thinking and research.[23]

Perhaps one of the most fitting examples of the hiatus is the minimal influence of the 1940 Pan-American Convention on Nature Protection, and the failure of half of regional countries to ratify or accept it initially.[24] Modeled on wildlife treaties such as the London Convention, the treaty for the western hemisphere had less influence in part because of the start of the war and the eagerness of politicians to demonstrate their independence from U.S. influence.[25] Another example of this bristling autonomy comes with the role of the U.S. scientist William Vogt. As one of the last biologists involved with the Guano Commission in Peru, Vogt had decades of experience in the region before he served as a Latin American representative to the Pan-American Union with a primary goal of convincing countries to create national parks.[26] A skilled and prolific writer, he nevertheless had little success with this. Vogt quickly shifted to issues such as erosion and soil conservation, and it was not until the late 1960s that countries responded to outside requests for national parks, largely by inviting consultants from international bodies.[27] If external demand and models for conservation had an outsized influence, this should have been a time when they expanded.

World War II and the successes of ISI had contradictory impacts that sidelined conservation. U.S. demand for resources shifted some communities back into export-driven activities, especially for woods, fibers, and fuels. Industrialization policies made workers the centerpiece of economic policies, pivoting attention away from the countryside. Food prices needed to be kept low in order to facilitate urban living, and state intervention in food production overrode conservation efforts. Population numbers surged in this period, increasing demands on land and shuffling political priorities. Conservation disappeared inside developmentalism, with a few exceptions that merit more research. For instance, the Guatemalan dictator Castillo Armas created thirty-two conservation areas to protect as many volcanoes in 1956. It is not clear if these were a hedge against land reform efforts, an expression of elite demands, or an attempted populist measure.

There is some evidence to suggest that although park creation lagged in this period, the actual use of conservation areas increased. Argentina, under President Juan Perón, developed a mass tourism industry that incorporated the national parks. Perón added a few new parks, but more importantly initiated state subsidies for excursions to Nahuel Huapi and Iguazú. In 1948, the peak year for this initiative, more than twenty-six hundred people, largely workers and students, participated in these trips. Nationalist celebrations—such as an official "National Park Day" on the day Moreno donated his lands—heightened campaigns for conservation, and Perón used the parks for international as well as internal prestige.[28] The defense of nature and the development of tourism seemed to nicely coexist by appealing to elite, middle, and working-class patrons in different ways, thus diversifying the panorama of protection.

The International Turn: Amplifying Global Trends, 1970–2000

By the early 1970s, international organizations took the initiative in organizing and standardizing conservation around the world. Conservation reemerged as a global priority and a reaction to the developmental mandate. For instance, the Nature Conservancy opened its "Latin America Desk" in 1966 at the behest of Maria Buchinger, a resident of Argentina born and educated in Europe. She argued that the desk would serve as an information clearinghouse, promote public understanding, and assist in the establishment of similar organizations.[29] The first international meeting of conservation in Latin America took place in late March of 1968 in the location where Moreno proposed Argentina's first national park. The conference was organized by the International Union for Conservation of Nature (IUCN) and sponsored by the United Nations Educational, Scientific and Cultural Organization (UNESCO) and the Food and Agriculture Organization (FAO).[30] One hundred and fifty-five people participated in the conference and this included representatives from thirteen Latin American countries. The United States, the United Kingdom, and the sponsoring institutions were well represented (it is hard to say how many of the FAO or IUCN representatives were from Latin American countries). Conference panels included prominent scientists, foresters, and wildlife specialists, many of whom would go on to have long and distinguished careers leading their national conservation systems. Topics ranged from conservation laws and education to landscape planning and tourism. Although forestry or tropical forests were notably absent, migratory birds, sea turtles, primates, vicuña, and vampire bats all made the program. The longest sessions were dedicated to forging a regional agreement for coordinating conservation efforts. If nations were reluctant to commit to conservation in 1940, by 1968 this new generation of conservationists was ready to act. Instead of a comprehensive agreement, most developed their own national strategies. Conference participants looked to the past and the future to set forth new relationships for conservation in the region, especially through the coordination of national managers and scientists with international institutions.

Meetings provide historical barometers—they are moments of exchange that allow insight into how people communicated over a particular topic. For instance, Ian Grimwood, a British scientist born in Kenya, represented the

U.K. at the meeting but was based at the time in Lima, Peru, where he served as a consultant to professors at the National Agrarian University, La Molina. One interpretation might see this relationship as colonial, given the context of conservation in Africa, but to men like Marc Dourojeanni—a professor at La Molina and Peru's director of general forestry and fauna, who administered national parks—Grimwood and other experts were welcome assets and sources of perspective that often simply confirmed what was already known for Peruvians.[31] A wealth of knowledge and experience within the region was represented at the conference: there were over one hundred professionals working in conservation. They likely met each other for the first time at this meeting and developed friendships that were reinforced at subsequent meetings and endeavors. Global bodies sponsored the conference, but in practice, the meeting helped forge intraregional connections.

The combination of international support, scientific interest, and national policy changes dramatically increased the footprint of conservation in this era. Costa Rica is the paradigmatic case. The Forestry Law of 1969 was the principal milestone in the country's conservation history, and it set into place a General Forestry Directorate and within it a system of national parks.[32] This law, despite changes in subsequent decades, formed the cornerstone of conservationist responses to environmental problems, and it outlined national parks as sites for protecting flora and fauna, but also promoting recreation, tourism, and scientific research. In 1970, Costa Rica had no national parks or protected areas, and six years later it was used by the United Nations as a model for how to create plans to protect flora and fauna. Nearly a hundred thousand people visited the first parks by 1972, and such popularity persists today, forming a key component of Costa Rican's identity and economic development strategy. Professional administrators, international consultants, and volunteers—from the Boy Scouts to the Peace Corps—contributed to a surge in parks that catapulted the country to a position on the cutting edge of conservation strategies. By 1980, it conserved a greater percentage of its territory than did the United States.[33]

Perhaps the largest ideological change affecting conservation was the creation of a scientific framework for studying tropical nature as a specific subfield of ecology. Since Humboldt, scientists had been interested in the prodigious forests of the tropics, but the notion of biodiversity, which generally refers to the range of life in a particular place, emerged formally in the late 1970s. This gave new force to the tropics as a place of value for scientific study. As David Takacs has noted, the generation and dissemination of the term "biodiversity" was intended to change the way people viewed nature.[34] The emergence of conservation biology as a field in the 1980s was designed to put biology at the service of conservation, and biodiversity became a rationale for creating conservation areas and a source of pride for promoting them. It seems this idea has had a more widespread influence in justifying conservation than did the older notion of wilderness.

More parks created a situation in which the vast majority of conservation areas have had people residing within them.[35] Who these people are and how they live is a matter of continual debate, but cohabitation rather than evictions have been the norm. To be sure, evictions have occurred—sometimes with land exchanges and more often without—but they appear to have happened less frequently than in other regions of the world. As much as tools of dispossession, parks must also be seen as tools of distribution. They provided an alternative for people (local and otherwise) who would rather throw in their lot with intact nature than oil drilling, gold mining, or lumbering. In some cases, people within conservation areas have made specific demands on the state, and some groups, such as rubber-tappers, have appropriated conservationist discourse to further their assertions of the right to livelihoods on vast tracts of land. In 1978, Peru declared the "Law of the Jungle," which recognized the rights of indigenous populations in the Peruvian Amazonia to use natural resources on their territories.[36] This led to the creation of communal reserves, a useful instrument of management similar to Brazil's extractive reserves.

By the 1970s, many countries began focusing on specific environments with special attention to the areas that had been left out of their earlier efforts. Peru adopted a plan to create a major national park that showcased each area of the country—the coast, the sierra, and the jungle. Under the leadership of the politically astute and ruthlessly committed María Tereza Jorge Pádua, Brazil set about a program of plotting conservation areas across the tropical Amazon that resulted in the protection of nearly eleven million hectares of territory in Amazonian parks.[37] The plan, which drew on the notion of Pleistocene Refuges, was likely the first worldwide to use a synthesis of biological literature to attempt to locate the most promising sites for conservation. In many ways, the plotting of these parks across the Amazon epitomized the new era of conservation informed by science and collaboration across bureaucratic institutions.

Economic reorganization and political shifts also influenced the distribution of conservation. Mexico had a surge of park creation during the presidency of Carlos Salinas de Gortari (1988–1994), a president whose term is otherwise recognized for the implementation of neoliberal policies and downsizing state-run agencies. Newly valued environments became protected in this era, including deserts of the north and tropical forests of the south. Neoliberalism has had a contradictory influence as many new parks were designed with foreign, rather than domestic, tourists in mind. Concerns over resident peoples and conservation grew in the 1990s, and various scholars from within the region critiqued the notion of pristine nature as a justification for conservation.[38]

The emergence of social concerns for park inhabitants, paired with the creation of a scientific concept for measuring the biological value of lands, occurred within a shifting neoliberal economic context. In some countries, such as Costa Rica, these forces could be melded into a national identity that capitalized on protecting nature. In others, conservation areas were little more than disparate patches camouflaging landscape conversion at a breakneck pace. When Latin American countries increased their conservation activities, many had a deep and long—if not coordinated and comprehensive—past to draw upon.

Conclusion

Latin America's conservation history is both connected to and distinct from other world regions. Scholar Rob Nixon contends that South African conservation systems have been driven by the lucrative, easy sell of spectacular megafauna rather than the "boring" aesthetics of biodiversity.[39] There, large animals oriented nature reserves toward the desires of foreign tourists, who saw domestic and familiar scenes as anathema. For much of Latin America, where the spectacular megafauna disappeared in the Pleistocene, biodiversity became one of the drivers of conservation by the 1980s. This influenced the location and persistence of nature protection and its relationship with local, national, and international communities. Biodiversity may seem dull compared to trophy hunting, but it has likely contributed to the cohabitation of people and other animals within parks, which remains the norm. The widespread experience of people living within parks and the close territorial and ideological relationship between parks and formally recognized indigenous territories have made parks part of landscapes layered with overlapping claims and meanings. Conservation is one of the most classic and conventional topics covered by environmental historians, yet we still lack a clear understanding of the social, cultural, and political contours of nature conservation for nearly all individual countries.[40] Nations have their own conservation culture, but regional perspectives matter in a comparative global context. Argentinean's approach to national parks was shaped by the fear of losing territory. Mexicans created parks as part of their defining social revolution.[41] Costa Ricans proudly boast of their "green republic," and Peruvians point out that their parks have the world's highest concentrations of biodiversity.[42] National parks often had charismatic founders, and it is clear that conservation's cadence is linked to the features that help define the region and the forces that circulate beyond it.

Emily Wakild is Professor of History at Boise State University in Idaho (USA) where she researches and teaches Latin American and Environmental History. She earned her B.A. from Willamette University in Salem, Oregon in 1999 and her Ph.D. in History from the University of Arizona in 2007. She is the author of *Revolutionary Parks: Conservation, Social Justice, and Mexico's National Parks,* (University of Arizona Press, 2011) and with Michelle K. Berry, *A Primer for Teaching Environmental History,* (Duke University Press 2018).

Notes

The author would like to thank the editors, Claudia Leal, José Augusto Pádua, and John Soluri for the invitation to contribute to this project. The editors, along with Lisa Brady and Nick Miller, gave helpful comments that improved the chapter. A National Science Foundation Scholars Award (#1230911) and a National Endowment for the Humanities Fellowship (2015–2016) provided research and writing support.

1. Mark Dowie, *Conservation's Refugees: The Hundred-Year Conflict between Global Conservation and Native Peoples* (Cambridge, MA: MIT Press, 2009), xxvii.

2. Brian Kemp et al., "Genetic Analysis of Early Holocene Skeletal Remains from Alaska and Its Implications for the Settlement of the Americas," *American Journal of Physical Anthropology* 132 (2007): 605–621; Antonio Arnaiz-Villena et al., "The Origin of Amerindians and the Peopling of the Americas according to HLA Genes: Admixture with Asian and Pacific People," *Current Genomics* 11, no. 2 (2010): 103–114.

3. The classic paper is Paul S. Martin, "Pleistocene Overkill," *Natural History* 76, no. 10 (1967). More recently, see A. D. Barnosky et al., "Assessing the Causes of Late Pleistocene Extinctions on the Continents," *Science* 306 (2004): 70–75. Megafauna decreased everywhere, but the rates of extinction were most dramatic in the Americas and Australia.

4. Jane Carruthers, *The Kruger National Park: A Social and Political History* (Pietermaritzburg: University of Natal Press, 1995); William M. Adams, *Against Extinction: The Story of Conservation* (London: Earthscan, 2004).

5. Jürgen Haffer, "Speciation in Amazonian Forest Birds," *Science* 165, no. 3889 (1969): 131–37. For a recent critique see Paul Colinvaux, *Amazon Expeditions: My Quest for the Ice-Age Equator* (New Haven: Yale University Press, 2007). See discussion by David Cleary, "Extractivists, Indigenes and Protected Areas: Science and Conservation Policy in the Amazon," in *Global Impact, Local Action: New Environmental Policy in Latin America,* ed. Anthony Hall, (London: Institute for the Study of the Americas, 2005), 199–216, especially 204–210.

6. For a useful summary of the literature, see Charles Mann, *1491: New Revelations of the Americas before Columbus* (New York: Vintage, 2005).

7. J. O. Kaplan et al., "Holocene Carbon Emissions as a Result of Anthropogenic Land Cover Change," *Holocene* 21 (2001): 775–791; and Simon L. Lewis and Mark A. Maslin, "Defining the Anthropocene," *Nature* 519, no. 7542 (2015): 171–180. The authors note that the decline in atmospheric carbon dioxide caused by this plant growth is the most prominent feature in preindustrial atmospheric records (Antarctic ice cores) over the past two thousand years.

8. Shawn William Miller, *Fruitless Trees: Portuguese Conservation and Brazil's Colonial Timber* (Palo Alto: Stanford University Press, 2000).

9. Dan Brockington, Rosaleen Duffy, and Jim Igoe, *Nature Unbound: Conservation, Capitalism and the Future of Protected Areas* (London: Earthscan, 2008), 32; Mac Chapin, "A Challenge to Conservationists," *World Watch Magazine* (2004).

10. John Charles Chasteen, *Americanos: Latin America's Struggle for Independence* (New York: Oxford University Press, 2009).

11. José Luiz de Andrade Franco and José Drummond, "Nature Protection: The FBCN and Conservation Initiatives in Brazil, 1958–1992," *HALAC* 2, no. 1 (2013): 338–367, and "História das preocupações com o mundo natural no Brasil: da proteção à natureza à conservação da biodiversidade," in *História ambiental: Fronteiras, recursos naturais e conservação da natureza,* ed. José Luiz de Andrade Franco, Sandro Dutra e Silva, José Augusto Drummond, and Giovana Galvão Tavares (Rio de Janeiro: Garamond, 2012), 333–366.

12. André Rebouças, "Notas e considerações geraes pelo engenheiro André Rebouças, Excursão ao Salto do Guayra," *Revista Trimensal de História e Geographia ou Jornal do Instituto Histórico e Geográphico Brasileiro* 61, no. 1 (1898 [1876]): 74–87.

13. José Augusto Pádua, *Um sopro de destrucáo: pensamento político e critica ambiental no Brasil escravista, 1786–1888* (Rio de Janiero: Jorge Zahar Editor, 2002), 270.

14. Emily Wakild, *Revolutionary Parks: Conservation, Social Justice, and Mexico's National Parks, 1910–1940* (Tucson: University of Arizona Press, 2011); Christopher R. Boyer, *Political Landscapes: Forests, Conservation, and Community in Mexico* (Durham: Duke University Press, 2015).

15. Eduardo V. Moreno, ed., *Reminiscencias de Francisco P. Moreno: Versión propia* (Buenos Aires: Plantié Talls. Gráficos, 1942); Eduardo Miguel E. Bessera, "La nacionalización de las fronteras patagónicas. Los Parques Nacionales como herramienta estatal de ocupación e integración territorial," in *Procesos históricos, transformaciones sociales y construcciones de fronteras: Aproximaciones a las relaciones interétnicas,* ed. Graciela Maragliano Sebastián Valverde, Marcelo Impemba, and Florencia Trentini (Buenos Aires: Universidad de Buenos Aires, 2012), 67–88.

16. Chile had previously created forestry reserves at Malleco (1908) and Villarica (1912) in the same general area as these national parks. Gary Bernard Wetterberg, "The History and Status of South American Parks and an Evaluation of Selected Management Options" (Ph.D. diss., University of Washington, 1974), 37–42; and Minsterio de Agricultura, Corporación Nacional Forestal, Décima Región de Los Lagos, *Plan de manejo parque nacional Vicente Pérez Rosales* (Chile: Ministerio de Agricultura, Gobierno de Chile, 1994), 217.

17. See www.protectedplanet.net, where information is available by country. There were 4,861 conservation units listed for the region at this writing (9 September 2016).

18. Salvador San Martin, "Preface," in Exequiel Bustillo, *El despertar de Bariloche,* 2nd ed. (Buenos Aires: Casa Pardo, 1971).

19. Thomas Klubock, *La Frontera: Forests and Ecological Conflict in Chile's Frontier Territory* (Durham: Duke University Press, 2014), 20, 72; Pablo Camus Gayán, *Ambiente, bosques, y gestión forestal en Chile 1541–2005* (Santiago: Dirección Bibliotecas, Archivos y Museos, 2006), 150.

20. Dirección de Parques Nacionales, *Ley de Parques Nacionales* (Buenos Aires, Argentina, 1935).

21. Frederico Freitas, "A Park for the Borderlands: The Creation of the Iguaçu National Park in Southern Brazil, 1880–1940," *Revista de Historia IberoAmericana* 7, no. 2 (2014).

22. Claudia Leal, "Behind the Scenes and Out in the Open: Making Colombian National Parks in the 1960s and 1970s," in *The Nature State: Rethinking the History of Conserva-*

tion, ed. Wilko Graf von Hardenberg, Matthew Kelly, Claudia Leal, and Emily Wakild (London: Routledge, 2017), 135–156.

23. The Instituto Nacional de Pesquisas da Amazônia was created in 1952, and the Instituto de la Patagonia in 1969.

24. The entire name was the Convention on Nature Protection and Wild Life Preservation in the Western Hemisphere. For treaty text and ratification dates see http://www .oas.org/juridico/english/sigs/c-8.html. For a brief discussion, see Marc Cioc, *Game of Conservation: International Treaties to Protect the World's Migratory Animals* (Athens, OH: Ohio University Press, 2009), 94–97.

25. Juan Carlos Godoy in IUCN, *Proceedings of the Latin American Conference on the Conservation of Renewable Natural Resources* (San Carlos de Bariloche, Argentina: International Union for Conservation of Nature and Natural Resources, 1968), 416.

26. William Vogt Papers and Annette L. Flugger Papers, Conservation Collection, Denver Public Library, Denver, CO, USA.

27. Kenton Miller and Gary Wetterberg are two examples of U.S. consultants for the UN-FAO who advised park creation in the 1970s in Colombia, Chile, and Brazil.

28. Eugenia Scarzanella, "Las bellezas naturales y la nación: Los parques nacionales en Argentina en la primera mitad del siglo XX," *Revista Europea de Estudios Latinoamericanos y del Caribe/ European Review of Latin American and Caribbean Studies* 73 (2002): 5–21, esp. 17.

29. Maria Buchinger, "Why the Latin American Desk" (1966), Conservation Collection, Denver Public Library.

30. IUCN, *Proceedings.*

31. Marc J. Dourojeanni, *Crónica forestal del Perú* (Lima: Universidad Nacional Agraria, La Molina: Editorial San Marcos, 2009), 136–141.

32. Sterling Evans, *Green Republic: A Conservation History of Costa Rica* (Austin: University of Texas Press, 1999), 71; The forestry law came after a long history of forest extraction. See Anthony Goebel McDermott, *Los Bosques del "Progreso" explotación forestal y régimen ambiental en Costa Rica: 1883–1955* (San José, Costa Rica: Editorial Nuevas Perspectivas, 2013).

33. Evans, *Green Republic,* 7. By 1999, Costa Rica conserved twenty-five percent of its territory.

34. David Takacs, *Idea of Biodiversity: Philosophies of Paradise* (Baltimore: Johns Hopkins University Press, 2003), 2 and 35.

35. Stephan Amend and Thora Amend, *National Parks without People? The South American Experience* (Gland, Switzerland: IUCN, 1995) estimated 85 percent; Janis B. Alcorn, "Noble Savage or Noble State?: Northern Myths and Southern Realities in Biodiversity Conservation," *Ethnoecologica* 2, no. 3 (1994). Others argue that displacement rarely occurs; see Alejandro Velazquez, and David Bray, "From Displacement-Based Conservation to Place-Based Conservation," *Conservation and Society* 7, no. 1 (2009): 11–14.

36. Peru, Decreto Ley No. 20653 of 24 July 1978, Ley de Comunidades Nativas y de Promoción Agropecuaria de las Regiones de Selva y Ceja de Selva. See discussion, Dourojeanni, *Crónica,* 108.

37. Gary Bernard Wetterberg, Maria Teresa Jorge Padua, Celso Soares de Castro, and José Manuel Carvalho de Vasconcelos, *An Analysis of Nature Conservation Priorities in the Amazon,* Technical Series No. 8 (Brasilia: Forestry Development and Research Project,

UNDP/FAO/IBDF/Bra-545, 1976); José Augusto Drummond, "From Randomness to Planning: The 1979 Plan for Brazilian National Parks," in *National Parks beyond the Nation: Global Perspectives on "America's Best Idea,"* ed. Adrian Howkins, Jared Orsi, and Mark Fiege (Norman: University of Oklahoma Press, 2016).

38. Antônio Carlos Sant'Ana Diegues, *O mito moderno da natureza intocada* (São Paulo: NUPAUB, 1994); Arturo Gómez-Pompa and Andrea Kaus, "From Pre-Hispanic to Future Conservation Alternatives: Lessons from Mexico," *Proceedings of the National Academy of Sciences of the United States of America* 96, no. 11 (1999): 5982–5986.

39. Rob Nixon, *Slow Violence and the Environmentalism of the Poor* (Cambridge, MA: Harvard University Press, 2011), 197.

40. Mark Carey, "Latin American Environmental History: Current Trends, Interdisciplinary Insights, and Future Directions," *Environmental History* 14, no. 2 (2009): 221–252.

41. Scarzanella, "Las bellezas naturales," 6.

42. Evans, *Green Republic*; Enrique Ortiz, "Una joya del mundo en el Perú," *El Comercio,* Lima, 26 May 2013.

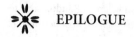 EPILOGUE

Latin American Environmental History in Global Perspective

J. R. McNeill

This book goes a long way toward making current scholarship in Latin American environmental history accessible in one handy volume. This concluding chapter goes a short way toward situating the Latin American experience in broader contexts. It does so in two allied ways. First, it draws on examples of environmental history research and perspectives concerning others parts of the world to offer a handful of suggestions for further work that could deepen or broaden Latin American environmental history. Second, it identifies, and to a limited extent explores, what I regard as the two most important eccentricities of Latin American environmental history, one each from the realms of the material world and the world of culture. Every region of the world is eccentric in some respects, Latin America no more or less so than others.[1] My aim here is to invite Latin American specialists to consider their subjects in relation to, and in comparison to, other parts of the world, and to invite readers just now coming to the study of Latin America to see how it might fit into patterns of history they are familiar with from prior study of somewhere else.

Opportunities

No single book, whether a compendium or an overview, can do everything.[2] On a subject as rich as Latin American environmental history, no book can touch on any but a few of the possible dimensions. This book offers four geographically bound overviews of subregions of Latin America. I suspect these will prove very useful especially for students. They might welcome similar compact overviews of Central America or the broad grasslands of the southern cone. The seven thematic chapters provide a wonderful set of cross-cutting essays that, in most cases, touch down widely throughout Latin America. Of the dozens of interesting themes not included here, I will mention only a few.

Modern history rests upon its premodern precedents. In the case of Latin America, the premodern is conventionally divided into a colonial era, roughly three centuries long, and a pre-Columbian era that is many millennia in duration. Several of the chapters here recognize the importance of colonial legacies for modern Latin American environmental history, and one of them, by Emily Wakild, delves into the meanings of the deep pre-Columbian past in interesting ways. In highlighting how the numerous extinctions of megafaunal species in the late Pleistocene affected the ecology of modern Latin America, and specifically how, why, when, and where parks were created, Wakild opens the door to other questions about the relevance of the deeper past. Among major world regions, Latin America was the last to acquire a human population (although it was not far behind North America in this respect). Thus flora and fauna have had less time here than anywhere else, excepting islands such as those of New Zealand or Madagascar, to learn[3] to live with human disturbances. Ecosystems of Latin America have had only about fourteen thousand years to adjust to the pressures brought by human actions, compared to fifty thousand years in Australia, sixty to one hundred thousand years in Eurasia, and more than two hundred thousand years in Africa.[4] Some parts of Latin America have much briefer histories of human occupation, such as Jamaica, which people left alone until about 600 C.E.[5] Wakild's example inspires one to ask, how has the deeper past, and the comparative brevity of human occupation in Latin America, affected the region's modern environmental history?

One might equally inquire into the modern implications of the environmental modifications undertaken by countless pre-Columbian peoples over the last ten millennia, from the first farmers roughly nine thousand years ago to the great empires in the Andes and Mexico.[6] Archaeologists in recent decades have uncovered remarkable fragments of Latin American's pre-Columbian history. The more they find, the stronger the impression that ancient peoples altered their environments both extensively and intensively. Wakild's example of hunting is matched by others involving farming. Some of these ancient impacts, dating back a millennium or several millennia, have affected more recent environmental history—the terracing of the Andes for example, or the formation of numerous patches of *terra preta* in Amazonia, or the vast network of water storage features built in the Maya lowlands (of Belize and Guatemala).[7]

A second opportunity, and one that does not require an education in paleoecology, is marine environmental history. In some settings, especially enclosed seas or bays, marine environmental history logically focuses on pollution issues. Thus environmental historians have found reason to explore the Baltic, Black, Yellow, and other seas suffering in modern times from heavy pollution loads. In Latin America, Guanabara Bay and Lake Maracaibo have won similar attention.[8] But in most settings, marine environmental history focuses on fish and marine mammals. The work of the historians collaborating in HMAP (History of Marine Animal Populations) is exemplary in this respect, as is the award-winning monograph on the Gulf of Maine by Jeffrey Bolster.[9] The last century or so has witnessed a vast human intrusion into marine ecosystems, mainly via fishing and whaling, the full effects of which are still working themselves out today. Here is a great opportunity in Latin American environmental history—more than one, in fact.

The Humboldt Current that caresses the western coast of South America sustains the world's richest fishing province. The chilly, upwelling waters carry extraordinary amounts of dissolved oxygen and are enriched by organic material tumbling out of the short, steep, erosive rivers of the Andes's western slope. These conditions give rise to a thick and tangled food web, anchored in phytoplankton and supporting uniquely dense fish populations. The coastal waters teem with jack mackerel, sardines, and anchovy, which for ages have fed marine mammals (notably seals) and seabirds making their homes on the coasts of Peru, Chile, and at times Ecuador.[10] In recent decades, roughly a fifth of the globe's marine fish catch has come from these waters. And of course fishermen and whalers have worked in other waters as well, such as the whaling grounds off Brazil and Argentina and those of the Caribbean.[11] The recent rise of salmonid (meaning mainly Atlantic salmon and rainbow trout) aquaculture in Chile is another dimension of marine environmental history that deserves more attention than it has received. Until a severe downturn brought on by the global financial crisis of 2008, Chile produced about one-third of the world's farmed salmonids.[12] Aside from a quick reference in McCook's chapter, and Santiago's discussion of guano mining, the possibilities of marine environmental history go unmentioned in this book—understandably enough because there is so little as yet to build on.[13]

A third opportunity is the environmental history of war. This is a particularly active subfield these days, but virtually none of the work in it considers Latin America.[14] That is not for a lack of warfare in the region: the nineteenth century was rife with it. And while the twentieth century was peaceful in comparison (with big exceptions such as Mexico), in Latin America, the big wars directly affected environments through accelerated resource extraction, notably of rubber, oil, bauxite, and copper. An environmental history of one or more of the more intense struggles of the modern era in Latin America, such as the War of the Triple Alliance (1865–70) or of the Mexican Revolution (1910–20) would be a fascinating, if grim, endeavor.[15]

The list of beckoning opportunities is long. I will mention only a few more without much attempt at elaboration. Gendered environmental history seems scarcer for Latin America than for some other regions. So does climate history. And while the introduction and the chapter by Sedrez and Duarte in this volume both draw attention to the history of informal settlements (favelas, barrios, etc.), this is a rich subject that—as far as I can tell—has not attracted much attention from environmental historians. Urban environmental history has flourished in North America and Europe, where informal settlements are rare, but it has scarcely taken account of the sprawling impromptu communities of the world, from the *gecekondu* of Istanbul to the *bidonvilles* of Antananarivo. Latin America is among the world regions that urbanized at breakneck speed in the last century, a circumstance that favored the emergence of informal settlements, and it still abounds in shantytowns.

Eccentricities

In this section I offer reflections on two of the distinctive features of Latin American environmental history, one of them drawn from the material realm and the other from the cultural. Both are broad subjects. First is the energy regime history of Latin America. Unlike most regions of the world, Latin America relied on biomass until it relied on oil and hydropower. A century ago, Mexico was Latin America's top coal producer, but its output came only to about 8 percent of India's (and 0.2 percent of the United States').[16] Today all of Latin America put together produces about 15 percent as much coal as India and 3 percent as much as China. Only Colombia is among the top twenty producers in the world. Due mainly to quirks of geology, there was no coal age, no "king coal" in Latin American history.

Biomass, such as fuelwood and crop residues, played a larger role in Latin American energy regimes until 1950 than in most parts of the world. Railways used fuelwood widely. A big slice of Brazil's steel industry used charcoal as fuel. Sugar boilers everywhere used biomass. When cheap oil came into play soon after 1900, there was little competition from coal-based constituencies. Latin America had minimal sunk investment in coal infrastructure and could pivot more readily to oil. And without a strong tradition of electricity generation from coal-burning plants, Latin America turned more readily to hydropower for electrification. (Today hydropower accounts for 50 to 65 percent of Latin America's electricity, and in Brazil 70 to 80 percent, compared to 16 percent for the world as a whole.)[17] Several Latin American countries had from the late nineteenth century *imported* modest amounts of coal, but were able to shift early and quickly when oil became available.

So an early and rapid embrace of oil characterized Latin America. The date at which oil permanently replaced coal as the dominant fossil fuel in Latin American countries came in all but two cases between 1896 and 1928. (The laggards were Brazil [1940] and Chile [1953]). This contrasts sharply with the experience of Europe and North America, where the dominance of oil came later (the dates at which it permanently outstripped coal are clustered between

1951 [U.S.] and 1971 [U.K.]) and the transitions, once begun, took decades longer than in Latin America. It also contrasts with the experiences of Africa and Asia, where coal accounted for upwards of 90 percent of primary energy consumption in 1925, while in Latin America at that date coal met only 37 percent of energy needs. In sum, Latin America never built its economy, society, and infrastructure around coal, and thus transitioned to oil early and quickly. In poorer countries of the region, such as Haiti or El Salvador, an energy transition took place directly from biomass to oil, an unusual trajectory by global standards.[18]

This distinctive energy path was not solely due to quirks of geology. Political considerations mattered as well. The great coal exporter of the nineteenth century was Britain. British authorities and businesses encouraged British colonies to become dependent on coal with greater effect than they could achieve in the independent countries of Latin America. The energy politics of World War I also played a role, because coal exports from Britain virtually ceased at a time when cheap oil from Mexico, the United States, and Venezuela became available.

Latin America's peculiar energy path had a number of consequences of which I will sketch two.[19] First, its urban air quality until 1950 (or so) was unusually good. Even cities that today are famous for air pollution, such as Mexico City, a century ago were famous for their clear air and long vistas. At a time when in Germany, Japan, the U.K., the United States, and dozens of other countries—all of which burned plenty of coal—urban residents suffered from lung ailments caused by pollution, Latin Americans breathed freely. In societies deeply invested in coal, such as Britain or Germany, every year tens of thousands of city people died from respiratory ailments either caused or exacerbated by coal smoke or coal dust. (Today coal smoke in China kills about 366 thousand people annually.)[20]

A second implication, if one accepts the analyses of Bruce Podobnik or Timothy Mitchell, is political.[21] Societies that converted deeply and widely to coal gradually created comparatively democratic institutions and political systems. They had little choice. Coal miners acquired too much power over national economies. If they chose to strike, economic life would grind to a halt in weeks. If they did so in time of war, national security teetered at the brink. Thus—the argument goes—elites had to share power with working classes lest coal miners flex their political muscles in disruptive ways.

This political dynamic did not extend to societies without much coal. Whereas in Britain (c.1880–1925) some 10 to 15 percent of families included coal miners who could stop British transport and industry any time they pleased, no one in Latin America's laboring populations had this sort of power. No one could compel business owners to pay higher wages or compel governments to allow the masses to vote.

Oil, in this analysis, held different political implications, less favorable for democracy. A society committed to oil as the heart of its energy regime was not hostage to mass unions. Oil workers were comparatively few in number, and if they threatened a national economy (or security) by going on strike, they could either be crushed or bought off without larger impacts. Only a few families would feel the effects. Other citizens would go about their daily lives as before. This was very different, according to Podobnik or Mitchell, from societies based on coal, in which elites had less latitude and were obliged to make far-reaching concessions that shaped national politics.

Whether this argument carries weight in understanding the political history of Latin America I leave to the reader. One can imagine objections to the argument—has coal-rich Poland been more democratic than oil-rich Norway? But if one accepts it, then the distinctive energy path of Latin America might hold some of the keys to its political experience. Whatever the range of its implications, the fact remains that Latin America pioneered an eccentric energy path by world standards.

Turning to the cultural realm, Latin American environmental history enjoys another distinction among world regions, matched perhaps only by Southeast Asia.[22] Latin American ideas and attitudes about nature, and the proper human relationship to nature, have been shaped by three broad cultural traditions: indigenous American, Iberian, and African (or to be more precise, Atlantic African). Each of these cultural traditions, of course, was in reality a composite of many more local traditions, which might be highly diverse themselves: Atlantic Africa, the home of the millions enslaved and bound for the Americas (c.1510–1860), stretched from Senegambia to Angola. Indeed, in the eighteenth century, slaves sent to Latin America also came from Mozambique and Madagascar. The indigenous American cultural traditions, from Patagonia to California, were probably equally diverse (not that there is any good way to measure such things). Those of Iberia probably were less diverse, because it is a smaller place and because of the cultural hegemony of the Roman Catholic Church in Spain and Portugal in recent centuries. But, against that, in the nineteenth and twentieth centuries, the Iberian strand broadened into a European one, thanks to sizeable influxes of immigrants from Italy, Germany, and elsewhere.

Naturally, all of the strands of cultural inheritance in modern Latin America have been evolving over time, many of them in response to one another. And from the late nineteenth century onwards, in Cuba, Peru, Brazil, and to a smaller extent elsewhere, the cultural mosaic grew more complex still as a result of immigration from China and Japan. Admittedly in some countries, Bolivia for example, the African component is tiny, and in others, Barbados for example, the indigenous or Amerindian is insignificant. But for the region as a whole, the cultural complexity is extreme, and has been extreme since colonial times, in a way that is very rare around the world.[23]

Moreover, ideas and practice concerning the environment received further influences independent of migration in the form of the spread of certain scientific outlooks. In the nineteenth century, the influential ones typically emanated from Germany, France, or Britain and included, for example, the germ theory of disease and the accompanying priority of sanitation.[24] After 1900, the prestige and influence of U.S. science and technology added to the mix, reflected in the quick transition to oil, and, later in the twentieth century, the package of genetically engineered seeds, pesticides, chemical fertilizers, and agricultural equipment known as the Green Revolution.[25]

All this—the migrations of peoples and ideas to and within Latin America—brought a rich menu of ideas and practices with respect to the environment from which generations of Latin Americans could pick and choose, combine and recombine, to suit their particular opportunities and constraints.

Environmental ideas and practices typically reflect, to greater or lesser extents, religious backgrounds and sensibilities.[26] In Latin America, that chiefly meant varieties of Catholicism, including liberation theology, which at times took on environmentalist hues, and—perhaps especially in the Andes—indigenous religion with strong notions of the sacred. Catholicism alone contained multitudes of contradictory positions.[27]

Cultural traditions relevant to environmental thought and action also include sharply different conceptions of property, and who has the right to do what with a forest, a river, or a hill full of tin. The range of views extended from reverent environmentalism to enthusiastic rapaciousness, the latter tradition lately bolstered by the strong Chinese market for Latin American commodities.[28]

The extraordinary cultural diversity of Latin American traditions, then, helped to flavor the myriad concepts informing environmental thought and ethics in the region. Just how it did so varied from place to place. In the Andes, the percolating mix of indigenous traditions, Catholic social thought, and the overlays of class identities with ethnic ones gave rise to what Joan Martinez Alier influentially termed "the environmentalism of the poor."[29] In Brazil, where strands of environmentalist thought emerged from the late eighteenth century onward, the cultural and social equations were sharply different, involving more often than not elite landowners with educations that gave them access to European science, which they creatively applied to their own settings. Their environmental thinking was often imbued with concerns about social issues such as slavery.[30] These two examples represent a drop in the bucket of what might be done with the cultural, ideological, political, and social history of environmentalism in Latin America.

Other eccentricities exist and other challenges abound in Latin American environmental history. This book touches on several, and this epilogue discusses only a few. Latin American environmental history is still young, vigorous, and vivacious so that many of these challenges, and others as yet unimagined, will soon be met. And then we will have a firmer notion of the respects in which Latin America's environmental history was eccentric and the respects in which it conformed to global patterns. The living past has a lively future.

J. R. McNeill has held two Fulbright awards and fellowships from Guggenheim, MacArthur, and the Woodrow Wilson Center. His books include *The Mountains of the Mediterranean* (1992); *Something New Under the Sun* (2000), winner of two prizes, listed by the London *Times* among the ten best science books ever written (despite not being a science book), and translated into nine languages; *The Human Web* (2003), translated into seven languages; *Mosquito Empires* (2010), which won the Beveridge Prize from the American Historical Association; and *The Great Acceleration* (2016). In 2010 he was awarded the Toynbee Prize for "academic and public contributions to humanity."

Notes

1. In other works I have tried to address the environmental historical eccentricities of China and of the Middle East. "The Eccentricity of the Middle East and North Africa's Environmental History," in *Water on Sand: Environmental Histories of the Middle East and North Africa,* ed. Alan Mikhail (New York: Oxford University Press, 2012), 27–50; and "Chinese Environmental History in World Perspective," in *Sediments of Time: Environment and Society in Chinese History,* ed. Mark Elvin and Ts'ui-jung Liu (New York: Cambridge University Press, 1998), 31–52.

2. The most accessible overview, already a decade old, is Shawn Miller, *An Environmental History of Latin America* (New York: Cambridge University Press, 2007).

3. By "learn" I mean both adaption over time via selection for genes embodying traits useful for survival and reproduction and, for those few species capable of it, learning in the cultural sense.

4. If one counts prehuman hominins, the time of adjustment in Asia and Africa is much longer still. All these spans of time are approximations.

5. Funes includes Jamaica in his chapter so I do so as well, even if its history since conquest by England in 1655 is often seen as part of the Caribbean but not part of Latin America.

6. An inspiration here is the work of Australian environmental historians attuned to modern impacts of the deeper past: Tim Flannery, *The Future Eaters: An Ecological History of the Australasian Lands and People* (Chatswood, NSW: Reed Books, 1994); Bill Gammage, *The Biggest Estate on Earth: How Aborigines Made Australia* (Sydney: Allen & Unwin, 2011); Tom Griffiths, "Environmental History, Australian Style," *Australian Historical Studies* 46, no. 2 (2015): 157–173. For a sense of how long-term fire regimes and forest exploitation have constrained modern uses of forests in Chile, see Juan J. Armesto, Daniela Ilona Manuschevich, Alejandra Mora, and Pablo A. Marquet, "From the Holocene to the Anthropocene: A Historical Framework for Land Cover

Change in Southwestern South America in the Past 15,000 Years," *Land Use Policy* 27 (2010): 148–160.

7. C. Cagnato, "Underground Pits (*Chultunes*) in the Southern Maya Lowlands: Excavation Results from Classic Period Maya Sites in Northwestern Petén," *Ancient Mesoamerica* 28, no. 1 (2017): 75–94.

8. Lise Sedrez, "'The Bay of All Beauties': State and Environment in Guanabara Bay, Rio de Janeiro, Brazil, 1875–1975" (Ph.D. diss., Stanford University, 2004); Nikolas Kozloff," Maracaibo Black Gold: Venezuelan Oil and Environment during the Juan Vicente Gómez Period, 1908–1935" (Ph.D. diss., Oxford University, 2002).

9. For HMAP, visit http://www.comlsecretariat.org/research-activities/history-of-marine-animal-populations-hmap/; Jeffrey Bolster, *The Mortal Sea: Fishing the Atlantic in the Age of Sail* (Cambridge, MA: Harvard University Press, 2012). See also Micah Muscolino, *Fishing Wars and Ecological Change in Late Imperial and Modern China* (Cambridge, MA: Harvard University Press, 2009); Ryan T. Jones, *Empire of Extinction: Russians and the North Pacific's Strange Beasts of the Sea* (New York: Oxford University Press, 2015). On a crayfish fishery that collapsed around 1990, see Françoise Pencalet-Kerivel, *Histoire de la pêche langoustière: Les "Mauritaniens" dans la tourmente du second XXe siècle* (Rennes: Presses Universitaires de Rennes, 2008).

10. The exploitation of guano inspired an imaginative and prize-winning environmental history: Greg Cushman, *Guano and the Opening of the Pacific World: A Global Ecological History* (New York: Cambridge University Press, 2013).

11. On modern Argentine fisheries there is a history, although not attentive to environmental considerations: Raúl Ricardo Fermepín and Juan Pedro Villemur, *155 Años de la pesca en el mar argentino* (Buenos Aires: Instituto de Publicaciones Navales, 2004). On whaling in Argentine waters in the early twentieth century, see M. L. Palomares, E. Mohammed, and D. Pauly, "European Expeditions as a Source of Historic Abundance Data on Marine Organisms," *Environmental History* 11 (2006): 835–847; Randall R. Reeves, J. A. Khan, R. R. Olsen, S. L. Swarz, and T. D. Smith, "History of Whaling in Trinidad and Tobago," *Journal of Cetacean Research and Management* 3 (2001): 45–54; Aldemaro Romero and Joel Creswell, "Deplete Locally, Impact Globally: Environmental History of Shore-Whaling in Barbados," *The Open Conservation Biology Journal* 4 (2010): 19–27. An older work, with minimal consideration of environmental perspectives, is Myriam Ellis, *A baleia no Brasil colonial* (São Paulo: Editora da USP/ Melhoramentos, 1969). I thank Bruno Biasetto for this last citation.

12. For a start, see Pablo Camus and Fabián Jaksic, *Piscicultura en Chile: entre la productividad y el deterioro ambiental, 1856–2008* (Santiago: Pontificia Universidad Católica de Chile, 2009); John Soluri, "Something Fishy: Chile's Blue Revolution, Commodity Diseases and the Problem of Sustainability," *Latin American Research Review* 46 (2011): 55–81.

13. Pearl diving and pearl production in Latin American waters has begun to inspire work: Micheline Cariño and Mario Monteforte, *El primer emporio perlero sustentable del mundo: la Compañía Criadora de Concha y Perla de la Baja California S.A., y sus perspectivas para Baja California Sur* (Mexico: UABCS, SEP, FONCA-CONACULTA, 1999); and for the colonial era, Molly A. Warsh, "A Political Ecology in the Early Spanish Caribbean," *The William and Mary Quarterly* 71 (2014): 517–548; Warsh, *American Baroque: Pearls and the Nature of Empire 1492–1700* (Chapel Hill: UNC Press, 2018).

14. The work of Richard Tucker is central to this subject: See Tucker and Edmund Russell, eds., *Natural Enemy, Natural Ally: Towards An Environmental History of Warfare* (Corvallis: Oregon State University Press, 2004); Simo Laakkonen, Richard Tucker, and Timo Vuorisalo, eds., *The Long Shadows: A Global Environmental History of the Second World War* (Corvallis: Oregon State University, 2017), which includes a chapter on Mexican forests by Chris Boyer. For a conceptually novel approach, see Micah Muscolino, *The Ecology of War in China: Henan Province, the Yellow River, and Beyond, 1938–1950* (New York: Cambridge University Press, 2014); for orientation see Chris Pearson, "Researching Militarized Landscapes: A Literature Review on War and the Militarization of the Environment," *Landscape Research* 37 (2012): 115–133.

15. For an example of how war, environment, and agriculture can come together in surprising ways, see , Erin Stewart Mauldin, *Unredeemed Land: Confronting the Ecological Legacies of War and Emancipation in the U.S. South, 1840–1880* (New York: Oxford University Press, 2018).

16. *Encyclopedia Britannica*, 1911 edition, as reported in Wikipedia, accessed 27 May 2017, https://en.wikipedia.org/wiki/History_of_coal_mining.

17. Ramón Espinasa and Carlos G. Sucre, "What Powers Latin America?" *ReVista: The Harvard Review of Latin America* (Fall 2015), accessed 27 May 2017, https://revista .drclas.harvard.edu/book/what-powers-latin-america; and the International Energy Agency, accessed 27 May 2017, https://www.iea.org/topics/renewables/subtopics/ hydropower/. As one might expect, the variability within Latin America is considerable. Paraguay gets almost every kilowatt from hydropower, while the small islands of the Caribbean or Belize get none.

18. The figures in this paragraph come from M. de Mar Rubio, César Yañez, Mauricio Folchi, and Albert Carreras, "Energy as an Indicator of Modernization in Latin America, 1890–1925," *Economic History Review* 63 (2010): 769–804; M. de Mar Rubio and Mauricio Folchi, "Will Small Energy Consumers Be Faster in Transition? Evidence from the Early Shift from Coal to Oil in Latin America," *Energy Policy* 50 (2012): 50–61. For a global perspective, see Vaclav Smil, *Energy and Civilization: A History* (Cambridge, MA: MIT Press, 2017). The figures given for Africa and Asia in 1925 reflect the low levels of energy use overall and the existence of pockets of coal-fired industry in places such as South Africa and Bengal.

19. Another, which I will not sketch, might be late industrialization in response to the absence of coal and dependence on imports from (mainly) Britain. In the mid twentieth century, several Latin American states, notably Mexico, Brazil, and Argentina, sponsored "import-substitution industrialization" (ISI), motivated by disappointment over the pace of free-market industrialization in the region.

20. Michael Brauer, "Poor Air Quality Kills 5.5 Million Worldwide Annually," Institute for Health Metrics and Evaluation (originally published by the University of British Columbia on 12 February 2016), accessed 27 May 2017, http://www.healthdata.org/ news-release/poor-air-quality-kills-55-million-worldwide-annually.

21. Bruce Podobnik, *Global Energy Shifts* (Philadelphia: Temple University Press, 2005); Timothy Mitchell, *Carbon Democracy: Political Power in the Age of Oil* (London: Verso, 2013).

22. Southeast Asia featured the cultural inheritances of diverse indigenous populations, influenced for many centuries by immigrant communities from South China and from

eastern India, from Bengal to Tamil Nadu. In religious terms, Southeast Asia included several strands of Hinduism, Buddhism, and Islam, with a sprinkling of Christian influence here and there, not to mention Confucian ideology (which some people also consider to be a religion).

23. On smaller scales, some cities have long been islands of cultural complexity by virtue of attracting migrants from near and far.

24. A useful overview is Marcos Cueto and Steven Palmer, *Medicine and Public Health in Latin America: A History* (New York: Cambridge University Press, 2014). See also Christopher Abel, "Health, Hygiene and Sanitation in Latin America, c. 1870 to 1950," University of London Institute of Latin American Studies Research Papers 42 (1996), http://sas-space.sas.ac.uk/3408/1/B24_Health,_Hygiene_and_Sanitation_in_Latin_America_c1870_to_c1950.pdf.

25. Northern Mexico served as the laboratory experiment of the early Green Revolution. See Pamela Matson, ed., *Seeds of Sustainability: Lessons from the Birthplace of the Green Revolution* (Washington, DC: Island Press, 2012). A vivid treatment of the health consequences is Angus Wright, *The Death of Ramon Gonzalez* (Austin: University of Texas Press, 1990). See also Noriega Orozco Blanca Rebeca, *Revolución Verde (1944–2008): Modernidad y tecnociencia en Sonora* (Madrid: Editorial Académica Española, 2013).

26. Mark Stoll, *Inherit the Holy Mountain: Religion and the Rise of American Environmentalism* (New York: Oxford University Press, 2015) makes this case for the United States, specifically examining the impacts of Protestant sensibilities.

27. It is a curious moment in the history of religion when both the spiritual heads of the Catholic and Orthodox Churches are publicly and vociferously in support of environmentalism. Pope Francis and Patriarch Batholomew find the consequences of capitalism an affront to God's creation and recommend, urgently, more environmentally respectful behavior in the tradition of St. Francis. Kevin Mongrain, "The Burden of Guilt and the Imperative of Reform: Pope Francis and Patriarch Bartholomew Take Up the Challenge of Re-Spiritualizing Christianity in the Anthropocene Age," *Horizons* 44 (2017): 80–107.

28. On the variety of environmental thought, see for example F. Estenssoro Saavedra, *Historia del debate ambiental en la política mundial 1945–1992: La perspectiva latinoamericana* (Santiago de Chile: Instituto de Estudios Avanzados, Universidad de Santiago de Chile, 2014); Joan Martinez-Alier, Michiel Baud, and Héctor Sejenovich, "Origins and Perspectives of Latin American Environmentalism," in *Environmental Governance in Latin America*, ed. Fábio de Castro, Barbara Hogenboom, Michiel Baud (Dordrecht: Springer, 2016), 29–57.

29. Joan Martinez-Alier, *The Environmentalism of the Poor: A Study of Ecological Conflicts and Valuation* (Cheltenham, UK: Edward Elgar, 2002).

30. José Augusto Pádua, *Um sopro de destruição: Pensamento político e crítica ambiental no Brasil escravista (1786–1888)* (Rio de Janeiro: Zahar, 2002).

Selected Bibliography

Aguilar, Luis Aboites. *El agua de la nación: Una historia política de México (1888–1946).* Mexico City: CIESAS, 1998.

Aide, T. Mitchell, Matthew L. Clark, H. Ricardo Grau, David López-Carr, Marc A. Levy, Daniel Redo, Martha Bonilla-Moheno, George Riner, María J. Andrade-Núñez, and María Muñiz. "Deforestation and Reforestation of Latin America and the Caribbean (2001–2010)." *Biotropica* 45, no. 2 (2013): 1–10.

Alier, Joan Martinez. *The Environmentalism of the Poor.* Cheltenhan, UK: Edward Elgar, 2003.

Alonso, Angela, and Débora Maciel. "From Protest to Professionalization: Brazilian Environmental Activism after Rio-92." *The Journal of Environment and Development* 19, no. 3 (2010): 300–317.

Amaral, Samuel. *The Rise of Capitalism on the Pampas: The Estancias of Buenos Aires, 1785–1870.* Cambridge: Cambridge University Press, 1998.

Anderson, Robert S., Richard Grove, and Karis Hiebert. *Islands, Forest and Gardens in the Caribbean: Conservation and Conflict in Environmental History.* Oxford: Macmillan Caribbean, 2006.

Ayuero, Javier and Débora Alejandra Swistun, *Flammable: Environmental Suffering in an Argentine Shantytown.* Oxford: Oxford University Press, 2009.

Barsky, Osvaldo, and Julio Djenderedjian. *Historia del capitalismo agrario pampeano.* Vol. 1, *La expansión ganadera hasta 1895.* Buenos Aires: Siglo XXI Editores Argentina, 2003.

Bauer, Sherrie L., Barbara Deutsch Lynch, and Marin Miller, eds. *Beyond Sun and Sand: Caribbean Environmentalisms.* New Brunswick, NJ: Rutgers University Press, 2006.

Bebbington, Anthony and Jeffrey Bury, eds. *Subterranean Struggles: New Dynamics of Mining, Oil, and Gas in Latin America.* Austin: University of Texas Press, 2013.

Bell, Stephen. *Campanha Gaúcha: A Brazilian Ranching System, 1850–1920.* Palo Alto, CA: Stanford University Press, 1998.

Bezerra, Joana. *The Brazilian Amazon: Politics, Science, and International Relations in the History of the Forest.* New York: Springer, 2015.

Boyer, Christopher R. *Political Landscapes: Forests, Conservation, and Community in Mexico.* Durham: Duke University Press, 2015.

———, ed. *A Land Between Waters: Environmental Histories of Modern Mexico.* Tucson: University of Arizona Press, 2012.

Brailovsky, Antonio, and Dina Foguelman. *Memoria Verde: Historia ecológica de la Argentina.* Buenos Aires: Editorial Sudamericana, 1991.

Brailovsky, Antonio. *Historia ecológica de la ciudad de Buenos Aires.* Buenos Aires: Kaicron, 2012.

Brannstrom, Christian. "Coffee Labor Regimes and Deforestation on a Brazilian Frontier, 1915–1965." *Economic Geography* 76 (2000): 326–346.

———. "Was Brazilian Industrialization Fuelled by Wood? Evaluating the Wood Hypothesis, 1900–1960." *Environment and History* 11, no. 4 (2005): 395–430.

———, ed. *Territories, Commodities, and Knowledges: Latin American Environmental History in the Nineteenth and Twentieth Century.* London: Institute of Latin American Studies, 2004.

Brothers, Timothy, Jeffrey S. Wilson, and Owen P. Dwyer, eds. *Caribbean Landscapes: An Interpretative Atlas.* Coconut Creek, FL: Caribbean Studies Press, 2008.

Brown, Kendall. *A History of Mining in Latin America from the Colonial Era to the Present.* Albuquerque: University of New Mexico Press, 2012.

Brush, Stephen B. *Farmers Bounty: Locating Crop Diversity in the Contemporary World.* New Haven: Yale University Press, 2004.

Cabral, Diogo. *Na presença da floresta: Mata Atlântica e história colonial.* Rio de Janeiro: Garamond, 2014.

Camus Gayán, Pablo. *Ambiente, bosques, y gestión forestal en Chile, 1541–2005.* Santiago: Dirección Bibliotecas, Archivos y Museos, 2006.

Camus, Pablo and Fabían Jaksic, *Piscicultura en Chile: Entre la productividad y el deterioro ambiental, 1856–2008.* Santiago: Pontificia Universidad Católica de Chile, 2009.

Candiani, Vera. *Dreaming of Dry Land: Environmental Transformation in Colonial Mexico City.* Palo Alto: Stanford University Press, 2014.

Carey, Mark. *In the Shadow of Melting Glaciers: Climate Change and Andean Society.* New York: Oxford University Press, 2010.

———. "Latin American Environmental History: Current Trends, Interdisciplinary Insights, and Future Directions." *Environmental History* 14, no. 2 (2009): 221–252.

Cariño, Micheline. *Historia de las relaciones hombre/naturaleza en Baja California Sur 1500–1940.* Mexico City: UABCS-SEP, 1996.

Carney, Judith, and Nicholas Rosomoff. *In the Shadow of Slavery: Africa's Botanical Legacy in the Atlantic World.* Berkeley: University of California Press, 2009.

Castro Herrera, Guillermo, *Los trabajos de ajuste y combate: Naturaleza y sociedad en la historia de América Latina.* La Habana/Bogotá: Casa de las Américas, Colcultura, 1995.

Clare, Patricia. "Un balance de la historia ambiental latinoamericana." *Revista de Historia* 59–60 (2009): 185–201.

Chacón, M. I., S. B. Pickersgill, and D. G. Debouck. "Domestication Patterns in Common Bean (*Phaseolus vulgaris L.*) and the Origin of the Mesoamerican and Andean Cultivated Races." *Theoretical and Applied Genetics* 110 (2005): 432–444.

Cleary, David. "An Environmental History of the Amazon: From Prehistory to the Nineteenth Century." *Latin American Research Review* 36, no. 2 (2001): 64–96.

Coates, Anthony G., comp. *Paseo pantera: Una historia de la naturaleza y cultura de Centroamérica.* Washington: Smithsonian Books, 2003.

Cook, Noble. *Born to Die: Disease and New World Conquest, 1492–1650.* Cambridge: Cambridge University Press, 1998.

Correa, Silvio, and Juliana Bublitz. *Terra de promissão: Uma introdução à eco-história do Rio Grande do Sul.* Passo Fundo: Editora da Universidade de Passo Fundo, 2006.

Cushman, Gregory T. *Guano and the Opening of the Pacific World: A Global Ecological History.* Cambridge: Cambridge University Press, 2012.

Cuvi, Nicolás. *Ciencia e imperialismo en América Latina: La Misión de Cinchona y las estaciones agrícolas cooperativas.* Saarbrücken: Editorial Académica Española, Ph.D. thesis, 2011. Available at http://www.tdx.cat/handle/10803/5182.

———. "The Cinchona Program (1940–1945): Science and Imperialism in the Exploitation of a Medicinal Plant." *Dynamis* 31 (2011): 183–206.

Dean, Warren. *Brazil and the Struggle for Rubber: A Study in Environmental History.* Cambridge: Cambridge University Press, 1987.

———. "The Green Wave of Coffee: Beginnings of Tropical Agricultural Research in Brazil (1885–1900)." *Hispanic American Historical Review* 69, no. 1 (1989): 91–115.

———. *With Broadax and Firebrand: The Destruction of Brazil's Atlantic Forest.* Berkeley: University of California Press, 1995.

Dematteis, Lou, and Kayana Szymczak. *Crude Reflections / Cruda Realidad: Oil, Ruin and Resistance in the Amazon Rainforest / Petróleo, Devastación y Resistencia en la Amazonía.* San Francisco: City Lights, 2008.

Denevan, William. "The Pristine Myth: The Landscape of the Americas in 1492." *Annals of the Association of American Geographers* 82 (1992): 369–385.

de Alcántara, Cynthia Hewitt. *Modernizing Mexican Agriculture: Socioeconomic Implications of Technological Change, 1940–1970.* Geneva: United Nations Research Institute for Social Development, 1976.

de Vos, Jan. *Oro verde: La conquista de la Selva Lacandona por los madereros tabasqueños, 1822–1949.* Mexico City: Fondo de Cultura Económica, 1988.

———. *Una tierra para sembrar sueños: Historia reciente de la Selva Lacandona, 1950–2000.* Mexico City: Fondo de Cultura Económica, 2002.

Diegues, Antônio Carlos Sant'Ana. *O mito moderno da natureza intocada.* São Paulo: NUPAUB, 1994.

Dollfus, Olivier, *El reto del espacio andino.* Lima: Instituto de Estudios Peruanos, 1981.

Dourojeanni, Marc J. *Crónica forestal del Perú.* Lima: Universidad Nacional Agraria, La Molina: Editorial San Marcos, 2009.

Droulers, Martine. *Brésil: Une Géohistoire.* Paris: PUF, 2001.

Drummond, José Augusto, and José Luiz Andrade Franco. *Proteção à natureza e identidade nacional no Brasil, anos 1920–1940.* Rio de Janeiro: Ed. Fiocruz, 2009.

Drummond, José, José Luiz Franco, and Daniele Oliveira. "Uma Análise sobre a História e a Situação das Unidades de conservação no Brasil." In *Conservação da Biodiversidade—Legislação e Políticas Públicas,* ed. Roseli Ganem. Brasília: Edições Câmara dos Deputados, 2011.

Duarte, Regina Horta. *Activist Biology: The National Museum, Politics and Nation Building in Brazil.* Tucson: The University of Arizona Press, 2016.

———. "It Does Not Even Seem Like We Are in Brazil: Country Clubs and Gated Communities in Belo Horizonte, Brazil, 1951–1964." *Journal of Latin American Studies* 44, no. 3 (2012): 435–466.

Duval, David T., ed. *Tourism in the Caribbean: Trends, Development, Prospects.* London: Routledge, 2004.

Earls, John. *La agricultura andina ante una globalización en desplome.* Lima: Pontificia Universidad Católica del Perú, 2006.

Edelman, Marc. *The Logic of the Latifundio: The Large Estates of Northwestern Costa Rica Since the Late Nineteenth Century.* Palo Alto: Stanford University Press, 1992.

Espinosa, Mariola. "Globalizing the History of Disease, Medicine, and Public Health in Latin America." *Isis* 104, no. 4 (December 2013): 798–806. doi:10.1086/674946.

Etter, Andrés, Clive McAlpine, and Hugh Possingham. "Historical Patterns and Drivers of Landscape Change in Colombia Since 1500: A Regionalized Spatial Approach." *Annals of the Association of American Geographers* 98 (2008): 2–23.

Evans, Sterling. *Bound in Twine: The History and Ecology of the Henequen-Wheat Complex for Mexico and the American and Canadian Plains, 1880–1950.* College Station: Texas A&M University Press, 2007.

———. *Green Republic: A Conservation History of Costa Rica.* Austin: University of Texas Press, 1999.

Fearnside, Philip. "Deforestation in Brazilian Amazonia: History, Rates, and Consequences." *Conservation Biology* 19, no. 3 (2005): 680–688.

Fernández Prieto, Leida. "Islands of Knowledge: Science and Agriculture in the History of Latin America and the Caribbean." *Isis* 104, no. 4 (1 December 2013): 788–797. doi:10.1086/674945.

Folchi, Mauricio. "Historia ambiental de las labores de beneficio en la minería del cobre en Chile, siglos XIX y XX." Ph.D. dissertation, Universidad Autónoma de Barcelona, 2006.

Franco, José Luiz de Andrade, Sandro Dutra e Silva, José Augusto Drummond, and Giovana Galvão Tavares, eds. *História ambiental: Fronteiras, recursos naturais e conservação da natureza.* Rio de Janeiro: Garamond, 2012.

Freyre, Gilberto. *Nordeste: Aspectos da influência da cana sobre a vida e a paisagem do Nordeste do Brasil.* Rio de Janeiro: Editora José Olympio, 1937.

Funes Monzote, Reinaldo. *From Rainforest to Cane Field in Cuba: An Environmental History since 1492.* Translated by Alex Martin. Chapel Hill: University of North Carolina Press, 2008.

———, ed. *Naturaleza en declive. Miradas a la historia ambiental de América Latina y el Caribe.* Valencia: Centro Francisco Tomás y Valiente, UNED Alzira-Valencia, Fundación Instituto de Historia Social, Colección Biblioteca Historia Social, 2008.

Gallini, Stefania. "Historia, ambiente, política: El camino de la historia ambiental en América Latina." *Nómadas* 30 (2009): 92–102.

———. *Una historia ambiental del café en Guatemala: La Costa Cuca entre 1830 y 1902.* Guatemala: Asociación para el Avance de las Ciencias Sociales en Guatemala, 2009.

García Martínez, Bernardo, and Alba González Jácome, eds. *Estudios sobre historia y ambiente en América.* Vol. 1, *Argentina, Bolivia, México, Paraguay.* Mexico City: Instituto Panamericana de Geografía e Historia, El Colegio de México, 1999.

García Martínez, Bernardo, and María del Rosario Prieto, eds. *Estudios sobre historia y ambiente en América.* Vol. 2, *Norteamérica, Sudamérica y el Pacífico.* Mexico: El Colegio de México, 2002.

Garfield, Seth. "A Nationalist Environment: Indians, Nature, and the Construction of the Xingu National Park in Brazil." *Luso-Brazilian Review* 41, no. 1 (2004): 139–167.

Gerhardt, Marcos. "Extrativismo e transformação na Mata Atlântica Meridional." In *Metamorfoses florestais: Culturas, Ecologias e as Transformações Históricas da Mata Atlântica,* edited by Diogo Cabral and Ana Bustamante. Curitiba: Primas, 2016.

Gligo, Nicolo, and Jorge Morello. "Notas para una historia ecológica de América Latina." In

Estilos de desarrollo y medioambiente en América Latina, edited by Nicolo Gligo and Osvaldo Sunkel. Mexico City: Fondo de Cultura Económica, 1980.

Grove, Richard. *Green Imperialism: Colonial Expansion, Tropical Island Edens and the Origins of Environmentalism, 1600–1860.* Cambridge: Cambridge University Press, 1995.

Guhl, Andrés. *Café y cambio del paisaje en Colombia, 1970–2005.* Medellín, Colombia: Fondo Editorial Universidad EAFIT, 2008.

Hall, Anthony, ed. *Global Impact, Local Action: New Environmental Policy in Latin America.* London: Institute for the Study of the Americas, 2005.

Hecht, Susanna. "Environment, Development, and Politics: Capital Accumulation and the Livestock Sector in Eastern Amazonia." *World Development* 13, no. 6 (1985): 663–684.

———. *The Scramble for the Amazon and the "Lost Paradise" of Euclides da Cunha.* Chicago: University of Chicago Press, 2013.

Hecht, Susanna, and Alexander Cockburn. *The Fate of the Forest: Developers, Destroyers and Defenders of the Amazon.* Chicago: Chicago University Press, 2010.

Hernández, Lucina, ed. *Historia ambiental de la ganadería en México.* Xalapa, Mexico City: Instituto de Ecología, A.C., and L'Institut de Recherche pour le Developppment, 2001.

Herrera, Alexander. *La recuperación de tecnologías indígenas: Arqueología, tecnología y desarrollo en los Andes.* Bogotá: Uniandes, CLACSO, Instituto de Estudios Peruanos y PUNKU, 2011.

Hochstetler, Kathryn, and Margaret Keck. *Greening Brazil: Environmental Activism in State and Society.* Durham: Duke University Press, 2007.

Holanda, Sérgio Buarque de. *Visão do paraíso: Os motivos edênicos no descobrimento e colonização do Brasil.* Rio de Janeiro: Editora José Olympio, 1959.

Kaimowitz, David. *Livestock and Deforestation in Central America in the 1980s and 1990s: A Policy Perspective.* Jakarta: Center for International Forestry Research, 1996.

Kimerling, Judith. "Oil, Lawlessness and Indigenous Struggles in Ecuador's Oriente." In *Green Guerrillas: Environmental Conflicts and Initiatives in Latin America and the Caribbean,* edited by Helen Collinson. London: The Latin American Bureau, 1996.

Klubock, Thomas Miller. *La Frontera: Forests and Ecological Conflict in Chile's Frontier Territory.* Durham: Duke University Press, 2014.

Kozloff, Nikolas. "Maracaibo Black Gold: Venezuelan Oil and Environment during the Juan Vicente Gómez Period, 1908–1935." Ph.D. dissertation, Oxford University, 2002.

Leal, Claudia. "Behind the Scenes and Out in the Open: Making Colombian National Parks in the 1960s and 1970s." In *The Nature State: Rethinking the History of Conservation,* edited by Wilko Graf von Hardenberg, Matthew Kelly, Claudia Leal, and Emily Walkid, 135–156. London: Routledge, 2017.

———. *Landscapes of Freedom: Building a Postemancipation Society in the Rainforests of Western Colombia.* Tucson: The University of Arizona Press, 2018.

———. "Un Puerto en la selva. Naturaleza y raza en la creación de la ciudad de Tumaco, 1860–1940." *Historia Crítica* 30 (2005): 39–66.

Leal, Claudia, and Eduardo Restrepo. *Unos bosques sembrados de aserríos. Historia de la extracción maderera en el Pacífico colombiano.* Medellín: Universidad de Antioquia, Universidad Nacional sede Medellín, Instituto Colombiano de Antropología e Historia, 2003.

Lehmann, Johannes, Dirse Kern, Bruno Glaser, and William Woods, eds. *Amazonian Dark Earths: Origin, Properties, Management.* Dordrecht: Kluwer Academic Publishers, 2003.

Lopes, Maria Margaret, and Irina Podgorny. "The Shaping of Latin American Museums of Natural History, 1850–1990." *Osiris* 15 (1 January 2000): 108–118. doi:10.2307/301943.

López, Rick A. "Nature as Subject and Citizen in the Mexican Botanical Garden, 1787–1829." In *A Land Between Waters: Environmental Histories of Modern Mexico,* edited by Christopher R. Boyer, 73–99. Tucson: University of Arizona Press, 2012.

McCook, Stuart. "The Neo-Columbian Exchange: The Second Conquest of the Greater Caribbean, 1720–1930." *Latin American Research Review* 46 (2011): 11–31.

———. *States of Nature: Science, Agriculture, and Environment in the Spanish Caribbean, 1760–1940.* 1st ed. Austin: University of Texas Press, 2002.

McNeill, John. *Mosquito Empires: Ecology and War in the Greater Caribbean, 1620–1914.* Cambridge: Cambridge University Press, 2010.

———. "The State of the Field of Environmental History," *The Annual Review of Environment and Resources* 35 (2010): 345–374.

Melillo, Edward. "The First Green Revolution: Debt Peonage and the Making of the Nitrogen Fertilizer Trade, 1840–1930." *American Historical Review* 117, no. 4 (October 2012): 1028–1060.

Melville, Elinor. *A Plague of Sheep: Environmental Consequences of the Conquest of Mexico.* Cambridge: Cambridge University Press, 1994.

Méndez, V. Ernesto, Christopher M. Bacon, Meryl Olson, Katlyn S. Morris, and Annie Shattuck. "Agrobiodiversity and Shade Coffee Smallholder Livelihoods: A Review and Synthesis of Ten Years' Research in Central America." *The Professional Geographer* 62 (2010): 357–376.

Miller, Shawn William. *An Environmental History of Latin America.* Cambridge: Cambridge University Press, 2007.

———. *Fruitless Trees: Portuguese Conservation and Brazil's Colonial Timber.* Palo Alto: Stanford University Press, 2000.

Mintz, Sidney. *Sweetness and Power: The Place of Sugar in Modern History.* New York: Penguin Books, 1986.

Mittermeier, Russell, Cristina Goettsch, and Patricio Robles. *Megadiversidad: los países biológicamente más ricos del mundo.* Mexico City: CEMEX, 1997.

Moraes, Antonio Robert de. *Geografia Histórica do Brasil.* São Paulo: Annablume, 2011.

Morlon, Pierre, ed. *Comprender la agricultura campesina en los Andes Centrales: Perú—Bolivia.* Lima: IFEA and CBC, 1996.

Morse, Richard. "The Development of Urban Systems in the Americas in the Nineteenth Century." *Journal of Interamerican Studies and World Affairs* 17, no. 1 (1975): 4–26.

Moya, Frank. *Historia del Caribe. Azúcar y plantaciones en el mundo atlántico.* Santo Domingo: Ediciones Ferilibro, 2008.

Murra, John. "El control vertical de un máximo de pisos ecológicos en la economía de las sociedades andinas." In *El mundo andino. Población, medio ambiente y economía,* edited by John V. Murra, 85–125. Lima: Pontificia Universidad Católica del Perú e Instituto de Estudios Peruanos, 2002/1972.

Murphy, Catherine. *Cultivating Havana: Urban Agriculture and Food Security in the Years of Crisis.* Oakland: Food First Institute, 1999.

Nodari, Eunice, and João Klug, eds. *História ambiental e migrações*. São Leopoldo: Oikos, 2012.

Nodari, Eunice, and Silvio Correa, eds. *Migrações e natureza*. São Leopoldo: Oikos, 2013.

North, Lisa, Timothy David Clark, and Viviana Patroni, eds. *Community Rights and Corporate Responsibility: Canadian Mining and Oil Companies in Latin America*. Toronto: Between the Lines, 2006.

Norton, Marcy. "The Chicken or the Iegue: Human-Animal Relationships and the Columbian Exchange." *The American Historical Review 120, no.* 1 (2015): 28–60.

Ochoa, Enrique C. *Feeding Mexico: The Political Uses of Food since 1910*. Wilmington, DE: Scholarly Resources, 2010.

Ohmstede, Antonio Escobar, Martín Sánchez Rodriguéz, and Ana María Gutiérrez Rivas, eds. *Agua y tierra en México, siglos XIX y XX*. 2 volumes. Zamora and San Luis Potosí: El Colegio de Michoacán/El Colegio de San Luis, 2008.

Ortiz Monasterio, Fernando. *Tierra profanada: Historia ambiental de México*. Mexico City: Instituto Nacional de Antropología e Historia, Secretaría de Desarrollo Urbano y Ecología, 1987.

Pádua, José Augusto. "As bases teóricas da história ambiental," *Estudos Avançados* 24 (2010): 81–101.

———. "Biosphere, History and Conjuncture in the Analysis of the Amazon Problem." In *The International Handbook of Environmental Sociology*, edited by Michael Redclift and Graham Woodgate. Cheltenham: Edward Elgar, 1997.

———. "Environmentalism in Brazil: A Historical Perspective." In *A Companion to Global Environmental History*, edited by John R. McNeill and Erin Mauldin. Oxford: Wiley-Blackwell, 2012.

———. "European Colonialism and Tropical Forest Destruction in Brazil." In *Environmental History—As If Nature Existed*, edited by John R. McNeill, José Augusto Pádua, and Mahesh Rangarajan. New Delhi: Oxford University Press, 2010.

———. "Tropical Forests in Brazilian Political Culture: From Economic Hindrance to Ecological Treasure." In *Endangerment, Biodiversity and Culture*, edited by Fernando Vidal and Nélia Dias. London: Routledge, 2015.

———. *Um sopro de destruição: pensamento político e crítica ambiental no Brasil escravista, 1786–1888*. Rio de Janiero: Jorge Zahar Editor, 2002.

Pajares, Erick, and Jaime Llosa. "Relational Knowledge Systems and Their Impact on Management of Mountain Ecosystems. Approaches to Understanding the Motivations and Expectations of Traditional Farmers in the Maintenance of Biodiversity Zones in the Andes." *Management of Environmental Quality: An International Journal* 22, no. 2 (2010): 213–232.

Palacio, Germán. "An Eco-Political Vision for an Environmental History: Toward a Latin American and North American Research Partnership." *Environmental History* 17, no. 4 (2012): 725–743.

———. *Fiebre de tierra caliente: Una historia ambiental de Colombia 1850–1930*. Bogotá: Universidad Nacional de Colombia, 2006.

———. "Urbanismo, naturaleza y territorio en la Bogotá Republicana, 1810–1910." In *Ciudad y naturaleza: tensiones ambientales en Latinoamérica, siglos XVIII–XXI*, edited by Rosalva Loreto, 165–187. Puebla: ICSyH/BUAP, 2012.

Palacios, Marco. *El café en Colombia, 1850–1870: Una historia económica, social y política.* Mexico City: El Colegio de México/El Ancora Editores, 1983 (1979).

Palmie, Stephan, and Francisco Scarano. *The Caribbean: A History of the Region and Its Peoples.* Chicago: University of Chicago Press, 2011.

Parsons, James J. "Spread of African Pasture Grasses to the American Tropics." *Journal of Range Management* 25, no. 1 (1972): 12–17.

Pérez, Leticia Merino. *Conservación o deterioro. El impacto de las políticas públicas en las instituciones comunitarias y en las prácticas de uso de los recursos forestales.* Mexico City: SEMARNAT/INE/CCMSS, 2004.

Patel, Raj. "The Long Green Revolution." *The Journal of Peasant Studies* 40, no. 1 (2012): 1–63.

Pérez, Louis A., Jr. *Winds of Change: Hurricanes and the Transformation of Nineteenth-Century Cuba.* Chapel Hill: North Carolina University Press, 2001.

Picado, Wilson. "Las buenas semillas: Plantas, capital genético y Revolución Verde en Costa Rica." *Historia Ambiental Latinoamericana y Caribeña* 2, no. 2 (March 2013): 308–337.

Picado, Wilson, Rafael Ledezma Díaz, and Roberto Granados Porras. "Territorio de Coyotes. Agroecosistemas y cambio tecnológico en una región cafetalera de Costa Rica." *Revista Historia* 59–60 (2009): 119–165.

Platt, Tristan. *Estado boliviano y ayllu andino: Tierra y tributo en el norte de Potosí.* Lima: Instituto de Estudios Peruanos, 1982.

Quintero, Camilo. *Birds of Empire, Birds of Nation: A Place for Science and Nature in U.S.-Colombia Relations.* Bogotá: Universidad Nacional de Colombia, 2012.

Rademaker, Kurt, Gregory Hodgins, Katherine Moore, Sonia Zarrillo, Christopher Miller, Gordon R. M. Bromley, Peter Leach, David A. Reid, Willy Yépez Álvarez, Daniel H. Sandweiss. "Paleoindian Settlement of the High-Altitude Peruvian Andes." *Science* 346, no. 6208 (2014): 466–469.

Radding, Cynthia. *Wandering Peoples: Colonialism, Ethnic Spaces, and Ecological Frontiers in Northwestern Mexico, 1700–1850.* Durham: Duke University Press, 1997.

Ramírez, Fernando. "Pestilencia, olores y hedores en el Santiago del Centenario." *HALAC* 1, supplement (2012): 58–59.

Richardson, Bonham. *The Caribbean in the Wider World, 1492–1992: A Regional Geography.* New York: Cambridge University Press, 1992.

Rivera, José Eustacio. *The Vortex.* Translated by E. K. James. Bogotá: Panamericana Editorial, 2001.

Robbins, Nicholas A. *Mercury, Mining, and Empire: The Human and Ecological Cost of Colonial Silver Mining in the Andes.* Bloomington: Indiana University Press, 2011.

Rogers, Thomas. *The Deepest Wounds: A Labor and Environmental History of Sugar in Northeast Brazil.* Chapel Hill: University of North Carolina Press, 2010.

Roseveare, G. M. *The Grasslands of Latin America.* Aberystwyth, UK: Imperial Bureau of Pastures and Field Crops, 1948.

Rudel, Thomas. *Tropical Deforestation: Small Farmers and Land Clearing in the Ecuadorian Amazon.* New York: Columbia University Press, 1993.

Sá, Lúcia. *Rain Forest Literatures: Amazonian Texts and Latin American Culture.* Minneapolis: University of Minnesota Press, 2004.

Sabin, Paul. "Searching for a Middle Ground: Native Communities and Oil Extraction in the Northern and Central Ecuadorian Amazon, 1967–1993." *Environmental History* 3, no. 2 (1998): 144–168.

Santiago, Myrna. *The Ecology of Oil: Environment, Labor, and the Mexican Revolution.* Cambridge: Cambridge University Press, 2006.

Santos, Milton, and Maria Laura Silveira. *O Brasil: Território e sociedade no início do século XXI.* Rio de Janeiro: Record, 2001.

Schmink Marianne, and Charles Wood. *Contested Frontiers in Amazonia.* New York: Columbia University Press, 1992.

Schwartz, Norman. *Forest Society: A Social History of Petén, Guatemala.* Philadelphia: University of Pennsylvania Press, 1990.

Schwartz, Stuart B. *Sea of Storms: A History of Hurricanes in the Greater Caribbean from Columbus to Katrina.* Princeton: Princeton University Press, 2015.

Sedrez, Lise. "Latin American Environmental History: A Shifting Old/New Field." In *The Environment and World History,* edited by Edmund Burke III and Edward Pomeranz. Berkeley: University of California Press, 2009.

———. "Rubber, Trees and Communities: Rubber Tappers in the Brazilian Amazon in the Twentieth Century." In *A History of Environmentalism,* edited by Marco Armiero and Lise Sedrez, 147–166. London: Bloomsbury Academic, 2014.

Sedrez, Lise, and Andréa Maia. "Narrativas de um dilúvio carioca: Memória e natureza na grande enchente de 1966." *Revista História Oral* 14, no. 2 (2011): 221–254.

Silva, Sandro Dutra, José Pietrafesa, José Luiz Franco, José Drummond, and Giovana Tavares, eds. *Fronteira cerrado: Sociedade e natureza no oeste do Brasil* (Goiânia: Editora da PUC-Goiás, 2013).

Silvestri, Graciela. *El color del río: Historia cultural del paisaje del Riachuelo.* Buenos Aires: Universidad Nacional de Quilmes, 2003.

Simonian, Lane. *Defending the Land of the Jaguar: A History of Conservation in Mexico.* Austin: University of Texas Press, 1995.

Slater, Candace. *Entangled Edens: Visions of the Amazon.* Berkeley: University of California Press, 2002.

Sluyter, Andrew. "Recentism in Environmental History on Latin America." *Environmental History* 10 (January 2005): 91–93.

Soluri, John. *Banana Cultures: Agriculture, Consumption, and Environmental Change in Honduras and the United States.* Austin: University of Texas Press, 2006.

———. "Seals and Seal Hunters along the Patagonian Littoral, 1780–1960." In *Centering Animals: Writing Animals into Latin American History,* edited by Martha Few and Zeb Totorici. 243–269. Durham: Duke University Press, 2013.

———. "Something Fishy: Chile's Blue Revolution, Commodity Diseases and the Problem of Sustainability." *Latin American Research Review* v. 46 (2011): 55–81.

———. "Tierras, montes y aguas: Apuntes sobre la energía, medio ambiente y justicia en las Américas." *Revista de Historia* (Costa Rica) 59–60 (2009): 169–184.

Steen, Harold, and Richard Tucker, eds. *Changing Tropical Forests: Historical Perspectives on Today's Challenges in Central and South America.* Durham: Forest History Society, 1992.

Stepan, Nancy. *Beginnings of Brazilian Science: Oswaldo Cruz, Medical Research and Policy, 1890–1920.* New York: Science History Publications, 1976.

———. *Picturing Tropical Nature*. London: Reaktion Books, 2002.

Studnicki-Gizbert, Daviken, and David Schecter. "The Environmental Dynamics of a Colonial Fuel-Rush: Silver Mining and Deforestation in New Spain, 1522 to 1810." *Environmental History* 15, no. 1 (2010): 94–119.

Super, John C., and Thomas C. Wright, eds. *Food, Politics and Society in Latin American History*. Lincoln: University of Nebraska Press, 1985.

Toledo, Víctor. *Ecología, espiritualidad y conocimiento: De la sociedad del riesgo a la sociedad sustentable*. Mexico City: PNUMA, Universidad Iberoamericana, 2003.

———. "¿Por qué los pueblos indígenas son la memoria de la especie?" *Papeles* 107 (2009).

Tortolero Villaseñor, Alejandro. *De la coa a la máquina de vapor: Actividad agrícola e innovación tecnológica en las haciendas de la región central de México, 1880–1914*. Mexico City: Siglo XXI, 1995.

———, ed. *Tierra, agua y bosques: historia y medio ambiente en el México Central*. Mexico City: Potrerillos Editores, 1996.

Tucker, Richard. *Insatiable Appetite: The United States and the Ecological Degradation of the Tropical World*. Berkeley: University of California Press, 2000.

United Nations. *Livestock in Latin America: Status, Problems and Prospects*. Vol. 1, *Colombia, Mexico, Uruguay and Venezuela*. New York: United Nations, 1962.

Vainer, Carlos. "Águas para a Vida, Não para a Morte: Notas para Uma História do Movimento de Atingidos por Barragens no Brasil." In *Justiça Ambiental e Cidadania*, edited by Henri Acserald, Selene Herculano and José Augusto Pádua. Rio de Janeiro: Relume-Dumará, 2004.

Van Ausdal, Shawn. "Pasture, Power, and Profit: An Environmental History of Cattle Ranching in Colombia, 1850–1950." *Geoforum* 40, no. 5 (2009): 707–719.

———. "Reimagining the Tropical Beef Frontier and the Nation in Early Twentieth-Century Colombia." In *Trading Environments: Frontiers, Commercial Knowledge and Environmental Transformation, 1820–1990*, edited by Gordon M. Winder and Andreas Dix, 166–192. New York: Routledge, 2016.

Vandermeer, John, and Ivette Perfecto. *Breakfast of Biodiversity: The Truth about Rain Forest Destruction*. Oakland: Food First Books, 1995.

Vessuri, Hebe M. C. "Academic Science in Twentieth-Century Latin America." In *Science in Latin America: A History*, edited by Juan José Saldaña, 197–230. Austin: University of Texas Press, 2006.

Viales Hurtado, Ronny J. and Anthony Goebel McDermott eds. *Costa Rica: Cuatro ensayos de historia ambiental*. Costa Rica: Sociedad Editora Alquimia 2000, 2011.

Vitz, Matthew. "La ciudad y sus bosques: La conservación forestal y los campesinos en el valle del México, 1900–1950." *Estudios de Historia Moderna y Contemporánea de México* 43 (2012): 135–172.

Warman, Arturo. *El campo mexicano en el siglo XX*. Mexico: Fondo de Cultura, 2001.

Weismantel, Mary. *Food, Gender and Poverty in the Ecuadorean Andes*. Philadelphia: University of Pennsylvania Press, 1988.

Wakild, Emily. "Parables of Chapultepec: Urban Parks, National Landscapes and Contradictory Conservation in Modern Mexico." In *A Land Between Waters: Environmental Histories of Modern Mexico*, edited by Christopher R. Boyer. Tucson: University of Arizona Press, 2014.

————. *Revolutionary Parks: Conservation, Social Justice, and Mexico's National Parks, 1910–1940*. Tucson: University of Arizona Press, 2011.

Walker, Robert , J. Browder, E. Arima, C. Simmons, R. Pereira, M. Caldas, R. Shirota, and S. de Zen. "Ranching and the New Global Range: Amazônia in the 21st century." *Geoforum* 40, no. 5 (2009): 732–745.

Walker, Charles. *Shaky Colonialism*. Durham: Duke University Press, 2008.

Watts, David. *The West Indies: Patterns: Development, Culture and Environment Change since 1492*. Cambridge: Cambridge University Press, 1987.

Wilcox, Robert W. *Cattle in the Backlands: Mato Grosso and the Evolution of Ranching in the Brazilian Tropics*. Austin: University of Texas Press, 2017.

————. "'The Law of the Least Effort': Cattle Ranching and the Environment in the Savanna of Mato Grosso, Brazil, 1900–1980." *Environmental History* 4, no. 3 (1999): 338–368.

Wilk, Richard, and Livia Barbosa. *Rice and Beans: A Unique Dish in a Hundred Places*. London: Bloomsbury Academic, 2012.

Wolfe, Mikael D. *Watering the Revolution: An Environmental and Technological History of Agrarian Reform in Mexico*. Durham: Duke University Press, 2017.

Zarrilli, Adrian Gustavo. "Capitalism, Ecology and Agrarian Expansion in the Pampean Region (1890–1950)." *Environmental History* 6 (2002): 560–583.

Zimmerer, Karl S. *Changing Fortunes: Biodiversity and Peasant Livelihood in the Peruvian Andes*. Berkeley: University of California Press, 1996.

Index

A

Africa
compared, 5–6, 9, 11, 115, 123, 128,
249, 250, 261, 267, 270
influence on Latin America, 5–6, 7,
9–10, 46, 47, 163
African grasses, 10, 52, 185–186, 191
agrarian reform, 34, 53, 126, 166, 169–170
agriculture. *See* agrodiversity; campesinos;
"commodity diseases"; plantations
agrodiversity
in Brazilian common beans, 171–173
and campesinos, 164, 176–177
in coffee farms, 173–176
defined, 163
and indigenous cultures, 80–81, 171,
176–177
in Mexican maize cultivation, 26,
165–168
in Peruvian tuber cultivation,
168–171
agroecology, 240
air pollution, 38, 84, 139, 154–156, 236,
238–239, 270
Albert, Federico, 252
Almaraz, Sergio, 205
Amazonia
and Brazilian territory, 92–93, 98, 117
and Colombian territory, 124
colonization of, 107, 124–126, 172
conservation of, 129–132, 196, 249
deforestation in, 13, 107, 109, 128,
191
and environmental activism, 108–109,
115, 131, 218
exploitation of, 102, 120, 123, 127,
146, 194, 217–219

indigenous presence in, 5, 131, 218,
260
as threatening, 116–119
Andes
agrodiversity in, 80–81, 168–171
biogeography of, 70–71
and cinchona bark, 76–78
deforestation of, 74–75
indigenous population of, 68
vertical landscapes of, 72–73
See also indigenous societies; mining
animals, 6, 10, 53, 68, 69, 82, 124–125, 129,
151, 183, 231, 249
See also extinction; hunting; ranching
aquaculture, 33, 268
Argentina
and conservation, 251–252, 255,
257–258
grasslands of, 22n48, 187–188, 240
immigration to, 143
indigenous population in, 68
and pollution, 1–2, 154–155
ranching in, 10, 186–188, 197
See also Buenos Aires; Mendoza;
Patagonia
Aridjis, Homero, 39
Asia, 46, 52, 76, 115, 120, 121, 122, 123,
128, 143, 250, 270
Atacama desert, 209–212
Atlantic Forest, 93, 95, 98, 99, 101, 104–
105, 107, 117, 121, 144
Ayllu. See *comuna*
Aymara, 67, 79, 84, 85, 169

B

Baja California, 26, 30, 33, 39
bananas, 12, 51–52, 54–55, 122, 175, 240

Bartra, Roger, 36
Bates, Henry Walter, 76, 129
bauxite, 56, 268
beans, 106, 163, 165, 171–173, 174–175,
176
beef consumption, 163, 174, 187, 190,
194–195, 196–197
    See *also* "hamburger connection"
Belo Horizonte, 156, 236
biocultural diversity, 9, 67, 85, 98
    See *also* agrodiversity
biodiversity, 49, 92, 119, 129, 191, 233, 234,
237, 259, 261
    See *also* Andes; tropical rainforests
biomass, 6, 28, 92, 106, 144, 150, 152, 269
biomes, 22n48, 25–26, 35, 92, 104, 108,
185, 269
Bogotá, 19n21, 72, 73, 84, 143, 144, 145,
148, 150
Bolivia
    biological diversity in, 71
    coca cultivation in, 75–76, 78
    indigenous influence on, 79–80, 84, 85
    indigenous population of, 68
    and *sumak kawsay*, 67
    See *also* Andes; mining; Potosí
borders (geopolitical), 7, 28, 94, 123, 246,
255
Brazil
    Atlantic Forest of, 98–101, 104–105
    ecological diversity of, 13, 92–93
    energy consumption in, 106, 269
    and "Great Acceleration," 105–106
    immigration to, 103
    and indigenous societies, 98, 115
    nature in national identity of, 7, 12,
97–98
    rubber boom in, 102
    slash-and-burn agriculture in, 99–100
    territorial formation of, 94–102
    See *also* agrodiversity; Amazonia;
        ranching; Rio de Janeiro; São Paulo;
        tropical rainforests
breeding (crops and livestock), 10, 52, 55,
82, 173, 176, 187–188, 190, 233, 235
"BRICS," 91

Buenos Aires, 1–2, 139, 143, 144, 148, 151,
154, 187–189
burning, 6, 49, 96, 99, 101, 105, 124, 189,
197
    See *also* slash-and-burn agriculture

C
cacao, 52, 99, 102, 120, 122, 230
camaima, 116
campesinos, 35–37, 106, 163–177, 178n9
    and agrodiversity, 176–177
    and cuisine, 163–165, 176
    use of cattle by, 193
Capanema, Guilherme, 100
Cárdenas, Lázaro, 8, 34–35, 166, 214–215
Caracas, 58, 148, 195
Caribbean
    colonial period, 46–47
    deforestation in, 6, 117
    ecosystems of, 45–46
    plantations in, 12, 47–55
    and tourism, 13, 58–60
    urbanization in, 57–58, 153, 157
    See *also* bananas; "conquest of the
        tropics"; slavery; sugarcane; timber
        industries
cartography, 97, 123, 130
Cebu. *See* Zebu
cenotes, 27
Central America
    banana industry in, 51–52, 54–55
    coffee production in, 175, 241
    conservation in, 62
    deforestation in, 117, 128, 190
    indigenous people in, 132
    ranching in, 127, 186, 190, 196
    timber industry in, 120, 127
    See *also* Costa Rica; Guatemala;
        Honduras; Panama
Cerrado, 93, 105, 107–108, 109, 188–189,
191, 192
Chachawayna, Manuel, 85
Chapultepec Park, 6, 150, 157
charcoal. *See* biomass
Chardón, Carlos, 230
Chiapas, 41, 116, 166–167

chicle, 120
Chile, 155–156, 208–213, 240, 246, 252, 255, 268
   See also mining; Santiago de Chile
Chincha Islands, 208, 230. *See also* guano
climate change, 197, 241, 248
coal, 30, 103, 269–271
Coatsworth, John, 30
coffee, 49–50, 99, 121, 172, 173–176, 235, 240
Colombia
   biodiversity in, 13, 70–71, 233
   coffee in, 163, 175, 176, 240, 241
   colonization of rainforests, 124–125, 126
   conservation in, 131–132
   deforestation in, 74, 122
   indigenous people in, 68, 131
   rainforests in, 116
   ranching in, 186, 187, 189, 194
   research centers in, 173, 176, 191, 233
   urban areas in, 143, 145–146
   See also Bogotá
colonato, 174–175
colonialism, 5–7
   in Brazil, 95, 98
   in the Caribbean, 46–48, 59
   and conservation, 249–251
   impacts of, compared, 6, 9, 12, 76, 120–125, 128, 250
   in Mexico (New Spain), 28
   and ranching, 183–184, 187–188
   and urbanization, 140–142
Columbian Exchange, 9–11
"commodity diseases," 10, 52, 122, 230, 240
comunas, 79, 84
"conquest of the tropics," 50, 51, 63
conservation, 3–4, 13
   chronology of, 253–254
   and colonialism, 250
   cosmopolitan roots of, 48, 251–253
   and developmentalism, 234, 256–258
   and nationalism, 35, 254–256
   and resident people, 260
   and scientists, 33, 236–240

   See also agrodiversity; conservation biology; protected areas
conservation biology, 129, 237, 259
consumption, 3, 105–106, 153, 194–197
Convention on Nature Protection and Wildlife Preservation, 234
Costa Rica
   agrodiversity in, 175
   banana industry in, 52, 55, 122, 123
   deforestation, 128
   environmental science in, 229, 233, 234, 237
   national parks in, 234, 259
   ranching, 127, 185–186
   tourism in, 60
   See also Central America
cowboys, 192–193
criollo culture, 68–70, 74, 85
Cruz, Oswaldo, 148, 231
Cuba
   agrodiversity in, 174
   cattle pasture in, 189
   and "conquest of tropics," 12, 50
   conservation in, 61
   environmental science in, 229, 230, 231
   mining in, 56
   sugarcane production in, 48–49, 50–51, 53–54, 228
   tourist trade in, 13, 58, 59
   See also Caribbean; Havana
Cubatão (Brazil), 155
cuisine, 37, 81, 83, 153, 163–177, 194–195
   See also food
Cusco, 80, 86, 169, 170

D
da Cunha, Euclides, 116–117
dams, 36, 57, 106, 129, 152
   See also Itaipu dam; Movement of People Affected by Dams
Darien National Park (Panama), 131
Dean, Warren, 2–3
de Andrada e Silva, José Bonifacio, 229
deforestation, 3, 6, 13
   in Brazil, 99–100, 102, 104, 107–108, 121, 128, 149, 175

in Caribbean, 48, 49, 52
in Mexico, 28, 33, 37, 38
in tropical Andes, 74–75
of tropical rainforests, 115–116, 122,
    124–125, 127–128, 133
See also plantations; petroleum
    industry; ranching
de Quevedo, Miguel Angel, 33, 234, 251
deserts, 4, 22n48, 26, 29, 93, 260
See also Atacama desert; Sonora desert
developmentalism, 226–227, 232, 256–257
in Brazil, 105–106
in Mexico, 35–40
disasters, 38, 46, 73, 141, 148, 211, 241
Dominican Republic, 49, 51, 52, 54, 56, 61
Dos Bocas (Mexico), 213–214
Dutilleux, Jean Pierre, 115, 129

E
Earls, John, 82
earthquakes. *See* disasters
"ecological revolutions," 23, 40
ecology. *See* environmental science
ecotourism, 60, 130, 239
Ecuador
cacao cultivation in, 122
criollo culture in, 68
deforestation in, 75, 126
indigenous population in, 68
indigenous practices in, 80
petroleum and conflict in, 217–219
population growth in, 125
and rights of nature, 67
See also Andes
El Cobre (Chile), 211
El Teniente (Chile), 211–212
energy, 106, 150, 152, 269
environmental conflicts:
in the Caribbean, 56
and conservation, 260
and extraction, 205–221
in Mexico, 28, 33–34, 40–41
and tropical rainforests, 108–109, 129,
    131, 132
in urban areas, 1–2, 146, 147 152,
    156–157

environmental history
defined, 2–5, 92
and indigenous ontologies, 85–87
of ranching, 183–184, 198
and social history, 221
of urban areas, 138, 158, 198
environmentalism, 39, 57, 109, 115, 129–
131, 218, 220, 272
See also Nature (ideas about)
environmental science
and agroexports, 228–230
and biodiversity, 233–234, 237–238
and developmentalism, 226–227
and disease epidemics, 230–231
and Green Revolution, 234–236
institutionalization of, 232–233
natural history as antecedent to,
    228–229
and sustainability, 238–240
See also scientists
epidemics, 46, 145, 147, 230–231
erosion, 29, 32, 48, 50, 99–100, 148, 175,
    252, 255, 257
eucalyptus, 75
export economies, 3, 8, 10–11, 37, 46–49,
    51–52, 75–78, 109–110, 173–174
extinction, 237, 248–249, 267
extractive economies, 119–121, 205–225
See also mining; petroleum industry;
    timber industries
extractive reserves, 260

F
farming. *See* agrodiversity; export
    economies; haciendas; indigenous
favelas, 107, 149, 153, 154, 268–269
See also shantytowns
feedlots, 197
Fernández Prieto, Leida, 228–229
fertilizers, 10, 36, 37, 99, 121, 170, 175,
    207–210, 230
Finlay, Carlos, 50, 231
flooding. *See* disasters
food, 55, 57, 81, 99, 156, 217, 257
See also cuisine
food sovereignty, 35, 39, 63

forests, 2–3, 6, 7, 8, 10, 11–13, 22n48, 25, 27, 184, 187, 246, 255
    See also deforestation; tropical rainforest
Funes, Reinaldo, 3

G
Gallegos, Rómulo, 12, 116
garbage, 150, 151, 156
    See also waste
García Márquez, Gabriel, 133
glaciers, 4, 13, 73, 241, 252, 255
Gorgas, William, 50
grasslands, 4, 11, 22n48, 93, 101, 183, 184, 187–188, 189, 249, 250
    See also Pampas; ranching; savanna
"Great Acceleration," 14, 103–109
Green Revolution, 9, 24, 26, 35–37, 78, 81, 177, 236, 272
Guanabara Bay (Brazil), 151, 267
guano, 207–209, 212, 230, 257
Guatemala, 29, 120, 246, 257
    See also Caribbean; Central America

H
haciendas, 26, 29, 31, 72, 79, 144, 165, 169
Haiti, 46, 47, 48, 49–50, 58, 61
"hamburger connection," 56, 127, 190
Havana, 46, 50, 58, 142, 151, 156
Hecht, Susanna, 194
henequen, 31
Honduras, 58, 122, 124, 130, 148, 201n51
    See also Caribbean; Central America
Huasteca, 35, 37, 122, 193, 213–214
Humboldt Current, 268
hunting, 10, 30, 101, 124–125, 127, 234
hurricanes. *See* disasters
hydroelectricity, 57, 103, 106, 129, 150, 152, 269
    See also dams

I
Iguazú/Iguaçu National Parks, 255
indigenous societies
    in Brazil, 94–95, 98, 104, 107
    in Caribbean, 46

    in Colombia, 132
    cosmovisions of, 67, 85–87, 272
    demographic collapse of, 5–6
    and environmental conflict, 41, 107, 115–116, 131, 199n6, 217–219
    and environmentalists, 129, 130–131, 132, 213, 220
    foodways of, 166, 169, 171, 176
    land use practices of, 83
    livelihoods of, 81, 119, 166–171, 176–177
    in Mexico, 26, 27, 29, 34, 41
    pre-columbian, 5, 71, 248–250
    in tropical Andes, 67–87, 260
    twenty-first-century population (Andes), 68
    water management practices of, 82–83, 141
industrialization, 53, 57, 103, 152, 164, 232, 256, 257, 275n19
industrial metabolism, 56–58
Inferno Verde, 116
influenza, 147
    See also epidemics
infrastructure. *See* dams; railways; roads; urban areas
intercropping, 174–176
    See also agrodiversity
International Center for Tropical Agriculture (ICTA), 173
Itaipu dam, 152

J
Jalisco, 166–167
Jamaica, 56, 58
jungle. *See* tropical rainforests

K
Kayapó, 115

L
labor. *See* workers
labor movements, 205–221
Lake Maracaibo, 53, 215–216
land colonization, 38, 75, 78, 101, 107, 124–129, 255

landraces (of crops), 26, 170, 173
land tenure, 24, 31, 33–34, 79, 80, 106, 126, 166
La Oroya (Peru), 78
livestock, 29, 31, 46, 52, 53, 55, 101, 127
　　See also ranching
logging. *See* mahogany; timber industries
Lomnitz, Claudio, 193
Love in the Time of Cholera, 133
Lulu da Silva, Luiz, 106
Luquillo Forest (Puerto Rico), 61

M
mahogany, 29, 46, 120, 127
maize. *See* agrodiversity; Mexico
Manaus, 119
mangroves, 46, 107, 148, 213
Manú National Park (Peru), 130
Mapocho River (Chile), 150
marine environments, 27, 35, 39, 46–47, 267–268
markets, 52, 168, 169, 171, 175, 177, 187, 190, 194–195
medicine. *See* public health
"megamining," 219–221
Melville, Elinor, 184
Mendes, Chico, 108–109, 131
Mendoza, 141
Merchant, Carolyn, 23
mestizaje, 12, 68–69
Mexican Revolution, 33–35, 166, 213, 214
Mexico
　　agrodiversity in, 26
　　biomes in, 24–28
　　community forestry in, 34, 41
　　deforestation in, 38
　　Green Revolution, 24, 35–37
　　liberal policies in, 23, 24, 27, 28, 29–33
　　maize cultivation in, 26, 37, 165–168
　　petroleum industry of, 35, 213–215
　　population change in, 24, 29, 35
　　ranching in, 193
　　water resources in, 31, 36, 38
　　See also agrodiversity; Mexican Revolution; Mexico City

Mexico City
　　air pollution in, 154–156, 236, 239
　　in colonial era, 6
　　food consumption in, 164, 167, 168, 195
　　as megacity, 27, 38, 139
　　parks in, 150, 157
　　shantytowns in, 153
　　ties to rural, 149–150, 255
migration, 59, 78, 84, 126, 143, 164, 231, 248, 272
mining, 3, 10–11, 28–29
　　bauxite, 56
　　copper, 26, 30, 210–212
　　gold, 127–128
　　guano in Peru, 207–209
　　nickel, 56
　　nitrates in Atacama desert, 209–210
　　and social conflict, 205–207
　　See also La Oroya (Peru), "megamining," Potosí
monoculture, 31, 48, 116, 121, 122, 174, 230
　　See also "commodity diseases"; plantations
Montalvo, Juan, 75
Montes Azul Biosphere Reserve, 130
Moreno, Francisco P., 251
Mosquitia, 130, 132
mountains, 27, 100, 149, 166, 177, 205, 211, 255
　　See also Andes; mining
Movement of People Affected by Dams, 106
museums of natural history, 97, 144, 228–229
Murra, John, 72
Myers, Norman, 190

N
Nahuel Huapi National Park, 251, 255, 257–258
nationalization (of resources), 9, 34–35, 56, 206, 212, 214–215, 217
national parks, 9, 35, 61, 66n41, 130, 131, 251, 255, 257–258
　　See also conservation; protected areas

nation-states, 7–9, 85, 94–97, 256
Nature (ideas about)
    as abundant, 7, 93, 96, 144, 228, 259
    commodification of, 23–24, 37, 59
    and cultural diversity, 7, 13, 67–69,
        85–86, 214, 241, 271–272
    as dangerous, 12, 123
    as endangered, 128–129, 234, 255
    as metaphor for social relations, 116,
        146, 192, 205–206
    and nationalism, 7–9, 35, 97–98, 144,
        229, 234
    See also environmental science;
        indigenous societies
Nature Conservancy, 258
naturalists, 4, 129, 144, 228–230, 233–234,
    237
neoliberalism, 1, 24, 39, 59, 218–221, 260
Nicaragua, 120, 132, 175, 186
    See also Caribbean; Central America
nitrate mines, 208–210, 212
Nixon, Rob, 261
non-governmental organizations (NGOs),
    115, 131, 234, 238, 258
North American Free Trade Agreement
    (NAFTA), 39
nuclear power, 152–153

O
oceans. *See* marine environments
oil. *See* petroleum industry

P
Pádua, José Augusto, 149
Pádua, María Tereza Jorge, 260
Pampas, 1, 194
    See also pasture; ranching
Panama, 52, 122, 132
    See also Caribbean; Central America
Panama Canal, 12, 47, 50, 123
Paraguay, 152, 188, 256, 275n17
páramos, 72, 74–75
Paraná River, 152
Parque Centenario, 150
Parque de la Papa, 86
Parsons, James, 183, 193

pasture, 52, 101, 126–127, 183, 185, 186–
    187, 188, 189, 190, 191, 194, 196
    See also ranching
Patagonia, 1, 10, 220, 255
Paucartambo, 80, 169–171
peasant. *See* campesino
Pereira Passos, Francisco, 148
Peru, 258–259, 260
    biodiversity in, 13, 71
    conservation in, 259–260
    export economies of, 75, 76–78
    indigenous agriculture in, 80–83,
        168–171
    indigenous population of, 68
    mining industries in, 207–209, 220
    research institutions in, 154, 176, 228,
        259
    See also Andes; guano; indigenous
        societies
pesticides, 36–37, 52, 75–76, 107, 235
petroleum industry, 1, 3, 8, 11, 78, 92, 206,
    220, 269–271
    in Ecuador, 217–219
    in Mexico, 35, 38, 122, 213–215, 217
    in Venezuela, 53, 57, 215–217
plantations, 3, 12, 45, 47–55, 58, 62, 83,
    99–100, 121–122, 173–174
Pleistocene, 248–249, 267
"Pleistocene Refuges," 249, 260, 261, 267
Pomeranz, Kenneth, 226
Population (human)
    of Brazil, 91, 94, 103, 109
    demographic collapse, 5, 46, 248, 250
    of Greater Caribbean, 57, 58, 61
    of Latin America, 125, 184, 195
    of Mexico, 29, 33, 35, 38
    in twentieth-century tropical forests,
        125–126
    in urban areas, 84, 107, 138, 139, 143,
        145, 164, 169, 236
potatoes, 80, 81, 83, 86, 233
    See also agrodiversity; Andes
Potosí, 6, 140
pre-Columbian, 139, 267
Primera Conferencia Latinoamericana de
    Contaminación del Aire, 154

protected areas, 3, 9, 24, 27, 60, 61–62, 109, 129–133, 250, 253–254
   See also conservation; national parks
public health, 1, 12–13, 33, 50, 148, 154, 155, 197, 217, 231, 236
Puerto Cortés (Honduras), 148
Puerto Rican Industrial Mission, 57
Puerto Rico, 50, 51, 53, 55, 57, 61, 175, 230

Q

qompis. See potatoes
Quechua, 67, 81, 84, 169
Quinine. *See* Andes: and cinchona bark
Quintero, Camilo, 233
Quito, 71, 84

R

railways 30, 34, 48, 51, 72–73, 100, 122, 123, 145–146, 151, 211, 269
Rainforest Alliance, 115
Rainforest Foundation, 115
Rainforests. *See* tropical rainforests
ranching, 10
   and beef consumption, 187, 190–191, 195, 196
   and biological advantages of livestock, 193–194
   cattle breeds used in, 55, 187, 192
   and climate change, 197
   in colonial Latin America, 183–184
   cultural aspects of, 192–193
   and deforestation, 74, 183, 185–186, 190–191
   geographical expansion of, 101, 126–127, 185–189, 196
   and planted pasture, 185–186, 188, 189, 191
   See also cowboys; livestock
ranchos, 29, 165
Rangel, Alfredo, 116
Rebouças, André, 251
reforestation, 61, 149, 196
research centers, 10, 52, 55, 173, 191, 230, 232, 233, 238, 257
   See also environmental science
"Revolta da Vacina," 148, 231

Riachuelo river (Argentina), 1, 148
Rio de Janeiro (city), 143, 144, 148, 149, 150–151, 231
Rio de Janeiro (state), 100, 121
Rio Plátano Biosphere Reserve, 130
Rivera, José Eustacio, 12, 116–117
rivers. *See under* water
roads, 35, 104, 126, 140, 153, 217
rubber, 12, 102, 115–116, 117, 120–121, 123, 146, 230
rubber-tappers, 102, 121, 260
   See also Mendes, Chico; rubber

S

sanitation, 109, 148, 231, 272
San Rafael National Forest (Haiti), 61
Santiago de Chile, 144, 150, 154, 156
Santos, Milton, 91
São Paulo (city), 124, 139, 144, 152, 153, 232, 233
São Paulo (state), 100, 148, 150, 172, 174–175, 186, 189
sarsaparilla, 120
savanna, 11, 93, 101, 188–189, 190
Schwartz, Stuart, 141
scientists, 4, 8, 10, 36, 49, 101, 257, 259
   See also conservation biology
seringais, 102
   See also rubber; rubber-tappers
sertões, 95–96
shantytowns, 1, 107, 147, 148–149, 153, 269
   See also favelas
sheep, 10, 184, 187
Sierra de Cristal National Park (Cuba), 61
slash-and-burn agriculture, 99–101
   See also burning
slaughterhouses, 1, 103, 187, 189–190, 200n35
slavery, 6, 9, 12, 31, 47–49, 95, 99–100, 116, 124, 174, 187, 252, 271
soils, 119–120, 121–122, 141, 165–166, 172, 175, 188, 210, 229–230, 257
Sonora desert, 26, 29, 30, 36
soybeans, 11, 107, 172–173, 197, 240
Stavenhagen, Rodolfo, 36

Sting (Gordon Sumner) 115, 129
subsidies (state), 36, 57, 107, 127, 166, 174, 191, 196, 257
sugarcane, 6, 12, 31, 45, 47–49, 50–51, 53, 54, 121, 230, 269
Sumak kawsay (*sumaq qamaña*), 7, 67
sustainability, 39–41, 60, 62, 86–87, 115, 130–132, 238–240

T
tagua, 146
Takacs, David, 259
tasajo. See beef consumption
"technification," 175, 235–236, 238
temporal lands, 165–166, 168
terras pretas, 120, 267
"territorial funds," 95–96
Texaco, 217–218
The Vortex, 12, 116–117
Tijuca Forest (Brazil), 149
timber industries, 3, 26, 29, 30, 34, 41, 46, 49, 116, 120, 127
    See also mahogany
tourism, 13, 58–60, 63, 157, 239, 246, 255, 257–258
Traven, B., 116
tropicality, 11–14, 58–63
tropical rainforests, 11–13
    changing meanings of, 115–133
    conservation of, 129–133
    distribution of in Latin America, 118
    extractive economies in, 120–121, 213, 217
    global rates of deforestation of, 128
    as "green hells," 116–117
    and plantation agriculture, 121–122
    post-1950 settlement of, 124–126
    See also Amazonia; plantations;
        ranching; rubber; timber industries
Tumaco (Colombia), 146

U
United Fruit Company, 51, 58, 122, 148
United Nations Conference on
    Environment and Development, 108, 238

United Nations Conference on the Human
    Environment (Stockholm), 155, 237
United States, 13, 28, 36, 40, 50–55, 58, 78, 122, 126, 168, 175, 190, 251
universities, 154, 228, 232
urban areas, 1–2, 4, 6
    colonial, 140–142
    and consumption, 153
    green spaces in, 150, 157
    indigenous presence in, 84–85
    and industrialization, 152
    infrastructure of, 144–146
    markets in, 168, 169–170
    pollution in, 103, 107, 236 (*see also* air
        pollution; waste)
    population growth of, 35, 38, 57–58, 143
    and privatization, 157
    and reforms, 143
    rural linkages to, 30, 38, 107, 145–146, 149–150, 164
    sanitation in, 148, 231
    and suburbanization, 145
    See also names of specific cities
Uruguay, 143, 187, 188, 197, 228

V
Vasconcelos, José, 119
Velasco, Juan, 169
Venezuela, 11, 13, 53, 58, 61–62, 68, 116, 188, 229, 233
    See also petroleum industry
"vertical archipelago," 72
Villa Inflamable (Argentina), 1–2
Vives, Gastón, 33
Vogt, William, 257
von Humboldt, Alexander, 11, 129, 195

W
Wakild, Emily, 234
waste
    cruise ship, 59
    industrial, 1, 31, 151, 189–190, 197, 211, 213–218
    radioactive, 153
    regulation of, 142

sewage, 141, 150, 151
  urban, 150, 151
water
  conflicts over, 31, 33, 157, 218, 220
  fresh, 92, 130, 141, 241
  irrigation, 31, 36, 74, 82–83
  pollution, 1–2, 6, 47, 128, 150–151,
    211, 213–214, 215–216, 217
  rivers, 1–2, 43n17, 73, 117, 133, 141,
    145, 188, 211
  urban, 1–2, 4, 31, 38, 109, 141, 144,
    149
  See also Amazonia; dams; marine
    environments
Werklé, Karl, 233
West, Robert C., 28–29, 125
wheat, 36, 69, 166, 169, 170, 255
  See also Green Revolution
women, 81, 166, 170, 200, 234
workers
  migrant, 12, 51, 54, 102, 208, 209
  in mining, 206–213, 220–221
  in petroleum industry, 35, 213–217
  political protests of, 8, 212, 213–214,
    216–217, 271

in rural livelihoods, 37, 52, 99, 103,
    163, 169, 174–175, 193
  in urban areas, 151, 166–167, 257
  and occupational health, 37, 50, 52,
    210–211, 213–214, 216
  See also colonato; labor movements;
    slavery
World Database of Protected Areas,
  253–254
  See also protected areas

X
Xingú, 129

Y
Yaque Aquatic Reserve (Dominican
  Republic), 61
Yaqui Valley, 36
  See also Green Revolution
Yellow Fever, 12, 13, 45, 50, 141, 145, 148,
  231

Z
Zebu, 52, 55, 192, 196–197